VOLUME ONE HUNDRED AND THIRTY FIVE

CURRENT TOPICS IN DEVELOPMENTAL BIOLOGY

The Immortal Germline

CURRENT TOPICS IN DEVELOPMENTAL BIOLOGY

"A meeting-ground for critical review and discussion of developmental processes"
A.A. Moscona and Alberto Monroy (Volume 1, 1966)

VOLUME ONE HUNDRED AND THIRTY FIVE

CURRENT TOPICS IN DEVELOPMENTAL BIOLOGY

The Immortal Germline

Edited by

RUTH LEHMANN
HHMI and Kimmel Center for Biology and Medicine of the Skirball Institute, Department of Cell Biology, New York University School of Medicine, New York, NY, United States

ELSEVIER

ACADEMIC PRESS
An imprint of Elsevier

Academic Press is an imprint of Elsevier
50 Hampshire Street, 5th Floor, Cambridge, MA 02139, United States
525 B Street, Suite 1650, San Diego, CA 92101, United States
The Boulevard, Langford Lane, Kidlington, Oxford OX5 1GB, United Kingdom
125 London Wall, London, EC2Y 5AS, United Kingdom

First edition 2019

ISBN: 978-0-12-809752-6
ISSN: 0070-2153

For information on all Academic Press publications
visit our website at https://www.elsevier.com/books-and-journals

Publisher: Zoe Kruze
Acquisition Editor: Zoe Kruze
Editorial Project Manager: Shellie Bryant
Production Project Manager: Denny Mansingh
Cover Designer: Greg Harris

Typeset by SPi Global, India

Contents

Contributors ix
Preface xi

1. The vertebrate Balbiani body, germ plasm, and oocyte polarity 1
Allison Jamieson-Lucy and Mary C. Mullins

1. Introduction 2
2. Inductive and inherited germ cell specification 3
3. Inheritance of germ plasm 6
4. The Balbiani body: A conserved oocyte organelle 8
5. Early oocyte development and the Balbiani body in mice 11
6. Early oocyte development in zebrafish 13
7. The Balbiani body in zebrafish and frogs 15
8. The role of mitochondria in the Balbiani body 17
9. Bucky ball: An essential Balbiani body protein 19
10. Macf1 and Balbiani body disassembly 21
11. Secondary pathway mRNA localization 23
12. Animal pole mRNA localization 23
13. Concluding remarks 24

Acknowledgments 25
References 25

2. Genetic basis for primordial germ cells specification in mouse and human: Conserved and divergent roles of PRDM and SOX transcription factors **35**
Anastasiya Sybirna, Frederick C.K. Wong, and M. Azim Surani

1. Introduction 36
2. Early developmental divergence in mammalian embryos 43
3. Modeling peri-implantation development and PGC specification 46
4. Gene regulatory networks for mouse and human PGC specification 53
5. Perspective 72

Acknowledgments 73
References 74

3. Germ cell reprogramming **91**
Kazuki Kurimoto and Mitinori Saitou

1. Introduction 92

2. Mouse germ cell reprogramming 93
3. Human germ cell reprogramming 111
4. Conclusion and perspectives 117
Acknowledgment 117
Funding 117
References 117

4. From worm to germ: Germ cell development and regeneration in planarians 127

Melanie Issigonis and Phillip A. Newmark

1. Introduction 128
2. Planarian biology: Some background about the "other" worm 129
3. Germ cell-intrinsic regulation 133
4. Extrinsic regulation via local cues 138
5. Extrinsic regulation via systemic cues 139
6. A blurred line between neoblasts and germ cells 143
References 145

5. Heterogeneity of primordial germ cells 155

Daniel H. Nguyen, Rebecca G. Jaszczak, and Diana J. Laird

1. Introduction 156
2. Heterogeneous phenotypes of primordial germ cells 157
3. Determinants of heterogeneity 173
4. Selection and heterogeneity 184
5. Conclusion 188
References 189

6. Germline stem cell homeostasis 203

Jonathan O. Nelson, Cuie Chen, and Yukiko M. Yamashita

1. Introduction 204
2. Specifying and maintaining germline stem cells 205
3. Asymmetric stem cell division 213
4. Transit-amplifying divisions to reduce the burden on stem cells 218
5. Dedifferentiation maintains GSC populations beyond the lifespan of individual cells 220
6. Cell death in GSC homeostasis 225
7. The disruption of germline homeostasis during aging 228
8. Maintaining the germline over successive generations 229
9. Conclusion 233
Acknowledgments 233
References 233

7. Heterogeneous, dynamic, and stochastic nature of mammalian spermatogenic stem cells **245**

Shosei Yoshida

1. Introduction 246
2. Overview of mouse spermatogenesis in adult testes 248
3. Advantages and available techniques in the study of SSCs 255
4. Identity and dynamics of SSCs 258
5. Microenvironmental regulation of SSCs 268
6. Conclusion 277

Acknowledgments 278

References 278

8. Sperm success and immunity **287**

Stuart Wigby, Susan S. Suarez, Brian P. Lazzaro, Tommaso Pizzari, and Mariana F. Wolfner

1. Introduction 288
2. Sperm success and the male immune system 288
3. Sperm success and the female immune system 294
4. Conclusions 304

Acknowledgments 306

References 306

9. A symbiont's guide to the germline **315**

Shelbi L. Russell, Laura Chappell, and William Sullivan

1. Background: Symbiont transmission modes maintain symbiotic associations 316
2. Soma-to-soma strategies of vertical symbiont transmission 319
3. Vertical transmission through the germline part 1: Migration from the soma to oocytes/embryos 322
4. Vertical transmission through the germline part 2: Migration from the soma to germ stem cells (GSCs) 330
5. Vertical transmission through the germline part 3: Continuous germ cell association throughout development 331
6. Parallels between symbiont transmission and the origin of the germline 339
7. Conclusions 342

References 343

Contributors

Laura Chappell
Department of Molecular, Cell, and Developmental Biology, University of California Santa Cruz, Santa Cruz, CA, United States

Cuie Chen
Life Sciences Institute, Department of Cell and Developmental Biology, Howard Hughes Medical Institute, University of Michigan, Ann Arbor, MI, United States

Melanie Issigonis
Regenerative Biology, Morgridge Institute for Research, Madison, WI, United States

Allison Jamieson-Lucy
Department of Cell and Developmental Biology, University of Pennsylvania Perelman School of Medicine, Philadelphia, PA, United States

Rebecca G. Jaszczak
Department of Obstetrics, Gynecology and Reproductive Science, Center for Reproductive Sciences, Eli and Edythe Broad Center for Regeneration Medicine and Stem Cell Research, University of California, San Francisco, CA, United States

Kazuki Kurimoto
Department of Embryology, Nara Medical University, Nara, Japan

Diana J. Laird
Department of Obstetrics, Gynecology and Reproductive Science, Center for Reproductive Sciences, Eli and Edythe Broad Center for Regeneration Medicine and Stem Cell Research, University of California, San Francisco, CA, United States

Brian P. Lazzaro
Departments of Biomedical Sciences (SSS), Entomology (BPL), and Molecular Biology and Genetics (MFW), Cornell University, Ithaca, NY, United States

Mary C. Mullins
Department of Cell and Developmental Biology, University of Pennsylvania Perelman School of Medicine, Philadelphia, PA, United States

Jonathan O. Nelson
Life Sciences Institute, Department of Cell and Developmental Biology, Howard Hughes Medical Institute, University of Michigan, Ann Arbor, MI, United States

Phillip A. Newmark
Regenerative Biology, Morgridge Institute for Research; Department of Integrative Biology, University of Wisconsin, Madison, WI; Howard Hughes Medical Institute, Chevy Chase, MD, United States

Daniel H. Nguyen
Department of Obstetrics, Gynecology and Reproductive Science, Center for Reproductive Sciences, Eli and Edythe Broad Center for Regeneration Medicine and Stem Cell Research, University of California, San Francisco, CA, United States

Tommaso Pizzari
Edward Grey Institute, Department of Zoology, University of Oxford, Oxford, United Kingdom

Shelbi L. Russell
Department of Molecular, Cell, and Developmental Biology, University of California Santa Cruz, Santa Cruz, CA, United States

Mitinori Saitou
Institute for the Advanced Study of Human Biology (ASHBi); Department of Anatomy and Cell Biology, Graduate School of Medicine; Center for iPS Cell Research and Application, Kyoto University, Kyoto, Japan

Susan S. Suarez
Departments of Biomedical Sciences (SSS), Entomology (BPL), and Molecular Biology and Genetics (MFW), Cornell University, Ithaca, NY, United States

William Sullivan
Department of Molecular, Cell, and Developmental Biology, University of California Santa Cruz, Santa Cruz, CA, United States

M. Azim Surani
Wellcome Trust Cancer Research UK Gurdon Institute; Physiology, Development and Neuroscience Department; Wellcome Trust/Medical Research Council Cambridge Stem Cell Institute, University of Cambridge, Cambridge, United Kingdom

Anastasiya Sybirna
Wellcome Trust Cancer Research UK Gurdon Institute; Physiology, Development and Neuroscience Department; Wellcome Trust/Medical Research Council Cambridge Stem Cell Institute, University of Cambridge, Cambridge, United Kingdom

Stuart Wigby
Edward Grey Institute, Department of Zoology, University of Oxford, Oxford, United Kingdom

Mariana F. Wolfner
Departments of Biomedical Sciences (SSS), Entomology (BPL), and Molecular Biology and Genetics (MFW), Cornell University, Ithaca, NY, United States

Frederick C.K. Wong
Wellcome Trust Cancer Research UK Gurdon Institute; Physiology, Development and Neuroscience Department, University of Cambridge, Cambridge, United Kingdom

Yukiko M. Yamashita
Life Sciences Institute, Department of Cell and Developmental Biology, Howard Hughes Medical Institute, University of Michigan, Ann Arbor, MI, United States

Shosei Yoshida
Division of Germ Cell Biology, National Institute for Basic Biology, National Institutes of Natural Sciences; Department of Basic Biology, School of Life Science, SOKENDAI (Graduate University for Advanced Studies), Okazaki, Aichi, Japan

Preface

Germ cells have the sole, yet fundamental task to generate the next generation. Distinct from all other embryonic cells in the body, which contribute to particular tissues or organs, germ cells, dependent on the sex of the organism, will differentiate into egg or sperm. But for germ cells, differentiation is not an endpoint but a new beginning: upon the union of egg and sperm an entirely new organism will emerge and the cycle can begin anew. The unique role of germ cells makes them at the same time the safekeepers of the genetic blueprint of the species to be passed on from generation and generation, but also the mediators and propagators of heritable change that is needed for organisms to adapt, evolve, and compete.

The essays in this book describe the intriguing biology of germ cells and their strategies for transgenerational survival. The germ cell life cycle has been described in great detail in a number of model organisms, such as flies, worms, fishes, frogs, and mice, and most recently in humans. In general, germ cells acquire their distinct fate during embryogenesis. This can occur either by preformation, as is found in flies, worms, fish, and frogs, whereby cells that inherit maternally synthesized germ plasm are destined to become germ cells, or by induction, as found in mouse, salamanders, and humans, whereby cell-to-cell signaling instills germ cell fate. Often specified at the margin of the embryo proper, germ cells have to migrate to a new destination within the embryo to meet up with the somatic part of the gonad, which provides important developmental signals and niche function to the maturing germ cells. Here, germ cells and their somatic support cells develop according to sexual differentiation cues into testis and ovaries. In most species, germline stem cells or immature precursor cells exist in the testis, which via amplification produce large amounts of sperm throughout adult life. The existence of germline stem cells in the ovary is rarer (observed and studied in *Drosophila* and *C. elegans*). More commonly (observed and studied in mice and humans), the pool of germ cells in females is limited. After a period of amplification, early germ cells enter meiosis and remain arrested at an early differentiated stage of oogenesis for years or decades to await final maturation. Upon fertilization, the life cycle begins anew.

Several excellent reviews and collections of germline-centered articles have been published. The essays in this book summarize some striking, emerging developments in our understanding of germ cell biology and to highlight some intriguing aspects of germline development, which I would like to briefly highlight below.

The oocyte plays a unique part in the development of the organism. In contrast to sperm, the oocyte does not only provide nuclear, genetic information to the offspring but also cytoplasm that contains cytoskeletal specializations and organelles, including mitochondria, which are vital for embryonic development. In Chapter 1, Mary Mullins and colleagues describe a granular structure, called the Balbiani body, that is found in one form or another in the oocytes of species as diverse as flies and mice. Multiple functions have been attributed to this still rather obscure structure, including roles in oocyte maturation and the establishment of embryonic polarity but also functions as an organizing center for the germ plasm and its associated RNAs in animals that specify germ cells by preformation. Despite its mystery, recent genetic, molecular, and functional dissection of components in the Balbiani body and other germline-specific membraneless granules is beginning to reveal common building principles that may lead to a more mechanistic understanding. Oocytes are also the preferred passage for endosymbionts to reach the next generation. In Chapter 9, William Sullivan and colleagues describe some of the fascinating strategies symbionts have chosen to adapt to their host and obtain a safe path for transmission.

Significant progress has been made in identifying the molecular mechanisms leading to germ cell specification, which requires the epigenetic resetting of the genome and repression of somatic genes. Indeed, significant insights have been gained recently through the development of culturing techniques to study germ cell specification during early human embryogenesis (Azim Surani and colleagues, Chapter 2). A comprehensive understanding of the molecular framework for germ cell specification has made it also possible to recapitulate the process in culture. Primordial germ cell-like cells can now be derived from human and mouse pluripotent stem cells and, together with gonadal somatic cells, can be prodded toward differentiation in culture (Chapter 3, Mitinori Saitou and colleagues). Fundamental principles of germ cells specification can also be gained by "newer" model systems such as planarians, which are well-known for their awesome potential to regenerate a complete flatworm from tiny cuttings of their former self. In the planarian *Schmidtea mediterranea*, both sexual and asexual strains exist. In Chapter 4, Phillip Newmark and colleagues describe conserved germ cell-intrinsic factors and somatically derived signals that coordinate germ cell development in the sexual strains. During regeneration, the "germ cell cycle" is broken and dividing neoblast not only regenerates every part of the body but also generates germ cells "de novo." Thus, similar to the emerging findings in mouse and human cultures, the distinction between pluripotency and germ cells may be more plastic than previously thought.

Of all primordial germ cells in the body only a tiny fraction will ever contribute to a germline stem cell or suitable egg or sperm for reproduction. In Chapter 5, Diana Laird and colleagues explore how heterogeneity and selection during different stages of the germline life cycles may influence the success of germ cells and ultimately benefit next generations. Selection, regeneration, and homeostasis are of central importance during spermatogenesis. In Chapters 6 and 7, some of the best-understood germline stem cell systems are discussed: the *Drosophila* and mouse testes. Progress in both systems has been made by linking genetic and molecular analysis with unprecedented advances in life imaging. These two systems exhibit mechanisms that maintain homeostasis between stem cells renewal and differentiating population, that replenish germ line stem cells upon aging or injury and that 'reset' the clock for rejuvenation during successive generation, such as telomere lengthening and rDNA repair that may account for rejuvenation. But these systems have strikingly different morphologies. The fly stem cell niche is a morphologically highly ordered system of germ cells and somatic support cells. At the tip of the testis (or ovary in flies) designated germline stem cells divide predominantly asymmetrically, under the control of signals from the somatic niche, to produce a new stem cell and a differentiating cell, which will go on to produce sperm (or eggs). However, as Yukiko Yamashita and colleagues (Chapter 6) describe, germ cell populations are germ cell populations are more plastic than previously thought and stem cells can be rejuvenated by dedifferentiation of already partially differentiated germ cells. Shosei Yoshida (Chapter 7) provides an in-depth look of spermiogenesis in the mouse that questions and expands some of the established views of this system. In contrast to the fly, the niche system in mouse is made up of a spatially open microenvironment, where sperm stem cells are motile and part of a flexible system of undifferentiated spermatogonia. These gradually commit toward differentiation possibly guided by their susceptibility for signals from the somatic microenvironment. With millions of sperm produced only one is needed to fertilize an egg. Mariana Wolfner and her colleagues discuss in Chapter 8 mechanisms of "sperm success" and the interdependence between female and male reproductive biology. In addition, the immune system plays an important role in reproduction. While needed to maintain a healthy body, its inert defense mechanisms against intruders can have antagonistic effects on sperm success that need to be overcome for successful reproduction.

In conclusion, I have tremendously enjoyed reading these exciting chapters and want to thank the authors not only for their scholarly work but also their courage to be provocative and speculative. With no doubt, enormous progress has been made in understanding germ cell biology. Yet, at the same time, it is humbling to realize that we still lack a firm grasp of understanding even the first principles that underlie immortality. I am indebted to Paul Wassarman, who twisted my arm to undergo this endeavor, to Shellie Bryant, who stood by me during the preparation of this book and to Kristen Dancel-Manning, who created the cover illustration. Finally, a big 'thank you' to my lab members, past and present, who share my love of everything germline.

Ruth Lehmann
HHMI and Kimmel Center for Biology and
Medicine of the Skirball Institute, Department of Cell Biology,
New York University School of Medicine,
New York, NY, United States

CHAPTER ONE

The vertebrate Balbiani body, germ plasm, and oocyte polarity

Allison Jamieson-Lucy, Mary C. Mullins*
Department of Cell and Developmental Biology, University of Pennsylvania Perelman School of Medicine, Philadelphia, PA, United States
*Corresponding author: e-mail address: mullins@pennmedicine.upenn.edu

Contents

1. Introduction 2
2. Inductive and inherited germ cell specification 3
3. Inheritance of germ plasm 6
4. The Balbiani body: A conserved oocyte organelle 8
5. Early oocyte development and the Balbiani body in mice 11
6. Early oocyte development in zebrafish 13
7. The Balbiani body in zebrafish and frogs 15
8. The role of mitochondria in the Balbiani body 17
9. Bucky ball: An essential Balbiani body protein 19
10. Macf1 and Balbiani body disassembly 21
11. Secondary pathway mRNA localization 23
12. Animal pole mRNA localization 23
13. Concluding remarks 24
Acknowledgments 25
References 25

Abstract

The fate of future generations depends on a high-quality germ line. For a female to successfully produce offspring, her oocytes must be successfully specified and their contents meticulously organized. Germ cells are specified by two general mechanisms: inductive and inherited. In the inductive mechanism, the primordial germ cells (PGCs) are induced by signals from the surrounding cells. In the inherited mechanism, PGCs are specified by passing localized germ plasm material from the oocyte to the future germ cells. The Balbiani body, a conserved oocyte aggregate, facilitates the organization of the oocyte into a polarized cell with discrete cytoplasmic domains, including localizing the germ plasm. In the mouse, the Balbiani body is implicated in oocyte survival, while in frogs and zebrafish the Balbiani body carries specific mRNAs to the vegetal pole. These asymmetric mRNAs form the foundation of the functionally polarized oocyte and play important roles in axial patterning and germ plasm formation of the embryo.

Current Topics in Developmental Biology, Volume 135
ISSN 0070-2153
https://doi.org/10.1016/bs.ctdb.2019.04.003

1. Introduction

The specification and development of the germline is a fundamental process across the animal kingdom. Germ cells have two common goals (Aguero, Kassmer, Alberio, Johnson, & King, 2017). The first is to block somatic identity and maintain pluripotency. Any somatic differentiation will destroy the germ cell's ability to generate a complete embryo. Germ cell identity is maintained by blocking translation and protecting pluripotency factors from degradation early in development, until long-term chromatin remodeling can firmly establish germ cell competency (Leitch & Smith, 2013). The second goal is to induce migration to the gonad, where the PGCs can settle as an immortal stem cell population and begin to differentiate and undergo meiosis.

The oocyte has its own additional goals (Cantu & Laird, 2017). Oocytes must arrest partway through meiosis and wait for an indeterminate amount of time, before completing meiosis I typically in the final stage of oogenesis (Tripathi, Kumar, & Chaube, 2010). The oocyte is responsible for passing nearly all of the cytoplasmic material on to the future offspring, including the mitochondria (Marlow, 2017). It controls the sperm entry point and guards against polyspermy, protecting the chromosome copy number (Berois, Arezo, & Papa, 2011; Escobar-Aguirre, Elkouby, & Mullins, 2017). Also the organization within the oocyte contributes in many organisms to patterning the future embryo and specifying the next generation of germ cells (Houston, 2013).

In this chapter, we will review how early stage oocytes develop into complex, highly organized cells. First, we will look at germ plasm, a material generated within the oocyte, that is responsible for producing the next generation of germ cells in many vertebrates. We will contrast that with the germ plasm-independent germ cell specification mechanism found in mammals (Houston & King, 2000b; Strome & Updike, 2015). Second, we will describe the processes of segregating the oocyte into distinct cytoplasmic regions. Organization of the oocyte into a polarized cell is initiated by an aggregate organelle called the Balbiani body (Bb), which sets up its initial asymmetry (Escobar-Aguirre, Elkouby, et al., 2017; Kloc, Jedrzejowska, Tworzydlo, & Bilinski, 2014). This asymmetry is built upon by secondary pathways to further localize essential elements within the oocyte (Micklem, 1995). The final result is an egg that can dictate the sperm entry point, the embryonic axis, and the future germ line.

2. Inductive and inherited germ cell specification

There are two general mechanisms for germ cell specification. The first is by inheritance of maternal factors (Fig. 1B and C). In this mechanism, a specific material called germ plasm is deposited within the oocyte. The cells that inherit this material eventually develop into the germ line (Houston & King, 2000b). Organisms that use this mechanism include insects such as Drosophila (Lehmann, 1992; Mahowald, 2001) and some vertebrates, where it is best studied in zebrafish and frogs (Escobar-Aguirre, Elkouby, et al., 2017; Kloc et al., 2001). The second specification mechanism is cell non-autonomous or inductive. These germ cells are not defined by maternally-deposited material, but instead are induced by the cells around them during embryonic development (Fig. 1A). Mice and axolotl are among the vertebrates using this mechanism (Chatfield et al., 2014; Kumar & DeFalco, 2017).

In organisms with the inductive mechanism, cells from the regions of the embryo that would not normally form germ cells can be forced to form PGCs by application of the correct signaling molecules. In the mouse embryo at the early primitive streak stage (E6.5), distal epiblast cells will normally not form germ cells, while a subset of proximal epiblast cells will differentiate into PGCs. Transplantation of distal cells into the proximal region of the embryo leads to the ability of some of the transplanted cells to contribute to the germ line. Conversely, proximal cells transplanted to the distal region of the embryo generally lose their ability to become PGCs (Tam & Zhou, 1996). Therefore, external signals from the surrounding cells are responsible for specifying the mouse germ line. These local factors have been identified as BMP and Wnt signaling pathway components (Fig. 1A) (Cantú & Laird, 2013). Consistent with this, mouse epiblast-like stem cells (EpiLCs) in culture can be induced to form PGCs with the addition of BMP4 (Hayashi, Ohta, Kurimoto, Aramaki, & Saitou, 2011). Additionally, a time-sensitive pulse of Wnt3a is required for cells to respond to Bmp4 (Ohinata et al., 2009). Importantly, such cultured EpiLCs when implanted into a mouse testis can produce healthy, fertile offspring, demonstrating that Wnt and Bmp together can produce normal and functional sperm from generic embryonic cells (Nagaoka & Saitou, 2017).

The inherited mechanism, conversely, is largely indifferent to the context in which the future germ cells are placed. The maternally-supplied germ plasm is sufficient to induce germ cell fate in any cell regardless of its position.

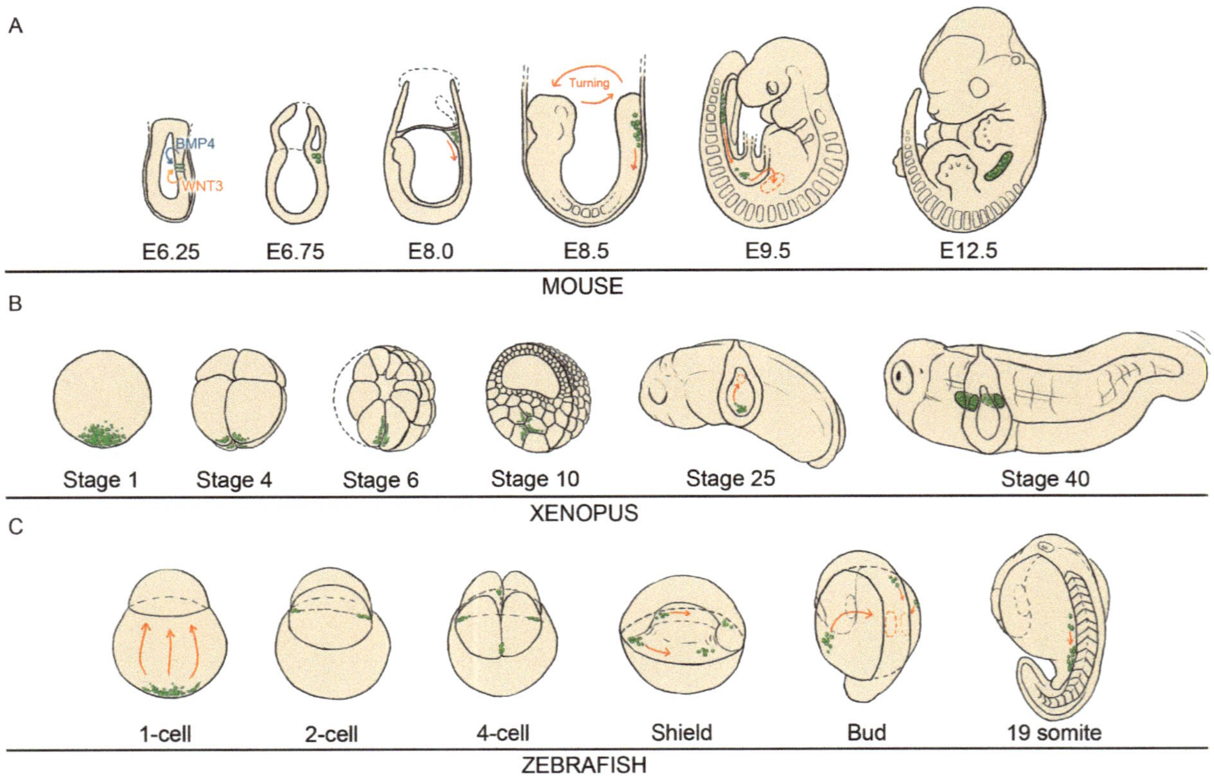

Fig. 1 See legend on opposite page.

The germ plasm (or pole plasm) in Drosophila has long been known to be sufficient to induce the formation of functional germ cells, shown by transplanting the pole plasm from its endogenous location at the posterior pole of the egg to the anterior pole and observing the formation of ectopic germ cells (Illmensee & Mahowald, 1974). In Xenopus embryos, germ plasm taken from the vegetal half of the eight-cell embryo and transplanted into the animal half, where no germ cells would usually form, is sufficient to generate ectopic germ cells. The ectopic germ cells express germ cell markers such as *Xdazl*. However, they do not migrate to the genital ridge. To test if the germ plasm-induced PGCs could form functional germ cells, the PGCs were transplanted a second time to the genital ridge, where they went on to produce functional sperm and eggs (Tada, Mochii, Orii, & Watanabe, 2012).

Over the course of evolutionary history, lineages have switched between inductive and inherited mechanisms of germ cell specification (Fig. 2) (Johnson, Richardson, Bachvarova, & Crother, 2011). Mice are the only mammal whose germ cell specification mechanism has been well-characterized, but it is assumed that the inductive mechanism is used across all mammals (De Felici, 2016). Among reptiles, turtles appear to use the inductive mechanism (Bachvarova, Crother, Manova, et al., 2009), while chickens have germ plasm (Tsunekawa, Naito, Sakai, Nishida, & Noce, 2000). Amphibians are similarly split: salamander germ cell development is inductive (Chatfield et al., 2014; Wakahara, 1996) while in frogs it is inherited

Fig. 1 Illustration of germ cell specification and migration during embryonic development. (A) Mouse: Beginning at embryonic day 6.25 (E6.25), external signals (BMP4, blue; WNT3, yellow) from the surrounding cells begin to specify the primordial germ cell (PGC) precursors, shown in green. The specified PGCs begin to migrate during E8.0 and E8.5. After the turning of the embryo, the PGCs migrate down the hindgut (E9.5). They then exit the endoderm and colonize the gonadal ridge by E12.5. (B) Xenopus: The germ plasm (green) is associated with the cortex at the vegetal pole of the egg (Stage 1), and remains there through the first several divisions (Stage 4). Beginning at Stage 6, the germ plasm moves animally, still associated with the cell margins, into the endodermal region (Stage 10), where the germ plasm specifies the PGCs. Over time, the PGCs migrate apically (Stage 25), before exiting the endoderm to colonize the future gonad. (C) Zebrafish: The germ plasm is initially localized to the vegetal cortex of the 1-cell embryo. It is carried by cytoplasmic streaming to the base of the blastodisc. The germ plasm forms two (2-cell stage), then four (4-cell stage) aggregates at the margin of the cleavage furrows. The four aggregates specify four clusters of PGCs. At shield stage (early gastrula), the PGCs migrate dorsally. The PGCs continue to migrate during gastrulation and somitogenesis until they reach the location of the future gonad at the base of the yolk extension.

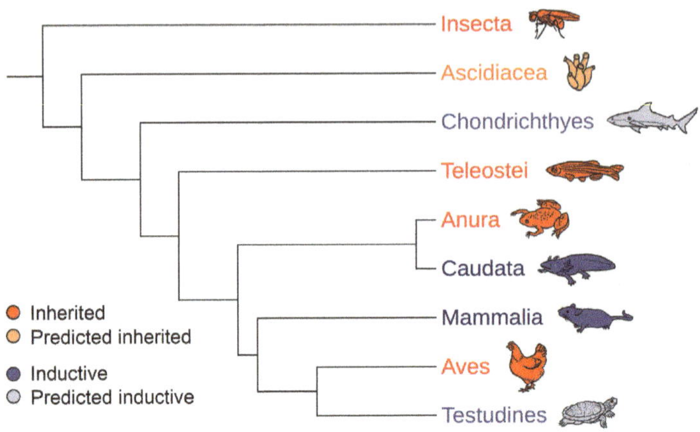

Fig. 2 Phylogenetic tree of inherited versus inductive germ cell specification. Tree generated using Taxonomy Browser (https://www.ncbi.nlm.nih.gov/taxonomy) and Interactive Tree of Life (https://itol.embl.de/).

(Elinson et al., 2011; Kloc et al., 2001). Teleosts such as zebrafish and sturgeon contain germ plasm (Dosch, 2015; Saito et al., 2014); basal fishes such as skates and sharks may use the inductive mechanism, but evidence is scarce overall in these animals (Bachvarova, Crother, & Johnson, 2009). Finally, ascidians, which are basal to chordates, have a segregated subset of cytoplasm called pole plasm, which is assumed to be germ plasm but also plays a role in somatic patterning (Shirae-Kurabayashi et al., 2006).

It is hypothesized that the inductive mechanism came first, and inherited maternal determinants have been derived multiple times in different evolutionary branches (Fig. 2). This is supported by the fact that inherited mechanisms of germ cell specification are relatively diverse in their embryonic location (Bachvarova, Crother, & Johnson, 2009; Johnson & Alberio, 2015). It is still, however, possible that germ plasm is basal, and has been lost several times during evolutionary history, giving way to the inductive mechanism. Ultimately, we may have to adopt a more complex view of germ cell determination as more organisms are studied and the binary of inductive/inherited is explored further.

3. Inheritance of germ plasm

Germ plasm can be first identified within the oocyte as an electron-dense material typically tucked between mitochondria or associated with the endoplasmic reticulum and nuclear envelope (Voronina, Seydoux, Sassone-Corsi, & Nagamori, 2011). It forms irregular granules or fibrillar aggregates, which are also called germ granules or nuage (Gao & Arkov, 2013; Reunov, 2006). This material contains both RNA and proteins: the RNAs include those involved in germline development, such as *vasa*, *dazl*, and *nanos*, as well as non-coding RNAs (Houston, Zhang, Maines, Wasserman, & King, 1998; Kloc et al., 2002; Knaut, Pelegri, Bohmann, Schwarz, & Nusslein-Volhard, 2000; MacArthur, Bubunenko, Houston, & King, 1999). The protein components are predominately members of the piRNA pathway (Schisa, 2012), which protects the integrity of the germline genome by attacking RNA generated by transposons (Arkov & Ramos, 2010; Juliano, Wang, & Lin, 2011). Mutations in elements of the piRNA pathway, such as Piwi and Pili, cause reduced fertility or failure of spermatogenesis (Bao et al., 2014; Houwing, Berezikov, & Ketting, 2008; Houwing et al., 2007).

Studies in flies, zebrafish and mouse have shown that Tudor Domain Containing proteins (Tud) are important for the assembly and structural integrity of the germ plasm (Anne, 2010; Boswell & Mahowald, 1985; Chuma et al., 2006; Roovers et al., 2018; Thomson & Lasko, 2005). Many Tud proteins contain multiple Tudor domains alongside RNA-binding elements. Tud proteins may form multivalent networks by linking multiple binding partners together, acting as nodes in the middle of a net of mRNA and proteins (Arkov, Wang, Ramos, & Lehmann, 2006; Huang et al., 2011).

Germ plasm confers germ cell identity by controlling translation: both by regulating the translation of maternally-supplied mRNAs and repressing the translation of mRNAs that drive somatic cell fate (Lai & King, 2013). Nanos is a particularly well-studied component of the germ plasm, conserved from insects to vertebrates. It is a key player in the maintenance of germ cell fate and is present in a wide range of organisms (Jaruzelska et al., 2003). Nanos functions by repressing the translation of maternally-deposited mRNAs (Lai, Zhou, Luo, Fox, & King, 2011). Nanos forms a complex with Pumilio, which targets mRNAs containing the *cis*-acting Pumilio Binding Element (PBE) in their 3′UTRs (Cheong & Hall, 2006; White, Moore-Jarrett, & Ruley, 2001). Morpholino knockdown of Nanos in Xenopus causes a severe reduction in the number of PGCs and a loss of germ cells from the mature gonad (Lai, Singh, & King, 2012). Mutation of *nanos3* in the zebrafish causes a loss of germline stem cells and fertility in adult homozygous mutant females (Beer & Draper, 2013; Draper, McCallum, & Moens, 2007).

In zebrafish and Xenopus, the germ plasm is asymmetrically distributed to the vegetal cortex of the egg (Fig. 1B and C) (Savage & Danilchik, 1993). In Xenopus, the germ plasm remains in place at the yolky vegetal pole of the zygote for the first cell divisions, and then is segregated into a subset of cells within the endodermal embryonic domain (Ressom & Dixon, 1988). The PGCs are specified before eventually migrating dorsally into the gut region (Fig. 1B) (Lai & King, 2013; Zhou & King, 2004). Zebrafish germ plasm is also vegetally localized in the zygote, but within 30 min of fertilization it is carried to the interface between the yolk and the blastodisc by a combination of parallel microtubule arrays and cytoplasmic streaming (Fig. 1C) (Welch & Pelegri, 2014). During the first embryonic division, the germ plasm coalesces into granules within the cleavage furrow in an F-Actin dependent manner (Eno & Pelegri, 2013, 2018; Knaut et al., 2000; Yoon, Kawakami, & Hopkins, 1997). After the second division, the germ plasm gathers into four points along the cleavage furrows, arranged in the four cardinal directions around the periphery of the embryo (Fig. 1C) (Knaut et al., 2000; Theusch, Brown, & Pelegri, 2006; Yoon et al., 1997). There the germ plasm is incorporated into the future germ cells. The PGCs remain in four clusters until 70% epiboly. At bud stage, the PGCs merge into two clusters and migrate actively through the mesoderm until they reach their destination, and form two rows on either side of the embryo at the base of the yolk extension, where they will eventually become part of the gonad (Lo, Hui, Yu, Wu, & Cheng, 2011; Tarbashevich & Raz, 2010; Wong & Collodi, 2013).

The PGCs will become the germ line stem cells (GSCs) of the testis and ovary. The GSCs perdure in the ovary for varying durations in vertebrates, from rather short periods in mammals to long time-scales in fish (Beer & Draper, 2013; Lei & Spradling, 2013a; Nakamura, Kobayashi, Nishimura, Higashijima, & Tanaka, 2010). The ovarian GSCs produce oogonia that divide multiple times to produce interconnected cystoblasts that give rise to oocytes, which are the focus of the rest of this review.

4. The Balbiani body: A conserved oocyte organelle

A prominent feature in the oocytes of many diverse organisms is the Balbiani body (Bb), a large, non-membrane bound organelle (Oh & Houston, 2017). The Bb has been called many names over the years, such as Balbiani's vesicle or the yolk body of Balbiani after the mentor of the student who discovered the structure. It has also historically been called the yolk nucleus, yolk nucleus complex, nuage body, Dotterkern, and mitochondrial cloud (MC) (Bradley, Estridge, Kloc, Wolfe, & Bilinski, 2001).

The Bb is an aggregate of proteins, germ plasm, mRNA and membrane-bound organelles. In particular, it is enriched for mitochondria, but can also include Golgi and endoplasmic reticulum (ER) (Oh & Houston, 2017). It is distinct from germ granules or P granules, both in contents and in structure. Instead of a liquid droplet composition such as in *C. elegans* P granules (Schuster et al., 2018), the Bb has slower dynamics consistent with a solid aggregate (Chang et al., 2004). Generally it is a single aggregate located on one side of the nucleus. The Bb is only present during the early stages of oogenesis, and disassembles well before oocyte maturation (Wallace & Selman, 1990).

The Bb exists across a wide range of organisms, but is not consistently present in the same form (Fig. 3). Among invertebrates, it has several morphologies. In the oocytes of the tiger beetle, the sea urchin and the black cricket, it is a familiar single round mass (Cakici, 2017; Jaglarz, Nowak, & Bilinski, 2003; Yakovlev, 2016). In the cellar spider, the Bb forms a crescent shape (Fig. 3D) (Jędrzejowska & Kubrakiewicz, 2007). A different species of cricket (*Acheta domesticus*) appears to develop two Balbiani bodies: first an anterior Bb and then a later second posterior Bb (Fig. 3E) (Bradley et al., 2001). In the Drosophila oocyte it is difficult to identify a single coherent Bb; however, the sponge bodies (multiple aggregates of electron-dense nuage and membrane-bound organelles) have similar contents to the Bb and may play a similar role (Cox & Spradling, 2003; Wilsch-Bräuninger, Schwarz, & Nüsslein-Volhard, 1997).

The best functional studies of the Bb have been performed in fish and frogs. Zebrafish and Xenopus oocytes use the Bb to carry germ plasm and localized maternal mRNAs to the cortex. The localization of the germ plasm to the vegetal pole via the Bb is important for the maternally-inherited specification of germ cells. Three dimensional reconstructions from electron microscope slices show no internal organization or subdomains within the Bb; in these organisms it appears to be a homogenous, spherical structure filled with mitochondria (Fig. 3A) (Kloc et al., 2002). Both mRNAs required for germ cell formation, such as *dazl*, and transcripts involved in embryonic patterning such as *syntabulin* and *grip2a* are carried by the Bb (Ge et al., 2014; Houston & King, 2000a; Houston et al., 1998; Nojima et al., 2010).

Several other non-mammalian vertebrates contain a Bb or analogous structure. The Bb has not been observed in sturgeon, but their oocytes may have a parallel structure. A dense ring of cytoplasm packed with organelles, called the Balbiani cytoplasm or granular ooplasm, surrounds the

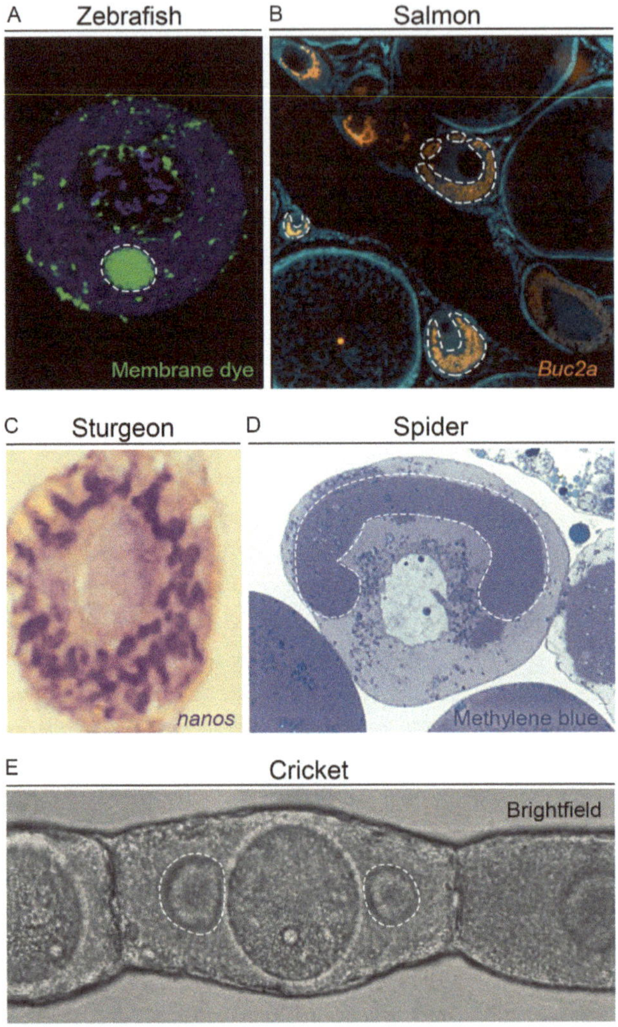

Fig. 3 Comparison of Balbiani body morphologies across species. (A) Zebrafish Balbiani body (dotted outline) stained with DAPI and DiOC6 membrane dye labeling mitochondria. (B) Salmon Balbiani body/granular cytoplasm (dotted outlines) marked by *Buc2a* mRNA (orange). DNA in blue (follicle cells). (C) Sturgeon granular cytoplasm labeled with *nanos* mRNA (purple). (D) Cellar Spider crescent shaped Balbiani body (dotted outline) stained with methylene blue. (E) Cricket oocyte with anterior and posterior Balbiani bodies (dotted outline). Images have been modified to remove original labels.

nucleus of the oocyte, separate from the smooth, organelle-free cytoplasm around the cortex (Fig. 3C). The organelle-rich Balbiani cytoplasm in sturgeon contains several classic Bb mRNAs: *nanos*, *xpat*, and possibly *dazl* (Zelazowska, Kilarski, Bilinski, Podder, & Kloc, 2007). Like zebrafish, sturgeon have material deposited in the vegetal region of the egg that is critical to creating the germ line: irradiating the vegetal pole of sturgeon eggs causes almost complete loss of PGCs in the resulting offspring (Saito, Güralp, Iegorova, Rodina, & Pšenicka, 2018). Salmon have both a Bb in early oocytes and granular ooplasm similar to that in sturgeon in later stages (Fig. 3B). The salmon Bb and granular ooplasm contain resident Bb mRNAs: *buc*, *dnd*, and *dazl* (Škugor, Tveiten, Johnsen, & Andersen, 2016). The Bb has further been observed in lizards, chickens, and other birds (Greenfield, 1966; Guraya, 1962, 1968).

The human Bb appears to have similar contents and structure to the Xenopus and zebrafish Bb (Hertig, 1968). In mice, the Bb does not contain mitochondria. Instead, the Bb is an electron-dense cloud of material surrounded by the Golgi body (Pepling, Wilhelm, O'Hara, Gephardt, & Spradling, 2007). The mitochondria in the mouse oocyte are often, but not always, enriched on the same side of the oocyte as the Golgi-wrapped Bb, but are largely not present within it. The mouse Bb likely influences oocyte quality control and survival (Lei & Spradling, 2016).

The diversity of Bb structures suggests that the Bb is a flexible oocyte organelle that different organisms utilize in different ways. This may explain how the inductive and inherited germ cell specification switched so many times during evolutionary history; the tools for germ plasm generation may be repurposed and maintained instead of lost.

5. Early oocyte development and the Balbiani body in mice

Oocyte development begins in the mouse once the PGCs have migrated to the genital ridge and sex determination has occurred, around E10.5. It begins with division of the GSCs to form oogonia, which then undergo several mitotic divisions with incomplete cytokinesis to form interconnected oocyte cysts, a hallmark of oogenesis (and spermatogenesis) (Fig. 4A) (Elkouby & Mullins, 2017). Over time, these oocyte cysts grow and fragment into clusters of different sizes (E10.5–E14.5) (Lei & Spradling, 2013b). Oocytes within the cysts are joined by intercellular bridges that are stabilized by the TEX14 protein (Greenbaum, Iwamori, Agno, & Matzuk, 2009; Greenbaum, Ma, & Matzuk, 2007). Oocytes often have multiple intercellular bridges, acting as branch points. Over time the number of oocytes decreases, with only about a third of the oocytes surviving in the mature ovary. Over the course of oocyte development,

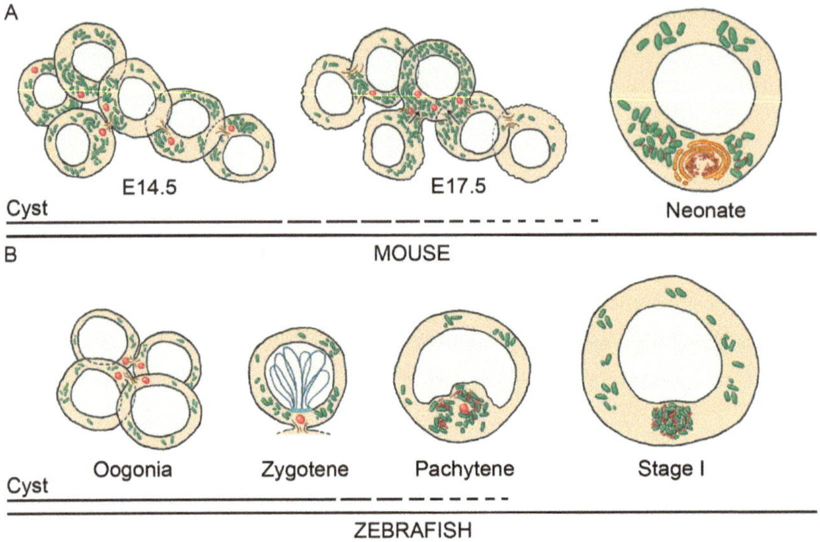

Fig. 4 Schematic comparing the oocyte cyst and Balbiani body of mouse and zebrafish oocytes. (A) Mouse: Early oocytes form cysts connected by intercellular bridges (E14.5). Organelles (green) and the centrosome (magenta) appear to pass through the bridges to enrich central oocytes with organelles, favoring them for survival (E17.5). In the neonatal oocyte, the Balbiani body is formed from the Golgi body (gold) and other organelles. Nuage (purple) is situated between the mitochondria (green). (B) Zebrafish: Oogonia divide to form a cyst of interconnected oocytes. At the zygotene stage, the telomeres in the chromosomal bouquet cluster on the same side of the oocyte as the centrosome (magenta). Organelles and Balbiani body precursors gather around the centrosome, and during the pachytene stage a nuclear cleft forms, within which the Balbiani body precursors aggregate and form the Balbiani body. The mature Balbiani body is present when stage I oocytes reach about 50 μm in diameter.

a widespread culling occurs, leaving a finite pool of high-quality oocytes within the ovary to serve the female's entire reproductive life span (Lei & Spradling, 2013b, 2016). This oocyte bottleneck is shared among mammals; for example, the human ovary also has a limited pool of eggs, which may or may not have regenerative properties (Bukovsky et al., 2005).

As the oocytes mature and begin to undergo meiosis, the percentage of oocytes with multiple intercellular bridges (more than three) increases (E14.5–E17.5) (Lei & Spradling, 2016). This indicates that oocytes at the terminal ends of the branched cyst, which would only have one to two intercellular bridges, are undergoing selective apoptosis, while central "hub" oocytes are preserved (Fig. 4A). As well as having multiple intercellular bridges, the surviving oocytes also have more than one centrosome, with the extra centrosomes thought to be donated by neighboring oocytes

(Lei & Spradling, 2016). Some oocytes are observed to shrink and lose their cytoplasmic contents, including mitochondria and centrosomes, which appear to pass through the cytoplasmic bridges into neighboring oocytes in a microtubule and dynein-dependent manner. The intercellular bridges present between oocytes within the cysts at this developmental stage are large and TEX14-negative. Female mice without TEX14 are fertile, which suggests the traditional TEX14 bridges are not crucial while these large bridges may be. A Balbiani body (Bb) coalesces in each oocyte containing multiple centrosomes (Fig. 4A). These Bb-containing cells lack expression of early apoptosis markers, indicating that they are being selected for survival. It is hypothesized that the dying oocytes are donating their organelles to the central oocytes, which form a Bb and survive to maturity (Lei & Spradling, 2016).

This echoes the way by which the oocyte and sister nurse cells are selected in Drosophila. Early in Drosophila oogenesis, a germline cell divides four times to form 16 cystoblasts, linked together by ring canals (intercellular bridges) and the fusome (McLaughlin & Bratu, 2015). The fusome is a cytoplasmic organelle made of densely packed ER-derived vesicles, situated between the centrioles of each dividing cystoblast. As the cells divide, it becomes densely branched. The fusome is essential to synchronize the divisions of the germ cells, initiate polarity, and to select which of the cystoblasts will become the future oocyte. Inheritance of the fusome contributes to future oocyte identity, possibly by mediating the inheritance of the centrosome (Megraw & Kaufman, 1999). The cell with the most fusome and centrosomes becomes the oocyte; this may mirror how mouse oocytes with multiple centrosomes are favored for survival.

In fish and frogs the Bb continues to play a role after the cyst fragments into individual oocytes, and contributes to the generation of oocyte polarity. It is an open question if the mouse Bb contributes to mouse oocyte development beyond organelle transfer. The murine follicle is a polarized structure, and the germinal vesicle in the mouse oocyte is asymmetrically positioned (Albertini & Barrett, 2004). It is possible that the mouse Bb influences oocyte polarity as development continues. If so, it would prove an exciting connection between the function of the Bb across vertebrates.

6. Early oocyte development in zebrafish

Oocyte development in zebrafish begins when a germline stem cell divides asymmetrically to produce an oogonium. Oogonia continue to divide mitotically, forming interconnected oocyte cysts. A cyst is surrounded by somatic follicle cells; inside, the oogonia are joined by cytoplasmic bridges formed by incomplete cytokinesis (Fig. 4B) (Elkouby & Mullins, 2017).

Unlike in mammals and Drosophila, the oocyte cyst in fish is not known to select oocytes for survival. In medaka there is little apoptosis within the ovarian cyst (Nakamura et al., 2010). More broadly, germ cell cysts are observed in a wide range of organisms where the majority of oocytes proceed to maturity. Intercellular connections between cystoblasts therefore likely have additional functions (Lu, Jensen, Lei, & Yamashita, 2017; Pepling, de Cuevas, & Spradling, 1999). It has been proposed that intercellular bridges maintain cell cycle synchrony throughout the cyst (Lilly, de Cuevas, & Spradling, 2000; Pepling & Spradling, 1998), and/or sensitize the connected germ cells to DNA damage (Lu & Yamashita, 2017).

In the zebrafish, the cytoplasmic bridges connecting oocytes may also provide the initial symmetry-breaking cue within the oocyte (Marlow, 2010). There are suggestions that the centrosome of the early oocyte tends to be oriented facing the cytoplasmic bridges between oocytes, although this has not been rigorously investigated. The cell division orients the centrosome, which subsequently may act as a first step in conferring polarity to the oocyte (Fig. 4B) (Elkouby, Jamieson-Lucy, & Mullins, 2016). It is unknown if there is a fusome-like organelle mediating this connection, as in Drosophila and suggested in Xenopus (Kloc, Bilinski, Dougherty, Brey, & Etkin, 2004; Snapp, Iida, Frescas, Lippincott-Schwartz, & Lilly, 2004).

After the mitotic divisions within the cyst, retinoic acid induces oocytes to enter meiosis I (Rodríguez-Marí et al., 2013). The oocytes develop through leptotene and zygotene meiotic stages within the cyst (Elkouby et al., 2016). During this process the homologous chromosomes must pair together and undergo recombination. This is facilitated by a transient, polarized nuclear architecture called the chromosomal bouquet (Blokhina, Nguyen, Draper, & Burgess, 2019). The telomeres anchor themselves to the nuclear envelope and then gather together on one side of the nucleus, gathering the ends of the chromosomes together like stems with the loops of chromosomes arranged like blooms (Elkouby et al., 2016; Harper, Golubovskaya, & Cande, 2004). The rearrangement of the telomeres shuffles the anchored chromosomes in a way that facilitates correct pairing (Blokhina et al., 2019; Golubovskaya, Harper, Pawlowski, Schichnes, & Cande, 2002). The side of the nucleus where the telomeres gather is the same as the location of the centrosome, which organizes microtubules that aid in the directed transport of the telomere-tethering complex.

Balbiani body (Bb) precursors (Buc protein and mitochondria) first become enriched at this stage, on the same side of the oocyte as the centrosome and the bouquet telomeres (Fig. 4B) (Elkouby et al., 2016).

The Bb begins to coalesce into a more coherent mass at the pachytene stage. At this point, the follicle cells begin to invade the cyst and surround individual oocytes, leading to cyst breakdown. The nucleus forms a pocket, or nuclear cleft, that partially surrounds the Bb components, which collect into a loose cloud around the centrosome. This persists from the pachytene stage to the beginning of the diplotene stage. As the oocyte develops to 30 μm in diameter, the centrosome is lost and the Bb persists, maintaining oocyte asymmetry. As the Bb components become more compacted in a single mass, the cleft gradually becomes less pronounced, and when the oocyte is 40–50 μm in diameter the Bb is mature and the nucleus is spherical again (Fig. 4B) (Elkouby et al., 2016).

7. The Balbiani body in zebrafish and frogs

In zebrafish and Xenopus, the mature Balbiani body (Bb) is initially closely associated with the nucleus. During the course of stage I of oogenesis, the Bb continues to grow, presumably by continuing to pick up its protein and mRNA cargo over time (Chang et al., 2004; Escobar-Aguirre, Zhang, Jamieson-Lucy, & Mullins, 2017; Heasman, Quarmby, & Wylie, 1984; Marlow & Mullins, 2008). The mature Bb is competent to acquire newly introduced cargo: injection of a tagged Bb-specific mRNA into Xenopus oocytes results in its slow enrichment within the Bb (Chang et al., 2004). In the cytoplasm, the tagged mRNA moves freely, but once within the Bb, the mRNA is almost completely immobilized. Photobleaching a region of the fluorescently tagged mRNAs within the Bb results in only about 20% recovery of fluorescence over the course of 6 h (Chang et al., 2004)! Therefore, mRNAs within the Bb move very slowly. These experiments suggest that the Bb acquires its cargo by a diffusion-entrapment mechanism. Interestingly, the injected mRNA becomes enriched in the Bb from the outside in, penetrating only a short way into the center of the aggregate even after 17 h of incubation (Chang et al., 2004). However, *in situs* show a relatively even distribution of mRNAs within the Bb (Kosaka, Kawakami, Sakamoto, & Inoue, 2007), although the high resolution analysis required to detect microdomains has not yet been performed. Endogenous Bb mRNAs permeate the Bb completely, without forming a gradient or a pattern like the growth rings on a tree. This indicates that in the natural context, mRNAs are either constitutively taken up by the Bb and/or have time to spread to the center of the Bb, despite their lethargic movement within the organelle.

The energy state of the oocyte affects the efficiency of mRNA localization to the Bb. Increasing the temperature or ATP concentration within the oocyte increases the speed of *nanos* localization to the Xenopus Bb. Conversely, depleting ATP, Kinesin II or lowering the temperature reduces mRNA accumulation (Heinrich & Deshler, 2009). This suggests that there could also be active transport involved in the localization of mRNAs to the Bb. Alternatively, changing the temperature, energy availability, or motor protein abundance could simply increase or decrease the general movement within the cytoplasm. Faster movement of mRNAs carried by the cytoplasm would give them more chances to encounter the Bb and become trapped, increasing localization speed without requiring directed transport.

The Bb entraps only the subset of mRNAs it will carry to the vegetal pole. Specific mRNAs are directed to the Bb by motifs in their 3′UTR. Several unrelated Bb-localization motifs have been characterized. *nanos* mRNA was the first Bb mRNA where a well-defined localization element (Mitochondrial Cloud Localization Element, MCLE) was described. Deletion analysis narrowed the necessary region to 250 nt at the beginning of the 3′UTR (Zhou & King, 1996). Further investigation revealed that six repeats of the motif UGCAC within this region are essential for localization. The MCLE is sufficient to localize a reporter mRNA to the Bb (Kloc, Bilinski, Pui-Yee Chan, & Etkin, 2000). A few mRNAs in the Xenopus Bb (*xpat* and *xnif*) have similar required localization motifs (Claussen, Horvay, & Pieler, 2004). However, other Bb mRNAs such as *dazl*, *vasa* and others do not have critical MCLE motifs. Deletion analysis of the *dazl* 3′UTR reveals a different 375 nt localization sequence that directs it to the Bb. The same deletion analysis showed that a second element within the 3′UTR is required to maintain the *dazl* mRNA at the oocyte cortex at later stages (Kosaka et al., 2007). Therefore, distinct mRNA sequences and possibly different RNA binding proteins are involved in Bb localization and cortical retention.

Long non-coding mRNAs have been implicated in the tethering of Bb cargo to the vegetal cortex after Bb disassembly. In Xenopus, *short interspersed repeat transcripts* (*xlsirts*) localize to the vegetal cortex (Kloc, Spohr, & Etkin, 1993). This family of transcripts include a well-conserved non-coding repeat region. The conservation of this region implies that it has a function within the oocyte (Zearfoss, Chan, Kloc, Allen, & Etkin, 2003). Destruction of xlsirt in the late Xenopus oocyte causes the vegetally-localized *vg1* transcript to become untethered from the cortex, suggesting that the *xlsirt* transcript itself is important for retaining cargo at the vegetal pole (Kloc & Etkin, 1994). Depleting xlsirt also perturbs the organization of cytokeratins at the oocyte cortex during maturation (Kloc, Bilinski, & Dougherty, 2007; Kloc et al., 2005). Possibly RNAs themselves play a role in the microfiber structure of the oocyte cortex, acting as a tether or crosslinker.

Another RNA-centric tethering mechanism for asymmetrically localized mRNAs is homotypic clustering. In the Drosophila oocyte, mRNAs at the posterior pole organize themselves into small puncta comprised of multiple transcripts of a single mRNA. The mRNAs themselves appear to interact with each other directly, in a novel mechanism of mRNA regulation and storage (Eagle, Yeboah-Kordieh, Niepielko, & Gavis, 2018; Trcek et al., 2015). Recently published data show that mRNA in the germ plasm of the early zebrafish embryo is also organized into homotypic clusters (Eno, Hansen, & Pelegri, 2019). Given the similarity between germ plasm, the Bb, and mRNA granules at the vegetal cortex, it seems likely that homotypic clustering also plays a role in the storage of vegetally-localized mRNAs.

The organization of maternal factors within the oocyte is the first step in both patterning the embryo and ensuring the future germ line. Once transported to the vegetal pole, the Bb mRNAs are retained throughout the rest of oocyte development. Following fertilization, they play critical roles in axial patterning and germ cell development (Escobar-Aguirre, Elkouby, et al., 2017). For example, *syntabulin* and *grip2a* are Bb mRNAs important for specifying dorsal identity in the embryo. In the egg, *syntabulin* is carried toward the prospective dorsal side of the blastodisc along a network of vegetally-localized parallel microtubules that require Grip2a to form (Ge et al., 2014; Nojima et al., 2010). Mutants that lack Syntabulin or Grip2a exhibit ventralized phenotypes due to a lack of dorsal axial tissue specification (Colozza & De Robertis, 2014; Ge et al., 2014; Nojima et al., 2010). Another Bb mRNA, *dazl*, is essential for germ cell specification (Houston & King, 2000a). Maternal mRNAs are therefore essential for the survival of both the first and second generations of offspring.

8. The role of mitochondria in the Balbiani body

Mitochondria are enriched in the Balbiani body (Bb) of most animals, including zebrafish and Xenopus. They are generally distributed evenly throughout the Bb. In zebrafish, the mitochondria are clustered in the Bb but are also present throughout the oocyte cytoplasm (Marlow & Mullins, 2008). Ultrastructural analysis of Xenopus Balbiani bodies shows both branched and bean-shaped mitochondria within the Bb (Kloc et al., 2002). In the mouse Bb, few if any mitochondria are present, but typically mitochondria are enriched on the same side of the oocyte as the Bb (Pepling et al., 2007). However, this does not hold true for all mammals; in the human oocyte, the Bb is strongly enriched with mitochondria (Hertig, 1968). Overall, mitochondria are a hallmark characteristic of the Bb.

Some have hypothesized that the Bb is involved in mitochondrial quality control (Bilinski, Kloc, & Tworzydlo, 2017; Zhou, Wang, Wang, Schatten, & Zhang, 2010). Oocytes play an important role in providing the mitochondria that will be passed along to their offspring, since sperm do not provide mitochondria to the zygote. If a significant number of defective mitochondria are present in the oocyte, a large proportion of the cells of the offspring may inherit low quality mitochondria. This can cause severe disorders in humans (Chiaratti, Garcia, Carvalho, Machado, & Macabelli, 2018). If the Bb is enriched with high quality mitochondria, it could protect or select them, protecting the offspring from mitochondrial disease.

However, the evidence for this hypothesis is contradictory. In silverfish, mitochondria within developing PGCs separate into two subpopulations with different morphologies: branched and bean-shaped. The branched mitochondria associate with nuage, and over time the mitochondrial network in the Balbiani-body-like structure becomes more branched and more complex. Electron microscopy shows that the un-networked mitochondria at the edge of the aggregate degenerate, possibly indicating a mitochondrial selection process (Tworzydlo, Kisiel, Jankowska, Witwicka, & Bilinski, 2016). However, in zebrafish, high activity mitochondria are not enriched within the Bb compared to the cytoplasm. Simultaneous staining of oocytes with a high-activity-staining red dye and low-activity-staining green dye shows an equal ratio of red/green fluorescence in Bb mitochondria and cytoplasmic mitochondria (Zhang et al., 2008). This suggests that the mitochondria within the Bb in the zebrafish are not selected for high activity, although could be selected for other advantageous properties.

It is also possible that the Bb is enriched for mitochondria for other reasons. For example, it may need extra mitochondria to power some energy-demanding process within the Bb. For instance, if there was active translation within the Bb, its energy requirements could partially explain the mitochondrial enrichment. Alas, it remains an open question if transcripts within the Bb are translationally active or repressed. Alternatively, the mitochondria or other membrane-bound organelles, such as the Golgi body in the mouse Bb, could be playing a structural role within the Bb. The mitochondria could act as a scaffold for components acting in Bb formation. In the case of the Golgi, it could act to enclose the Bb contents to keep them from diffusing away. In grasshoppers, small groupings of mitochondria encircle electron-dense filamentous nuage, as if they are forming a cage (Bilinski et al., 2017). In mice, the nuage can also be found cradled between mitochondria within the oocyte (Chuma et al., 2006). Mitochondria could also serve as a support or substrate for nuage aggregates, helping maintain its localization within the Bb. Ultimately, more studies are required to determine the complete function of the enriched mitochondria within the Bb.

9. Bucky ball: An essential Balbiani body protein

Maternal-effect mutants have been a key genetic tool to understand the functional elements of the Balbiani body (Bb). A forward genetic screen in zebrafish identified the *bucky ball* (*buc*) mutant, which produces oocytes completely lacking a Bb (Dosch et al., 2004; Marlow & Mullins, 2008). In these mutants the mRNAs that should be carried by the Bb become evenly distributed throughout the oocyte. Both vegetal and animal mRNAs are affected—without Bucky ball the oocyte effectively lacks all polarity and is radially symmetric (Bontems et al., 2009; Marlow & Mullins, 2008). Misexpression of Bucky ball leads to dorsal-ventral patterning defects and, in a portion of oocytes, the formation of multiple Balbiani bodies (Heim et al., 2014).

Zebrafish females homozygous mutant for *buc* lay eggs that are radially symmetric (Dosch et al., 2004). Instead of being separated into an animal pole where the blastodisc forms and a vegetal pole containing the yolk, these eggs display animal pole characteristics globally. The blastodisc forms all the way around the yolk ball in a thin layer. Maternal *buc*-deficient embryos display multiple sperm entry points (micropyles made by a specialized follicle cell), and are often fertilized multiple times (Marlow & Mullins, 2008). After fertilization the cells are able to divide a handful of times before the embryo lyses.

Both Buc protein and transcript localize to the Bb (Bontems et al., 2009; Heim et al., 2014). Furthermore, Buc is the most abundant protein within the Xenopus Bb (Boke et al., 2016). In the early embryo, it localizes to the cleavage furrows where the germ plasm aggregates, and then later resides within the PGCs as they form (Bontems et al., 2009). The Buc protein is predicted to be largely disordered and has no known structural domains. The only well-conserved regions are an approximately 100 amino acid stretch at the N-terminus of the protein and a short, centrally-located sequence (Bontems et al., 2009; Krishnakumar et al., 2018).

Buc has homologs in other vertebrates, including Xenopus, where the Buc homolog is named XVelo. In the frog, Buc has the same localization patterns as in zebrafish, and we expect plays the same role (Claussen & Pieler, 2004). The gene appears to have been lost in mice, but using alignments of the more conserved N-terminal region and synteny, versions of the gene have been identified in humans and other mammals (Bontems et al., 2009; Škugor et al., 2016).

Buc protein function in the Bb may stem from its ability to self-aggregate. In culture, the Buc analog XVelo can spontaneously form large aggregates (Boke et al., 2016). These aggregates stain positively with Thioflavin-T (Th-T), a dye that labels amyloid-like fibers. Furthermore, electron microscopy reveals that the Buc aggregates are formed of fibrous material. Th-T also stains the Bb in the whole oocyte (Boke et al., 2016). Interestingly, the conserved N-terminal domain of Buc, which has properties of a prion-like domain, is necessary and sufficient to form these aggregates (Boke et al., 2016). Amyloid-like fibers are capable of causing a phase transition to form hydrogels (Kato et al., 2012). In this case, the phase transition is liquid to solid. These results together support a role for Buc in aggregating the Bb via an amyloid-like fiber phase transition mechanism.

Amyloid-like fibers are formed from multiple proteins acting cooperatively. Proteins that would have no stable tertiary structure alone stack together, each contributing a single strand to a single long beta sheet (Kato & McKnight, 2017). These strands are stronger than actin or microtubules, and are linked to neurodegenerative disease (Fukuma, Mostaert, & Jarvis, 2006; Hartl, 2017). The presence of amyloid-like fibers in the Bb is significant because over the course of oocyte development the Bb naturally disassembles. In the degenerative disease context, these aggregates persist and are detrimental. Understanding the Bb may not only prove relevant to oocyte development and the generation of polarity, but also prove informative to disease states.

In addition to self-aggregating, Buc interacts with Tdrd6 and Rbpms2 (Heim et al., 2014; Nijjar & Woodland, 2013; Roovers et al., 2018). Tdrd6 is a multi-tudor domain-containing protein present in both the Bb and the germ plasm. Tudor proteins are known components of germ granules and germ plasm, and most are components of the Piwi pathway. However, Tdrd6 is not required for the piRNA pathway, and is instead important for PGC formation and the structural integrity of the Bb. It interacts with multiple germ plasm mRNAs, including *nanos*, *vasa* and *dazl*. In Tdrd6 mutants, the number of PGCs are reduced and the number of oocytes containing smaller than usual Balbiani bodies or Balbiani bodies that prematurely fragment increases. FRAP experiments assessing the motility of Buc-GFP within the Bb in Tdrd6 mutants show that Buc is less mobile in mutants compared to heterozygous siblings. This indicates that the mobility of Buc either in and out of the Bb or within the Bb may be important to its structural integrity, and that Tdrd6 is involved in Buc mobility (Roovers et al., 2018).

Buc's interaction with Tdrd6 is dependent on dimethylated arginines at the C-terminus of the Buc protein. Without these dimethylated arginines, the Bb does not form properly. The early Bb precursors briefly form puncta, but then disperse and do not coalesce into a mature Bb (Roovers et al., 2018), indicating the importance of this specific interaction.

Despite the importance of mRNA within the Bb, few Bb-localized RNA binding proteins have been identified. Rbpms2 is one of the handful of known resident RNA binding proteins within the Bb (Song et al., 2007; Zearfoss et al., 2003). Rbpms2 binds to Buc (Heim et al., 2014; Nijjar & Woodland, 2013), as well as to Bb mRNAs. In frogs, Rbpms interacts with the Bb mRNAs *nanos* and *Xvelo* (Xenopus *buc* homolog) (Aguero et al., 2016), while in zebrafish it has only been shown to interact with *buc* mRNA (Heim et al., 2014). Given the specific binding affinity of Rbpms2 to some Bb mRNAs, it could tether them to the fibrous Buc network that comprises the Bb structure, slowing their diffusion. Since Rbpms2 mutants develop exclusively into males in the zebrafish, it is difficult to directly test Rbpms2 function in selecting and maintaining cargo in the Bb (Kaufman, Lee, Martin, Rothhamel, & Marlow, 2018).

10. Macf1 and Balbiani body disassembly

While Buc is essential for the formation of the Balbiani body (Bb), Microtubule Actin Crosslinking Factor 1 (Macf1) is critical for disassembly of the Bb. Macf1 is a large protein with multiple domains: an actin-binding domain, a plakin domain, an intermediate-filament binding domain, a series of spectrin repeats, and finally a microtubule-binding domain (Karakesisoglou, Yang, & Fuchs, 2000; Kodama, Karakesisoglou, Wong, Vaezi, & Fuchs, 2003; Lin, Chen, Leung, Parry, & Liem, 2005; Wu, Kodama, & Fuchs, 2008). In zebrafish Macf1 mutants, the Bb does not disassemble (Escobar-Aguirre, Zhang, et al., 2017; Gupta et al., 2010). Instead, it remains in the cytoplasm, not associating with the oocyte cortex, and continues to grow. Additionally, the nucleus becomes asymmetrically localized and frequently the cytoplasm of stage II oocytes appears detached from the cortex. Like in Buc mutants, females lacking Macf1 are healthy, but their offspring display radial animal character (Gupta et al., 2010). In the oocyte, Macf1 protein localizes to the Bb and around the nucleus (Escobar-Aguirre, Zhang, et al., 2017).

The Bb's transition from associated with the nucleus to anchoring to the oocyte cortex is interesting in that it appears to have little to do with directed transport by the cytoskeleton, but still requires the presence of Macf1, a cytoskeletal crosslinker. The Bb is enriched for cytokeratins, but not for actin or microtubules (Gupta et al., 2010). Destabilizing microtubules has little effect on the Bb macrostructure, at least in the short term (Chang et al., 2004; Escobar-Aguirre, Zhang, et al., 2017). Furthermore, the localization of mRNA and mitochondria is also independent of actin and microtubules. However, the requirement of Macf1 for Bb disassembly indicates that the cytoskeleton is involved in that final step (Escobar-Aguirre, Zhang, et al., 2017).

The critical cytoskeletal component for Bb disassembly is actin. The intermediate-filament binding domain of Macf1 is dispensable for Bb disassembly, but deletion of the actin binding domain alone recapitulates the mutant phenotype (Escobar-Aguirre, Zhang, et al., 2017). The depolymerization of actin causes a similar phenotype to the Macf1 mutant. Macf1 likely acts as a crosslinker between the Bb and the actin cortex. Once anchored to the oocyte cortex, the Bb is capable of dissociating. The Bb disassembles into large granules, as if chunks or flakes of the Bb are being pulled away. Bb mRNAs, Buc protein, and mitochondria colocalize within these granules (Escobar-Aguirre, Zhang, et al., 2017). The pattern of mRNA localization at the vegetal pole is also often punctate (Marlow & Mullins, 2008). One possible mechanism of Bb disassembly is that pieces of the Bb are crosslinked to the actin network at the oocyte cortex in a Macf1-dependent manner, and then as the oocyte increases in size these anchor points grow farther apart, physically pulling the Bb apart into pieces that are attached to the vegetal cortex of the oocyte.

11. Secondary pathway mRNA localization

In Xenopus, after the Balbiani body deposits mRNAs at the vegetal cortex, a second, microtubule-mediated mRNA localization pathway kicks in. The mRNAs are carried along a microtubule array that stretches in a wedge from the vegetal nuclear envelope to the vegetal cortex. The dynamics of mRNAs traveling along this pathway are consistent with active transport by motor proteins (Gagnon, Kreiling, Powrie, Wood, & Mowry, 2013; Messitt et al., 2008; Zhou & King, 1996).

This secondary (or late) pathway localizes multiple specific transcripts to the vegetal pole, including *VegT*, a mesoderm-inducing factor of the embryo. The transcript contains a 240-nt localization element (LE) in its 3′UTR, which contains repeated binding motifs for Igf2bp3 (Vera) and Ptbp1. Deleting these motifs disrupts the localization of *VegT* (Bubunenko, Kress, Vempati, Mowry, & King, 2002; Kwon et al., 2002). *VegT* is also bound by Staufen, which was first described in Drosophila as localizing the *oskar* transcript to the posterior pole of the oocyte (St Johnston, Beuchle, & Nusslein-Volhard, 1991). Staufen also binds Kinesin I, which has led to the hypothesis that Staufen functions by cross linking the VegT RNP complex to motor proteins that then carry the mRNA to the vegetal pole (Yoon & Mowry, 2004). At this stage, a subpopulation of microtubules exists with plus-ends directed toward the vegetal pole, which could carry mRNAs (Messitt et al., 2008). Kinesin II is an additional motor protein candidate for the late pathway, as the transcript colocalizes with VegT mRNA during late pathway stages, and depletion of Kinesin II significantly reduces VegT localization (Betley et al., 2004).

12. Animal pole mRNA localization

In zebrafish, Xenopus secondary pathway mRNAs *VegT* and *Vg1* are localized animally instead of vegetally (Bally-Cuif, Schatz, & Ho, 1998; Bontems et al., 2009). Furthermore, Igf2bp3, Stau1 (Staufen), and Kinesin I are involved in the animal pole localization of *cyclin B1*, indicating that there may be parallels between the secondary mRNA localization pathway in Xenopus and the animal pole mRNA localization pathway in zebrafish (Takahashi, Ishii, & Yamashita, 2018; Takahashi, Kotani, Katsu, & Yamashita, 2014).

cyclin B1 is a canonical animal-pole mRNA that is asymmetrically localized in zebrafish, frog and mouse oocytes. The mechanisms for localizing, tethering and storing *cyclin B1* mRNA have all been investigated. *cyclin B1* mRNA forms an RNA-dependent complex with Stau1 along with Pumilio and Igf2bp3 (Kotani, Yasuda, Ota, & Yamashita, 2013; Takahashi et al., 2014). Like the secondary vegetal-localization pathway in Xenopus, the animal localization of cyclin B1 depends on microtubules. In stage IV oocytes, the plus ends of microtubules are anchored at the animal pole and treating oocytes with nocodazole abolishes the localization of a *cyclin B1* reporter mRNA to the animal pole. Injecting anti-Kinesin I or anti-Stau1 antibodies also mitigates the localization of *cyclin B1*. This suggests a possible mechanism where Pumilio, Stau1 and Igf2bp3 form a complex with *cyclin B1* mRNA that then interacts with Kinesin I to actively transport the mRNA to the animal pole along active microtubules (Takahashi et al., 2018). This leaves open the question of how the microtubules become polarized, given that the centrosome has been lost by this stage of oocyte development, and in previous stages microtubules are distributed uniformly.

The maintenance of *cyclin B1* at the animal pole and its translational repression requires two *cis*-acting sequences that Pumilio binds. While the cell cycle is arrested, *cyclin B1* is packed into dense granules that keep it translationally repressed. This repression requires the Pumilio binding sites that are involved in tethering the mRNA to the animal cortex. Mutations in these two elements cause the failure of *cyclin B1* mRNA to remain at the animal pole, and allow precocious translation. These studies show how RNP granules can function in the localization and translational repression of their RNA cargos.

13. Concluding remarks

In comparing oocyte development across vertebrates, it is striking how opposing mechanisms—inductive versus inherited germ cell specification, animal versus vegetal mRNA localization—have more complex relationships than are immediately obvious. Inductive and inherited mechanisms of germ cell specification have switched multiple times in evolution, with the ancestral mechanism still unclear. RNAs with highly defined molecular pathways directing them to the vegetal pole in Xenopus are instead localized to the animal pole in zebrafish, with the zebrafish mechanism bearing significant similarities to the Xenopus pathway despite targeting the opposite pole.

Evolution has repurposed its limited tool box to accomplish many new goals. The Balbiani body represents one such structure that comes in a wide variety of sizes and shapes, with varying functions. Most of its functions remain to be fully investigated, whether it be oocyte survival, polarity, the harboring of germ plasm or embryonic axis determinants. Considering its varying morphologies, it may form, grow, and dissociate by multiple mechanisms or by variations on a common theme. Further studies are clearly warranted to understand this conserved and still rather mysterious oocyte structure.

Acknowledgments

We would like to acknowledge grants from the National Institutes of Health: R01GM117981 to M.C.M. and T32 GM008216 and 1F31GM115066 to A.J-.L.

References

Aguero, T., Kassmer, S., Alberio, R., Johnson, A., & King, M. L. (2017). Mechanisms of vertebrate germ cell determination. *Advances in Experimental Medicine and Biology*, *953*, 383–440.

Aguero, T., Zhou, Y., Kloc, M., Chang, P., Houliston, E., & King, M. L. (2016). Hermes (Rbpms) is a critical component of RNP complexes that sequester germline RNAs during oogenesis. *Journal of Developmental Biology*, *4*(1), 2.

Albertini, D. F., & Barrett, S. L. (2004). The developmental origins of mammalian oocyte polarity. *Seminars in Cell & Developmental Biology*, *15*(5), 599–606.

Anne, J. (2010). Targeting and anchoring Tudor in the pole plasm of the Drosophila oocyte. *PLoS One*, *5*(12), e14362.

Arkov, A. L., & Ramos, A. (2010). Building RNA-protein granules: Insight from the germline. *Trends in Cell Biology*, *20*(8), 482–490.

Arkov, A. L., Wang, J. Y., Ramos, A., & Lehmann, R. (2006). The role of Tudor domains in germline development and polar granule architecture. *Development*, *133*(20), 4053–4062.

Bachvarova, R. F., Crother, B. I., & Johnson, A. D. (2009). Evolution of germ cell development in tetrapods: Comparison of urodeles and amniotes. *Evolution & Development*, *11*(5), 603–609.

Bachvarova, R. F., Crother, B. I., Manova, K., Chatfield, J., Shoemaker, C. M., Crews, D. P., et al. (2009). Expression of Dazl and Vasa in turtle embryos and ovaries: Evidence for inductive specification of germ cells. *Evolution & Development*, *11*(5), 525–534.

Bally-Cuif, L., Schatz, W. J., & Ho, R. K. (1998). Characterization of the zebrafish Orb/CPEB-related RNA binding protein and localization of maternal components in the zebrafish oocyte. *Mechanisms of Development*, 77(1), 31–47.

Bao, J., Zhang, Y., Schuster, A. S., Ortogero, N., Nilsson, E. E., Skinner, M. K., et al. (2014). Conditional inactivation of Miwi2 reveals that MIWI2 is only essential for prospermatogonial development in mice. *Cell Death and Differentiation*, *21*(5), 783–796.

Beer, R. L., & Draper, B. W. (2013). nanos3 maintains germline stem cells and expression of the conserved germline stem cell gene nanos2 in the zebrafish ovary. *Developmental Biology*, *374*(2), 308–318.

Berois, N., Arezo, M. J., & Papa, N. G. (2011). Gamete interactions in teleost fish: The egg envelope. Basic studies and perspectives as environmental biomonitor. *Biological Research*, *44*, 119–124.
Betley, J. N., Heinrich, B., Vernos, I., Sardet, C., Prodon, F., & Deshler, J. O. (2004). Kinesin II mediates Vg1 mRNA transport in Xenopus oocytes. *Current Biology*, *14*(3), 219–224.
Bilinski, S. M., Kloc, M., & Tworzydlo, W. (2017). Selection of mitochondria in female germline cells: Is Balbiani body implicated in this process? *Journal of Assisted Reproduction and Genetics*, *34*(11), 1405–1412.
Blokhina, Y. P., Nguyen, A. D., Draper, B. W., & Burgess, S. M. (2019). The telomere bouquet is a hub where meiotic double-strand breaks, synapsis, and stable homolog juxtaposition are coordinated in the zebrafish, Danio rerio. *PLoS Genetics*, *15*(1), e1007730.
Boke, E., Ruer, M., Wühr, M., Coughlin, M., Lemaitre, R., Gygi, S. P., et al. (2016). Amyloid-like self-assembly of a cellular compartment. *Cell*, *166*(3), 637–650.
Bontems, F., Stein, A., Marlow, F., Lyautey, J., Gupta, T., Mullins, M. C., et al. (2009). Bucky ball organizes germ plasm assembly in zebrafish. *Current Biology*, *19*(5), 414–422.
Boswell, R. E., & Mahowald, A. P. (1985). tudor, a gene required for assembly of the germ plasm in Drosophila melanogaster. *Cell*, *43*(1), 97–104.
Bradley, J. T., Estridge, B. H., Kloc, M., Wolfe, K. G., & Bilinski, S. M. (2001). Balbiani bodies in cricket oocytes: Development, ultrastructure, and presence of localized RNAs. *Differentiation*, *67*(4), 117–127.
Bubunenko, M., Kress, T. L., Vempati, U. D., Mowry, K. L., & King, M. L. (2002). A consensus RNA signal that directs germ layer determinants to the vegetal cortex of Xenopus oocytes. *Developmental Biology*, *248*(1), 82–92.
Bukovsky, A., Caudle, M. R., Svetlikova, M., Wimalasena, J., Ayala, M. E., & Dominguez, R. (2005). Oogenesis in adult mammals, including humans: A review. *Endocrine*, *26*(3), 301–316.
Cakici, O. (2017). Oocyte development in Melanogryllus desertus (Pallas, 1771) (Orthoptera: Gryllidae): Presence of Balbiani body. *Turkish Journal of Zoology*, *41*, 203–208.
Cantú, A. V., & Laird, D. J. (2013). Wnt and Bmp fit germ cells to a T. *Developmental Cell*, *27*(5), 485–487.
Cantu, A. V., & Laird, D. J. (2017). A pilgrim's progress: Seeking meaning in primordial germ cell migration. *Stem Cell Research*, *24*, 181–187.
Chang, P., Torres, J., Lewis, R. A., Mowry, K. L., Houliston, E., & King, M. L. (2004). Localization of RNAs to the mitochondrial cloud in Xenopus oocytes through entrapment and association with endoplasmic reticulum. *Molecular Biology of the Cell*, *15*(10), 4669–4681.
Chatfield, J., O'Reilly, M.-A., Bachvarova, R. F., Ferjentsik, Z., Redwood, C., Walmsley, M., et al. (2014). Stochastic specification of primordial germ cells from mesoderm precursors in axolotl embryos. *Development (Cambridge, England)*, *141*(12), 2429–2440.
Cheong, C. G., & Hall, T. M. (2006). Engineering RNA sequence specificity of Pumilio repeats. *Proceedings of the National Academy of Sciences of the United States of America*, *103*(37), 13635–13639.
Chiaratti, M. R., Garcia, B. M., Carvalho, K. F., Machado, T. S., Ribeiro, F. K. D. S., & Macabelli, C. H. (2018). The role of mitochondria in the female germline: Implications to fertility and inheritance of mitochondrial diseases. *Cell Biology International*, *42*(6), 711–724.
Chuma, S., Hosokawa, M., Kitamura, K., Kasai, S., Fujioka, M., Hiyoshi, M., et al. (2006). Tdrd1/Mtr-1, a tudor-related gene, is essential for male germ-cell differentiation and nuage/germinal granule formation in mice. *Proceedings of the National Academy of Sciences of the United States of America*, *103*(43), 15894–15899.

Claussen, M., Horvay, K., & Pieler, T. (2004). Evidence for overlapping, but not identical, protein machineries operating in vegetal RNA localization along early and late pathways in Xenopus oocytes. *Development, 131*(17), 4263–4273.
Claussen, M., & Pieler, T. (2004). Xvelo1 uses a novel 75-nucleotide signal sequence that drives vegetal localization along the late pathway in Xenopus oocytes. *Developmental Biology, 266*(2), 270–284.
Colozza, G., & De Robertis, E. M. (2014). Maternal syntabulin is required for dorsal axis formation and is a germ plasm component in Xenopus. *Differentiation, 88*, 17–26.
Cox, R. T., & Spradling, A. C. (2003). A Balbiani body and the fusome mediate mitochondrial inheritance during Drosophila oogenesis. *Development, 130*(8), 1579–1590.
De Felici, M. (2016). The formation and migration of primordial germ cells in mouse and man. In R. P. Piprek (Ed.), *Molecular mechanisms of cell differentiation in gonad development* (pp. 23–46). Cham: Springer International Publishing.
Dosch, R. (2015). Next generation mothers: Maternal control of germline development in zebrafish. *Critical Reviews in Biochemistry and Molecular Biology, 50*(1), 54–68.
Dosch, R., Wagner, D. S., Mintzer, K. A., Runke, G., Wiemelt, A. P., & Mullins, M. C. (2004). Maternal control of vertebrate development before the mid blastula transition: Mutants from the zebrafish I. *Developmental Cell, 6*(6), 771–780.
Draper, B. W., McCallum, C. M., & Moens, C. B. (2007). nanos1 is required to maintain oocyte production in adult zebrafish. *Developmental Biology, 305*(2), 589–598.
Eagle, W. V. I., Yeboah-Kordieh, D. K., Niepielko, M. G., & Gavis, E. R. (2018). Distinct cis-acting elements mediate targeting and clustering of Drosophila polar granule mRNAs. *Development, 145*(22), pii: dev.164657.
Elinson, R. P., Sabo, M. C., Fisher, C., Yamaguchi, T., Orii, H., & Nath, K. (2011). Germ plasm in *Eleutherodactylus coqui*, a direct developing frog with large eggs. *EvoDevo, 2*, 20.
Elkouby, Y. M., Jamieson-Lucy, A., & Mullins, M. C. (2016). Oocyte polarization is coupled to the chromosomal bouquet, a conserved polarized nuclear configuration in meiosis. *PLoS Biology, 14*(1), e1002335.
Elkouby, Y. M., & Mullins, M. C. (2017). Coordination of cellular differentiation, polarity, mitosis and meiosis—New findings from early vertebrate oogenesis. *Developmental Biology, 430*(2), 275–287.
Eno, C., Hansen, C. L., & Pelegri, F. (2019). Aggregation, segregation and dispersal of homotypic germ plasm RNPs in the early zebrafish embryo. *Developmental Dynamics, 248*(4), 306–318.
Eno, C., & Pelegri, F. (2013). Gradual recruitment and selective clearing generate germ plasm aggregates in the zebrafish embryo. *Bioarchitecture, 3*(4), 125–132.
Eno, C., & Pelegri, F. (2018). Modulation of F-actin dynamics by maternal mid1ip1L controls germ plasm aggregation and furrow recruitment in the zebrafish embryo. *Development, 145*(10), pii: dev.156596.
Escobar-Aguirre, M., Elkouby, Y. M., & Mullins, M. C. (2017). Localization in oogenesis of maternal regulators of embryonic development. *Advances in Experimental Medicine and Biology, 953*, 173–207.
Escobar-Aguirre, M., Zhang, H., Jamieson-Lucy, A., & Mullins, M. C. (2017). Microtubule-actin crosslinking factor 1 (Macf1) domain function in Balbiani body dissociation and nuclear positioning. *PLoS Genetics, 13*(9), e1006983.
Fukuma, T., Mostaert, A. S., & Jarvis, S. P. (2006). Explanation for the mechanical strength of amyloid fibrils. *Tribology Letters, 22*(3), 233–237.
Gagnon, J. A., Kreiling, J. A., Powrie, E. A., Wood, T. R., & Mowry, K. L. (2013). Directional transport is mediated by a Dynein-dependent step in an RNA localization pathway. *PLoS Biology, 11*(4), e1001551.
Gao, M., & Arkov, A. L. (2013). Next generation organelles: Structure and role of germ granules in the germline. *Molecular Reproduction and Development, 80*(8), 610–623.

Ge, X., Grotjahn, D., Welch, E., Lyman-Gingerich, J., Holguin, C., Dimitrova, E., et al. (2014). Hecate/Grip2a acts to reorganize the cytoskeleton in the symmetry-breaking event of embryonic axis induction. *PLoS Genetics*, *10*(6), e1004422.
Golubovskaya, I. N., Harper, L. C., Pawlowski, W. P., Schichnes, D., & Cande, W. Z. (2002). The pam1 gene is required for meiotic bouquet formation and efficient homologous synapsis in maize (Zea mays L.). *Genetics*, *162*(4), 1979.
Greenbaum, M. P., Iwamori, N., Agno, J. E., & Matzuk, M. M. (2009). Mouse TEX14 is required for embryonic germ cell intercellular bridges but not female fertility. *Biology of Reproduction*, *80*(3), 449–457.
Greenbaum, M. P., Ma, L., & Matzuk, M. M. (2007). Conversion of midbodies into germ cell intercellular bridges. *Developmental Biology*, *305*(2), 389–396.
Greenfield, M. L. (1966). The oocyte of the domestic chicken shortly after hatching, studied by electron microscopy. *Journal of Embryology and Experimental Morphology*, *15*(3), 297.
Gupta, T., Marlow, F. L., Ferriola, D., Mackiewicz, K., Dapprich, J., Monos, D., et al. (2010). Microtubule actin crosslinking factor 1 regulates the Balbiani body and animal-vegetal polarity of the zebrafish oocyte. *PLoS Genetics*, *6*(8), e1001073.
Guraya, S. S. (1962). The structure and function of the so-called yolk-nucleus in the oogenesis of birds. *The Quarterly Journal of Microscopical Science*, *103*(64), 411.
Guraya, S. S. (1968). Further morphological and histochemical studies on the yolk nucleus and associated cell components in the developing oocyte of the Indian wall lizard. *Journal of Morphology*, *124*(3), 283–294.
Harper, L., Golubovskaya, I., & Cande, W. Z. (2004). A bouquet of chromosomes. *Journal of Cell Science*, *117*(18), 4025.
Hartl, F. U. (2017). Protein misfolding diseases. *Annual Review of Biochemistry*, *86*(1), 21–26.
Hayashi, K., Ohta, H., Kurimoto, K., Aramaki, S., & Saitou, M. (2011). Reconstitution of the mouse germ cell specification pathway in culture by pluripotent stem cells. *Cell*, *146*(4), 519–532.
Heasman, J., Quarmby, J., & Wylie, C. C. (1984). The mitochondrial cloud of Xenopus oocytes: The source of germinal granule material. *Developmental Biology*, *105*(2), 458–469.
Heim, A. E., Hartung, O., Rothhamel, S., Ferreira, E., Jenny, A., & Marlow, F. L. (2014). Oocyte polarity requires a Bucky ball-dependent feedback amplification loop. *Development*, *141*(4), 842–854.
Heinrich, B., & Deshler, J. O. (2009). RNA localization to the Balbiani body in Xenopus oocytes is regulated by the energy state of the cell and is facilitated by kinesin II. *RNA (New York, NY)*, *15*(4), 524–536.
Hertig, A. T. (1968). The primary human oocyte: Some observations on the fine structure of Balbiani's vitelline body and the origin of the annulate lamellae. *The American Journal of Anatomy*, *122*(1), 107–137.
Houston, D. W. (2013). Regulation of cell polarity and RNA localization in vertebrate oocytes. *International Review of Cell and Molecular Biology*, *306*, 127–185.
Houston, D. W., & King, M. L. (2000a). A critical role for Xdazl, a germ plasm-localized RNA, in the differentiation of primordial germ cells in Xenopus. *Development*, *127*(3), 447.
Houston, D. W., & King, M. L. (2000b). Germ plasm and molecular determinants of germ cell fate. *Current Topics in Developmental Biology*, *50*, 155–181.
Houston, D. W., Zhang, J., Maines, J. Z., Wasserman, S. A., & King, M. L. (1998). A Xenopus DAZ-like gene encodes an RNA component of germ plasm and is a functional homologue of Drosophila boule. *Development*, *125*(2), 171–180.
Houwing, S., Berezikov, E., & Ketting, R. F. (2008). Zili is required for germ cell differentiation and meiosis in zebrafish. *The EMBO Journal*, *27*(20), 2702–2711.

Houwing, S., Kamminga, L. M., Berezikov, E., Cronembold, D., Girard, A., van den Elst, H., et al. (2007). A role for Piwi and piRNAs in germ cell maintenance and transposon silencing in zebrafish. *Cell*, *129*(1), 69–82.

Huang, H. Y., Houwing, S., Kaaij, L. J., Meppelink, A., Redl, S., Gauci, S., et al. (2011). Tdrd1 acts as a molecular scaffold for Piwi proteins and piRNA targets in zebrafish. *The EMBO Journal*, *30*(16), 3298–3308.

Illmensee, K., & Mahowald, A. P. (1974). Transplantation of posterior polar plasm in Drosophila. Induction of germ cells at the anterior pole of the egg. *Proceedings of the National Academy of Sciences of the United States of America*, *71*(4), 1016–1020.

Jaglarz, M. K., Nowak, Z., & Bilinski, S. M. (2003). The Balbiani body and generation of early asymmetry in the oocyte of a tiger beetle. *Differentiation*, *71*(2), 142–151.

Jaruzelska, J., Kotecki, M., Kusz, K., Spik, A., Firpo, M., & Reijo Pera, R. A. (2003). Conservation of a Pumilio-Nanos complex from Drosophila germ plasm to human germ cells. *Development Genes and Evolution*, *213*(3), 120–126.

Jędrzejowska, I., & Kubrakiewicz, J. (2007). The Balbiani body in the oocytes of a common cellar spider, *Pholcus phalangioides* (Araneae: Pholcidae). *Arthropod Structure & Development*, *36*(3), 317–326.

Johnson, A. D., & Alberio, R. (2015). Primordial germ cells: The first cell lineage or the last cells standing? *Development (Cambridge, England)*, *142*(16), 2730–2739.

Johnson, A. D., Richardson, E., Bachvarova, R. F., & Crother, B. I. (2011). Evolution of the germ line-soma relationship in vertebrate embryos. *Reproduction*, *141*(3), 291–300.

Juliano, C., Wang, J., & Lin, H. (2011). Uniting germline and stem cells: The function of Piwi proteins and the piRNA pathway in diverse organisms. *Annual Review of Genetics*, *45*, 447–469.

Karakesisoglou, I., Yang, Y., & Fuchs, E. (2000). An epidermal plakin that integrates actin and microtubule networks at cellular junctions. *The Journal of Cell Biology*, *149*(1), 195–208.

Kato, M., Han, T. W., Xie, S., Shi, K., Du, X., Wu, L. C., et al. (2012). Cell-free formation of RNA granules: Low complexity sequence domains form dynamic fibers within hydrogels. *Cell*, *149*(4), 753–767.

Kato, M., & McKnight, S. L. (2017). Cross-beta polymerization of low complexity sequence domains. *Cold Spring Harbor Perspectives in Biology*, *9*(3), pii: a023598.

Kaufman, O. H., Lee, K., Martin, M., Rothhamel, S., & Marlow, F. L. (2018). rbpms2 functions in Balbiani body architecture and ovary fate. *PLoS Genet*, *14*(7), e1007489.

Kloc, M., Bilinski, S., Chan, A. P., Allen, L. H., Zearfoss, N. R., & Etkin, L. D. (2001). RNA localization and germ cell determination in Xenopus. *International Review of Cytology*, *203*, 63–91.

Kloc, M., Bilinski, S., & Dougherty, M. T. (2007). Organization of cytokeratin cytoskeleton and germ plasm in the vegetal cortex of *Xenopus laevis* oocytes depends on coding and non-coding RNAs: Three-dimensional and ultrastructural analysis. *Experimental Cell Research*, *313*(8), 1639–1651.

Kloc, M., Bilinski, S., Dougherty, M. T., Brey, E. M., & Etkin, L. D. (2004). Formation, architecture and polarity of female germline cyst in Xenopus. *Developmental Biology*, *266*(1), 43–61.

Kloc, M., Bilinski, S., Pui-Yee Chan, A., & Etkin, L. D. (2000). The targeting of Xcat2 mRNA to the germinal granules depends on a cis-acting germinal granule localization element within the 3'UTR. *Developmental Biology*, *217*(2), 221–229.

Kloc, M., Dougherty, M. T., Bilinski, S., Chan, A. P., Brey, E., King, M. L., et al. (2002). Three-dimensional ultrastructural analysis of RNA distribution within germinal granules of Xenopus. *Developmental Biology*, *241*(1), 79–93.

Kloc, M., & Etkin, L. D. (1994). Delocalization of Vg1 mRNA from the vegetal cortex in Xenopus oocytes after destruction of Xlsirt RNA. *Science*, *265*(5175), 1101–1103.

Kloc, M., Jedrzejowska, I., Tworzydlo, W., & Bilinski, S. M. (2014). Balbiani body, nuage and sponge bodies—The germ plasm pathway players. *Arthropod Structure & Development, 43*(4), 341–348.

Kloc, M., Spohr, G., & Etkin, L. D. (1993). Translocation of repetitive RNA sequences with the germ plasm in Xenopus oocytes. *Science, 262*(5140), 1712–1714.

Kloc, M., Wilk, K., Vargas, D., Shirato, Y., Bilinski, S., & Etkin, L. D. (2005). Potential structural role of non-coding and coding RNAs in the organization of the cytoskeleton at the vegetal cortex of Xenopus oocytes. *Development, 132*(15), 3445–3457.

Knaut, H., Pelegri, F., Bohmann, K., Schwarz, H., & Nusslein-Volhard, C. (2000). Zebrafish vasa RNA but not its protein is a component of the germ plasm and segregates asymmetrically before germline specification. *The Journal of Cell Biology, 149*(4), 875–888.

Kodama, A., Karakesisoglou, I., Wong, E., Vaezi, A., & Fuchs, E. (2003). ACF7: An essential integrator of microtubule dynamics. *Cell, 115*(3), 343–354.

Kosaka, K., Kawakami, K., Sakamoto, H., & Inoue, K. (2007). Spatiotemporal localization of germ plasm RNAs during zebrafish oogenesis. *Mechanisms of Development, 124*(4), 279–289.

Kotani, T., Yasuda, K., Ota, R., & Yamashita, M. (2013). Cyclin B1 mRNA translation is temporally controlled through formation and disassembly of RNA granules. *The Journal of Cell Biology, 202*(7), 1041–1055.

Krishnakumar, P., Riemer, S., Perera, R., Lingner, T., Goloborodko, A., Khalifa, H., et al. (2018). Functional equivalence of germ plasm organizers. *PLoS Genetics, 14*(11), e1007696.

Kumar, D. L., & DeFalco, T. (2017). Of mice and men: In vivo and in vitro studies of primordial germ cell specification. *Seminars in Reproductive Medicine, 35*(2), 139–146.

Kwon, S., Abramson, T., Munro, T. P., John, C. M., Kohrmann, M., & Schnapp, B. J. (2002). UUCAC- and vera-dependent localization of VegT RNA in Xenopus oocytes. *Current Biology, 12*(7), 558–564.

Lai, F., & King, M. L. (2013). Repressive translational control in germ cells. *Molecular Reproduction and Development, 80*(8), 665–676.

Lai, F., Singh, A., & King, M. L. (2012). Xenopus Nanos1 is required to prevent endoderm gene expression and apoptosis in primordial germ cells. *Development, 139*(8), 1476–1486.

Lai, F., Zhou, Y., Luo, X., Fox, J., & King, M. L. (2011). Nanos1 functions as a translational repressor in the Xenopus germline. *Mechanisms of Development, 128*(1–2), 153–163.

Lehmann, R. (1992). Germ-plasm formation and germ-cell determination in Drosophila. *Current Opinion in Genetics & Development, 2*(4), 543–549.

Lei, L., & Spradling, A. C. (2013a). Female mice lack adult germ-line stem cells but sustain oogenesis using stable primordial follicles. *Proceedings of the National Academy of Sciences of the United States of America, 110*(21), 8585–8590.

Lei, L., & Spradling, A. C. (2013b). Mouse primordial germ cells produce cysts that partially fragment prior to meiosis. *Development, 140*(10), 2075–2081.

Lei, L., & Spradling, A. C. (2016). Mouse oocytes differentiate through organelle enrichment from sister cyst germ cells. *Science, 352*(6281), 95.

Leitch, H. G., & Smith, A. (2013). The mammalian germline as a pluripotency cycle. *Development, 140*(12), 2495–2501.

Lilly, M. A., de Cuevas, M., & Spradling, A. C. (2000). Cyclin A associates with the fusome during germline cyst formation in the Drosophila ovary. *Developmental Biology, 218*(1), 53–63.

Lin, C.-M., Chen, H.-J., Leung, C. L., Parry, D. A. D., & Liem, R. K. H. (2005). Microtubule actin crosslinking factor 1b: A novel plakin that localizes to the Golgi complex. *Journal of Cell Science, 118*(16), 3727.

Lo, K. H., Hui, M. N., Yu, R. M., Wu, R. S., & Cheng, S. H. (2011). Hypoxia impairs primordial germ cell migration in zebrafish (*Danio rerio*) embryos. *PLoS One, 6*(9), e24540.

Lu, K., Jensen, L., Lei, L., & Yamashita, Y. M. (2017). Stay connected: A germ cell strategy. *Trends in Genetics, 33*(12), 971–978.

Lu, K. L., & Yamashita, Y. M. (2017). Germ cell connectivity enhances cell death in response to DNA damage in the Drosophila testis. *eLife, 6*, e27960.

MacArthur, H., Bubunenko, M., Houston, D. W., & King, M. L. (1999). Xcat2 RNA is a translationally sequestered germ plasm component in Xenopus. *Mechanisms of Development, 84*(1–2), 75–88.

Mahowald, A. P. (2001). Assembly of the Drosophila germ plasm. *International Review of Cytology, 203*, 187–213. Academic Press.

Marlow, F. L. (2010). Developmental biology. In *Maternal control of development in vertebrates: My mother made me do it!*. San Rafael, CA: Morgan & Claypool Life Sciences Copyright (c) 2010 by Morgan & Claypool Life Sciences.

Marlow, F. L. (2017). Mitochondrial matters: Mitochondrial bottlenecks, self-assembling structures, and entrapment in the female germline. *Stem Cell Research, 21*, 178–186.

Marlow, F. L., & Mullins, M. C. (2008). Bucky ball functions in Balbiani body assembly and animal-vegetal polarity in the oocyte and follicle cell layer in zebrafish. *Developmental Biology, 321*(1), 40–50.

McLaughlin, J. M., & Bratu, D. P. (2015). Drosophila melanogaster oogenesis: An overview. *Methods in Molecular Biology, 1328*, 1–20.

Megraw, T. L., & Kaufman, T. C. (1999). The centrosome in Drosophila oocyte development. In R. E. Palazzo & G. P. Schatten (Eds.), *49. Current topics in developmental biology* (pp. 385–407): Academic Press.

Messitt, T. J., Gagnon, J. A., Kreiling, J. A., Pratt, C. A., Yoon, Y. J., & Mowry, K. L. (2008). Multiple kinesin motors coordinate cytoplasmic RNA transport on a subpopulation of microtubules in Xenopus oocytes. *Developmental Cell, 15*(3), 426–436.

Micklem, D. R. (1995). mRNA localisation during development. *Developmental Biology, 172*(2), 377–395.

Nagaoka, S. I., & Saitou, M. (2017). Reconstitution of female germ cell fate determination and meiotic initiation in mammals. *Cold Spring Harbor Symposia on Quantitative Biology, 82*, 213–222.

Nakamura, S., Kobayashi, K., Nishimura, T., Higashijima, S.-I., & Tanaka, M. (2010). Identification of germline stem cells in the ovary of the teleost medaka. *Science, 328*(5985), 1561.

Nijjar, S., & Woodland, H. R. (2013). Protein interactions in Xenopus germ plasm RNP particles. *PLoS One, 8*(11), e80077.

Nojima, H., Rothhamel, S., Shimizu, T., Kim, C. H., Yonemura, S., Marlow, F. L., et al. (2010). Syntabulin, a motor protein linker, controls dorsal determination. *Development, 137*(6), 923–933.

Oh, D., & Houston, D. W. (2017). RNA localization in the vertebrate oocyte: Establishment of oocyte polarity and localized mRNA assemblages. In M. Kloc (Ed.), *Oocytes: Maternal information and functions* (pp. 189–208). Cham: Springer International Publishing.

Ohinata, Y., Ohta, H., Shigeta, M., Yamanaka, K., Wakayama, T., & Saitou, M. (2009). A signaling principle for the specification of the germ cell lineage in mice. *Cell, 137*(3), 571–584.

Pepling, M. E., de Cuevas, M., & Spradling, A. C. (1999). Germline cysts: A conserved phase of germ cell development? *Trends in Cell Biology, 9*(7), 257–262.

Pepling, M. E., & Spradling, A. C. (1998). Female mouse germ cells form synchronously dividing cysts. *Development, 125*(17), 3323.

Pepling, M. E., Wilhelm, J. E., O'Hara, A. L., Gephardt, G. W., & Spradling, A. C. (2007). Mouse oocytes within germ cell cysts and primordial follicles contain a Balbiani body. *Proceedings of the National Academy of Sciences of the United States of America, 104*(1), 187–192.

Ressom, R. E., & Dixon, K. E. (1988). Relocation and reorganization of germ plasm in Xenopus embryos after fertilization. *Development*, *103*(3), 507–518.

Reunov, A. (2006). Structures related to the germ plasm in mouse. *Zygote*, *14*(3), 231–238.

Rodríguez-Marí, A., Cañestro, C., BreMiller, R. A., Catchen, J. M., Yan, Y.-L., & Postlethwait, J. H. (2013). Retinoic acid metabolic genes, meiosis, and gonadal sex differentiation in zebrafish. *PLoS One*, *8*(9), e73951.

Roovers, E. F., Kaaij, L. J. T., Redl, S., Bronkhorst, A. W., Wiebrands, K., de Jesus Domingues, A. M., et al. (2018). Tdrd6a regulates the aggregation of Buc into functional subcellular compartments that drive germ cell specification. *Developmental Cell*, *46*(3) 285–301.e9.

Saito, T., Güralp, H., Iegorova, V., Rodina, M., & Pšenicka, M. (2018). Elimination of primordial germ cells in sturgeon embryos by ultraviolet irradiation. *Biology of Reproduction*, *99*(3), 556–564.

Saito, T., Psenicka, M., Goto, R., Adachi, S., Inoue, K., Arai, K., et al. (2014). The origin and migration of primordial germ cells in sturgeons. *PLoS One*, *9*(2), e86861.

Savage, R. M., & Danilchik, M. V. (1993). Dynamics of germ plasm localization and its inhibition by ultraviolet irradiation in early cleavage Xenopus embryos. *Developmental Biology*, *157*(2), 371–382.

Schisa, J. A. (2012). New insights into the regulation of RNP granule assembly in oocytes. *International Review of Cell and Molecular Biology*, *295*, 233–289.

Schuster, B. S., Reed, E. H., Parthasarathy, R., Jahnke, C. N., Caldwell, R. M., Bermudez, J. G., et al. (2018). Controllable protein phase separation and modular recruitment to form responsive membraneless organelles. *Nature Communications*, *9*(1), 2985.

Shirae-Kurabayashi, M., Nishikata, T., Takamura, K., Tanaka, K. J., Nakamoto, C., & Nakamura, A. (2006). Dynamic redistribution of vasa homolog and exclusion of somatic cell determinants during germ cell specification in *Ciona intestinalis*. *Development*, *133*(14), 2683–2693.

Škugor, A., Tveiten, H., Johnsen, H., & Andersen, Ø. (2016). Multiplicity of Buc copies in Atlantic salmon contrasts with loss of the germ cell determinant in primates, rodents and axolotl. *BMC Evolutionary Biology*, *16*(1), 232.

Snapp, E. L., Iida, T., Frescas, D., Lippincott-Schwartz, J., & Lilly, M. A. (2004). The fusome mediates intercellular endoplasmic reticulum connectivity in Drosophila ovarian cysts. *Molecular Biology of the Cell*, *15*(10), 4512–4521.

Song, H.-W., Cauffman, K., Chan, A. P., Zhou, Y., King, M. L., Etkin, L. D., et al. (2007). Hermes RNA-binding protein targets RNAs-encoding proteins involved in meiotic maturation, early cleavage, and germline development. *Differentiation*, *75*(6), 519–528.

St Johnston, D., Beuchle, D., & Nusslein-Volhard, C. (1991). Staufen, a gene required to localize maternal RNAs in the Drosophila egg. *Cell*, *66*(1), 51–63.

Strome, S., & Updike, D. (2015). Specifying and protecting germ cell fate. *Nature Reviews. Molecular Cell Biology*, *16*(7), 406–416.

Tada, H., Mochii, M., Orii, H., & Watanabe, K. (2012). Ectopic formation of primordial germ cells by transplantation of the germ plasm: Direct evidence for germ cell determinant in Xenopus. *Developmental Biology*, *371*(1), 86–93.

Takahashi, K., Ishii, K., & Yamashita, M. (2018). Staufen1, kinesin1 and microtubule function in cyclin B1 mRNA transport to the animal polar cytoplasm of zebrafish oocytes. *Biochemical and Biophysical Research Communications*, *503*(4), 2778–2783.

Takahashi, K., Kotani, T., Katsu, Y., & Yamashita, M. (2014). Possible involvement of insulin-like growth factor 2 mRNA-binding protein 3 in zebrafish oocyte maturation as a novel cyclin B1 mRNA-binding protein that represses the translation in immature oocytes. *Biochemical and Biophysical Research Communications*, *448*(1), 22–27.

Tam, P. P. L., & Zhou, S. X. (1996). The allocation of epiblast cells to ectodermal and germ-line lineages is influenced by the position of the cells in the gastrulating mouse embryo. *Developmental Biology, 178*(1), 124–132.

Tarbashevich, K., & Raz, E. (2010). The nuts and bolts of germ-cell migration. *Current Opinion in Cell Biology, 22*(6), 715–721.

Theusch, E. V., Brown, K. J., & Pelegri, F. (2006). Separate pathways of RNA recruitment lead to the compartmentalization of the zebrafish germ plasm. *Developmental Biology, 292*(1), 129–141.

Thomson, T., & Lasko, P. (2005). Tudor and its domains: Germ cell formation from a Tudor perspective. *Cell Research, 15*, 281.

Trcek, T., Grosch, M., York, A., Shroff, H., Lionnet, T., & Lehmann, R. (2015). Drosophila germ granules are structured and contain homotypic mRNA clusters. *Nature Communications, 6*, 7962.

Tripathi, A., Kumar, K. V., & Chaube, S. K. (2010). Meiotic cell cycle arrest in mammalian oocytes. *Journal of Cellular Physiology, 223*(3), 592–600.

Tsunekawa, N., Naito, M., Sakai, Y., Nishida, T., & Noce, T. (2000). Isolation of chicken vasa homolog gene and tracing the origin of primordial germ cells. *Development, 127*(12), 2741.

Tworzydlo, W., Kisiel, E., Jankowska, W., Witwicka, A., & Bilinski, S. M. (2016). Exclusion of dysfunctional mitochondria from Balbiani body during early oogenesis of Thermobia. *Cell and Tissue Research, 366*(1), 191–201.

Voronina, E., Seydoux, G., Sassone-Corsi, P., & Nagamori, I. (2011). RNA granules in germ cells. *Cold Spring Harbor Perspectives in Biology, 3*(12), pii: a002774.

Wakahara, M. (1996). Primordial germ cell development: Is the urodele pattern closer to mammals than to anurans? *The International Journal of Developmental Biology, 40*, 653–659.

Wallace, R. A., & Selman, K. (1990). Ultrastructural aspects of oogenesis and oocyte growth in fish and amphibians. *Journal of Electron Microscopy Technique, 16*(3), 175–201.

Welch, E., & Pelegri, F. (2014). Cortical depth and differential transport of vegetally localized dorsal and germ line determinants in the zebrafish embryo. *Bioarchitecture, 5*(1–2), 13–26.

White, E. K., Moore-Jarrett, T., & Ruley, H. E. (2001). PUM2, a novel murine puf protein, and its consensus RNA-binding site. *RNA, 7*(12), 1855–1866.

Wilsch-Bräuninger, M., Schwarz, H., & Nüsslein-Volhard, C. (1997). A sponge-like structure involved in the association and transport of maternal products during Drosophila oogenesis. *The Journal of Cell Biology, 139*(3), 817–829.

Wong, T. T., & Collodi, P. (2013). Inducible sterilization of zebrafish by disruption of primordial germ cell migration. *PLoS One, 8*(6), e68455.

Wu, X., Kodama, A., & Fuchs, E. (2008). ACF7 regulates cytoskeletal-focal adhesion dynamics and migration and has ATPase activity. *Cell, 135*(1), 137–148.

Yakovlev, K. V. (2016). Localization of germ plasm-related structures during sea urchin oogenesis. *Developmental Dynamics, 245*(1), 56–66.

Yoon, C., Kawakami, K., & Hopkins, N. (1997). Zebrafish vasa homologue RNA is localized to the cleavage planes of 2- and 4-cell-stage embryos and is expressed in the primordial germ cells. *Development, 124*(16), 3157.

Yoon, Y. J., & Mowry, K. L. (2004). Xenopus Staufen is a component of a ribonucleoprotein complex containing Vg1 RNA and kinesin. *Development, 131*(13), 3035–3045.

Zearfoss, N. R., Chan, A. P., Kloc, M., Allen, L. H., & Etkin, L. D. (2003). Identification of new Xlsirt family members in the *Xenopus laevis* oocyte. *Mechanisms of Development, 120*(4), 503–509.

Zelazowska, M., Kilarski, W., Bilinski, S. M., Podder, D. D., & Kloc, M. (2007). Balbiani cytoplasm in oocytes of a primitive fish, the sturgeon *Acipenser gueldenstaedtii*, and its potential homology to the Balbiani body (mitochondrial cloud) of *Xenopus laevis* oocytes. *Cell and Tissue Research, 329*(1), 137–145.

Zhang, Y.-Z., Ouyang, Y.-C., Hou, Y., Schatten, H., Chen, D.-Y., & Sun, Q.-Y. (2008). Mitochondrial behavior during oogenesis in zebrafish: A confocal microscopy analysis. *Development, Growth & Differentiation*, *50*(3), 189–201.
Zhou, Y., & King, M. L. (1996). RNA transport to the vegetal cortex of Xenopus oocytes. *Developmental Biology*, *179*(1), 173–183.
Zhou, Y., & King, M. L. (2004). Sending RNAs into the future: RNA localization and germ cell fate. *IUBMB Life*, *56*(1), 19–27.
Zhou, R. R., Wang, B., Wang, J., Schatten, H., & Zhang, Y. Z. (2010). Is the mitochondrial cloud the selection machinery for preferentially transmitting wild-type mtDNA between generations? Rewinding Muller's ratchet efficiently. *Current Genetics*, *56*(2), 101–107.

CHAPTER TWO

Genetic basis for primordial germ cells specification in mouse and human: Conserved and divergent roles of PRDM and SOX transcription factors

Anastasiya Sybirna[a,b,c,*,†], Frederick C.K. Wong[a,b,†], M. Azim Surani[a,b,c,*]

[a]Wellcome Trust Cancer Research UK Gurdon Institute, University of Cambridge, Cambridge, United Kingdom
[b]Physiology, Development and Neuroscience Department, University of Cambridge, Cambridge, United Kingdom
[c]Wellcome Trust/Medical Research Council Cambridge Stem Cell Institute, University of Cambridge, Cambridge, United Kingdom
*Corresponding authors: e-mail address: as2439@cam.ac.uk; a.surani@gurdon.cam.ac.uk

Contents

1. Introduction 36
2. Early developmental divergence in mammalian embryos 43
3. Modeling peri-implantation development and PGC specification 46
   3.1 PGC-like cell specification from pluripotent cells *in vitro* 46
   3.2 Developmental models based on self-organizing properties of pluripotent cells 51
4. Gene regulatory networks for mouse and human PGC specification 53
   4.1 The genetic basis for mouse PGC specification 56
   4.2 Distinct network for human PGC development 57
   4.3 Differential roles of SOX factors in mouse and human PGC specification 59
   4.4 PRDM family genes in mouse and human PGC specification 65
5. Perspective 72
Acknowledgments 73
References 74

Abstract

Primordial germ cells (PGCs) are embryonic precursors of sperm and egg that pass on genetic and epigenetic information from one generation to the next. In mammals, they

† Equal contribution.

Current Topics in Developmental Biology, Volume 135
ISSN 0070-2153
https://doi.org/10.1016/bs.ctdb.2019.04.004

are induced from a subset of cells in peri-implantation epiblast by BMP signaling from the surrounding tissues. PGCs then initiate a unique developmental program that involves comprehensive epigenetic resetting and repression of somatic genes. This is orchestrated by a set of signaling molecules and transcription factors that promote germ cell identity. Here we review significant findings on mammalian PGC biology, in particular, the genetic basis for PGC specification in mice and human, which has revealed an evolutionary divergence between the two species. We discuss the importance and potential basis for these differences and focus on several examples to illustrate the conserved and divergent roles of critical transcription factors in mouse and human germline.

1. Introduction

In sexually reproducing metazoans genetic and epigenetic information is transmitted by primordial germ cells (PGCs). PGCs are diploid embryonic precursors of gametes that segregate from somatic cells during early embryogenesis via two distinct mechanisms: preformation and epigenesis (Extavour & Akam, 2003). The former is seen in teleost fish, frogs, the fruit fly *Drosophila melanogaster* and the nematode *Caenorhabditis elegans* and relies on maternal inheritance of the germ plasm that contains specific mRNAs and proteins (Extavour & Akam, 2003; Johnson & Alberio, 2015; Strome & Updike, 2015). Epigenesis, seen in mammals, urodele amphibians and the cricket *Gryllus bimaculatus*, involves inductive signaling from the surrounding tissues (Chatfield et al., 2014; Magnúsdóttir & Surani, 2014; Strome & Updike, 2015). Sponges, hydrozoans and planarians display a specific case of epigenesis, where germ cells originate from adult multipotent stem cells (Muller, 2006; Watanabe, Hoang, Mattner, & Holstein, 2009). Nevertheless, PGC development requires a conserved subset of genes, including *Nanos*, *Vasa* and *Piwi* (Juliano, Swartz, & Wessel, 2010).

Direct studies of human PGCs (hPGCs) are precluded by ethical and practical considerations. However, *ex vivo* hPGCs from Week 4 of development (Wk 4, migrating hPGCs), up to Wk 26 (gonadal hPGCs) have been examined (Gkountela et al., 2015; Guo et al., 2015; Li et al., 2017; Tang et al., 2015). Specification of hPGCs likely occurs around Wks 2–3 (Carnegie stage 7) in peri-gastrulation stage embryos (Irie, Sybirna, & Surani, 2018). Historic Carnegie collections contain samples of Wks 2–3 human embryos with putative PGCs identified morphologically (Hill, 2018; Hill & Florian, 1931; Politzer, 1933). Several studies suggested that hPGCs migrated to the primitive gonadal fold from the yolk sac

(Anderson, Copeland, Scholer, Heasman, & Wylie, 2000; De Felici, 2013; Felix, 1912; Fuss, 1911; Politzer, 1930, 1933; Witschi, 1948).

Similar to other lineages, mammalian PGCs have been extensively studied using mouse models. Initially, mouse PGCs (mPGCs) and hPGCs were distinguished from the surrounding somatic tissues as larger cells with prominent nucleoli and darker stained plasma membrane and nucleus (Chiquoine, 1954; Politzer, 1933; Witschi, 1948). Histochemical experiments then showed that mPGCs could be identified by alkaline phosphatase (AP) activity (Chiquoine, 1954). Later, clonal analysis showed that AP-positive cells originated from the proximal epiblast in pre-gastrulation and early-streak stage embryos (Lawson & Hage, 1994). At the end of gastrulation on embryonic day 7.5 (E7.5), a cluster of ~30–40 mPGCs were detected at the extraembryonic mesoderm, which expanded to 61–88 cells by the head-fold stage prior the onset of somitogenesis (Downs & Davies, 1993; Ginsburg, Snow, & McLaren, 1990; Lawson & Hage, 1994). Crucially, the removal of this cluster at E7.5 led to complete absence of mPGCs after a 48-hour (h) culture, suggesting that mPGC expansion observed later is solely due to proliferation (Ginsburg et al., 1990).

Although cell fate decisions occur in the regionalized postimplantation epiblast (Lawson, 1999; Lawson, Meneses, & Pedersen, 1991; Tam & Behringer, 1997), the developmental potency of epiblast cells is not yet determined in pre-somite stage embryos, as heterotopic cell transplantation resulted in adoption of fates typical for the host region (Beddington, 1982; Tam & Zhou, 1996). For example, when cells from the distal epiblast, normally not destined to become mPGCs, were grafted into the proximal epiblast at E6.5, they formed AP-positive mPGCs (Tam & Zhou, 1996). Alternatively, distal epiblast cells could become mPGCs when extraembryonic ectoderm (ExE) was placed adjacent them (Saitou, Barton, & Surani, 2002).

Knockout (KO) studies and *ex vivo* culture revealed that combined action of Wingless/Int-3 (WNT3), bone morphogenetic protein 4 (BMP4), BMP2 and BMP8b signaling is necessary to induce mPGCs (Lawson, 1999; Lawson & Hage, 1994; Ohinata et al., 2009; Ying, Liu, Marble, Lawson, & Zhao, 2000; Ying, Qi, & Zhao, 2001) (Table 1). WNT3/β-catenin pathway is activated in the posterior visceral endoderm at E5.5 and in the epiblast at E5.75, which is required for primitive streak formation (Rivera-Pérez & Magnuson, 2005). WNT3/β-catenin signaling also renders the epiblast cells competent to respond to BMP signaling and form mPGCs (Aramaki et al., 2013; Ohinata et al., 2009). BMP4 is the key PGC-inducing signal emanating from the ExE from E6.0. Its action is strengthened by BMP2

Table 1 Knockout studies in mouse embryos showing genes involved in mPGC development.

Gene	Mutation	Allele	Phenotype in mouse embryo	Germline phenotype	Reference
Bmp2	Homozygous null	*Bmp2*Δ(ex2)	Short allantois, amnion/chorion fusion, delayed and ectopic cardiac development, lethal.	Reduced number at E7.5	Ying and Zhao (2001)
Bmp4	Homozygous null	$Bmp4^{tm1}$	No allantois	Absent at E7.5	Lawson et al. (1999)
	Homozygous null	$Bmp4^{tm1blh}$ Δ(ex3)	Indistinguishable before E5.5, variable retardation in development from E9.5–11.5	Absent at E6.5–7.5	Ohinata et al. (2009) and Winnier, Blessing, Labosky, and Hogan (1995)
Bmp8b	Homozygous null	$Bmp8b^{tm}$	Short or missing allantois, spermatogenesis defects, small testis or infertile	Absent at E7.75	Ying et al. (2000)
	Homozygous null	$Bmp8b^{tm1blh}$ Δ(ex4)	Males and females sterile	Absent at E6.5, very few in E7.5	Ohinata et al. (2009) and Zhao, Deng, Labosky, Liaw, and Hogan (1996)
Smad1	Homozygous null	*Smad1*Δ(MH1)	Allantois development delayed	Absent at E6.5, reduced in E7.5	Hayashi et al. (2002) and Ohinata et al. (2009)
Smad2	Homozygous null	*Smad2*Δ(MH1)	Early embryonic lethal (E8.5), severely morphological defects at E7.5, lacked AP axis	Expansion of Blimp1 + domain at E6.75, abundant PGCs at E8.5	Nomura and Li (1998) and Ohinata et al. (2009)

Smad5	Homozygous null	*Smad5*Δ(ex2)	Enlarged base at Allantois, abberant amnion formation at E8.0, embryonic lethal (E9.5–11.5)	Absent at E8.0–8.5	Chang et al. (1999) and Chang and Matzuk (2001)
Wnt3	Homozygous null	*Wnt3*Δ(ex3–4)	No primitive streak, mesoderm or node, no AP neural patterning	No Blimp1/Prdm14 induction	Aramaki et al. (2013) and Liu et al. (1999)
Ctnnb1	Homozygous null	*Ctnnb1*Δ(ex2–6)	No primitive streak, mesoderm. No AP axis.	No Blimp1 induction	Aramaki et al. (2013) and Grigoryan, Wend, Klaus, and Birchmeier (2008)
FoxH1	Homozygous null	*FoxH1*Δ(ex1)	Embryonic lethal, lack anterior structures, node, severe defects of AP patterning	Expansion of Blimp1 + domain at E6.75	Ohinata et al. (2009) and Yamamoto et al. (2001)
Prdm1 *(Blimp1)*	Homozygous null	*Blimp1*Δ(ex5)	Embryonic lethal (E10.5), defects in caudal branchial arch and labyrinthine layer of placenta	Few abberant AP + cells at E7.5, no repression of Hox genes, no migration	Ohinata et al. (2005) and Vincent et al. (2005)
Prdm14	Homozygous null	*Prdm14*Δ(Venus)	Gross morphology indistinguishable, infertile	Absent at E12.5, reduced numbers at E7.5, no epigenetic reprogramming	Yamaji et al. (2008)

Continued

Table 1 Knockout studies in mouse embryos showing genes involved in mPGC development.—cont'd

Gene	Mutation	Allele	Phenotype in mouse embryo	Germline phenotype	Reference
Tfap2c	Homozygous null	*Mox1*-Cre;*Tfap2c*Δ/Δ or *Alpl*-Cre; *Tfap2c*Δ/Δ or *Sox2*-Cre; *Tfap2c*Δ/Δ	Gross morphology indistinguishable, infertile	Reduced number at E8.0, no migration, smaller testes/ovaries without germ cells	Weber et al. (2010)
Nanog	Homozygous null	*Prdm1*-Cre;*Nanog*Δ/−	Gross morphology indistinguishable	Reduced number at E11.5–12.5	Zhang, Leitch, et al. (2018)
	Conditional KD	*TNAP*-Cre;NRi-Tg or ER-Cre;NRi-Tg	Gross morphology indistinguishable	Increased TUNEL + at E10.5 then reduced number at E12.5	Yamaguchi et al. (2009)
	ESC chimera	RCNβH-B(t)	Gross morphology indistinguishable	No Oct4 +/Mvh + at E12.5	Chambers et al. (2007)
Esrrb	Knockin substitution	$Nanog^{Esrrb\ KI}$	Gross morphology indistinguishable	Restores number at E11.5–12.5	Zhang, Leitch, et al. (2018)
Pou5f1	Homozygous null	*TNAP*-Cre; *Pou5f1*Δ/Δ	Gross morphology indistinguishable, infertile	Reduced number at E10.5–12.5	Kehler et al. (2004)
(Oct4)	ESC chimera	ZHBTc4	Gross morphology indistinguishable	Absent at E7.25	Okamura, Tokitake, Niwa, and Matsui (2008)

Sox2	Homozygous null	*Blimp1*-Cre;*Sox2*Δ/−	Microphthalmia, lethal postnatal within hours	Reduced number at E7.5	Campolo et al. (2013)
		TNAP-Cre;*Sox2*Δ/−	Lethal postnatal within 2 weeks	Reduced number at E11.5, absent of oocyte, follicles and pro-spermatogonia	
		Spo11-Cre;Sox2Δ/−	Gross morphology indistinguishable. fertile, no effect on zygotes	No effect on spermatocyte or oocyte	
Sox17	Homozygous null	*Sox17*Δ(ex2–4)	Defective formation of definitive endoderm, no axis turning at E9.5, poor posterior trunk development, embryonic lethal before E10.5	No effect at E8.0, abberant migration to yolk sac mesoderm	Hara et al. (2009), Kanai-Azuma et al. (2002), and Sakamoto et al. (2007)
Otx2	Homozygous null	*Otx2*Δ(ex1–2)	Severe morphologiical defects at E7.5, delayed gastrulation. Absent of forebrain and midbrain region at E9.5, embryonic lethal	Increased percentage at E7.5	Acampora et al. (1995) and Zhang, Zhang, et al. (2018)

Alleles shown describe mutation strategies in mouse embryos or in knockout cell lines used in chimeric studies. Superscripts denote the nomenclature of the knockout loci. Δ denotes a deletion of the indicated exon(s)/protein domain or, in the case of Cre-containing strains, Cre-mediated excision. In studies where allelic series were described, only homozygous null phenotype is included.

produced in the visceral endoderm (VE) from around E5.5 (Ying & Zhao, 2001). BMP8b is also secreted by the ExE from E6.0, but its role is to restrict the growth of anterior visceral endoderm (AVE) that produces WNT and BMP signaling antagonists, DKK1 and CER1, respectively (Ohinata et al., 2009; Ying et al., 2000; Ying & Zhao, 2001). BMP4 and BMP2, but not BMP8b, are sufficient to induce mouse and human PGC-like cells (PGCLCs) *in vitro* (Hayashi, Ohta, Kurimoto, Aramaki, & Saitou, 2011; Irie et al., 2015; Ohinata et al., 2009). BMP2, BMP4 and BMP8B all display a dose-dependent effect, as the number of mPGCs decreases proportionally to the loss of respective gene alleles (Lawson et al., 1999; Ying et al., 2000, 2001; Ying & Zhao, 2001). BMPs act via SMAD1 and SMAD5 in this context, and deletion of corresponding genes leads to abrogation of mPGC specification (Table 1) (Chang & Matzuk, 2001; Hayashi et al., 2002; Ohinata et al., 2009).

Specified mPGCs migrate along the hindgut to the genital ridges and undergo global epigenetic reprogramming involving genome-wide DNA demethylation (including imprint erasure), progressive reduction in H3K9me2 and increase in H3K27me3 (Hajkova et al., 2008; Seki et al., 2005, 2007), which altogether is thought to be prerequisite for totipotency (Leitch et al., 2013; Tang et al., 2015). Sexual differentiation of both PGCs and the gonadal cells commences soon after colonization (Kim & Capel, 2006). Male gonadal PGCs (also known as gonocytes) become quiescent in G0/G1 phases of the cell cycle after several mitotic divisions (Western, Miles, van den Bergen, Burton, & Sinclair, 2008). After birth, these cells migrate to the basement membrane of the seminiferous tubule, resume proliferation and become spermatogonial stem cells (SSC), responsible for replenishing spermatogenesis (Brinster, 2002). Female gonadal PGCs arrest in the diplotene of prophase I of meiosis (Mehlmann, 2005); Meiosis I resumes at puberty and meiosis II is completed upon fertilization.

PGC specification and ensuing molecular events are orchestrated by a set of transcription factors (TFs), among which PRDM14, BLIMP1 and TFAP2C are the most critical in mice (Kurimoto, Yamaji, Seki, & Saitou, 2008; Magnúsdóttir et al., 2013; Ohinata et al., 2005; Weber et al., 2010). Together they are necessary and sufficient to induce mPGC fate *in vivo* and *in vitro* (Magnúsdóttir et al., 2013; Nakaki et al., 2013). In contrast, acquisition of hPGC fate requires SOX17 but not SOX2 expression. It has been hypothesized that perigastrulation embryo morphology might influence the spatio-temporal signaling events that lead to PGCs specification (Irie et al., 2018; Tang, Kobayashi, Irie, Dietmann, & Surani, 2016). Indeed, rodents typically form a cup-shaped egg cylinder, unlike other mammals (including primates) that feature a flat bilaminar disc epiblast (Boroviak & Nichols, 2017).

2. Early developmental divergence in mammalian embryos

While early embryonic development in many mammalian species follows a similar sequence of events from blastocyst formation to gastrulation, the timing, structural, and molecular mechanisms have diverged (Boroviak & Nichols, 2017; Kobayashi & Surani, 2018; Tang et al., 2016). In humans and other primates, developmental events are protracted when compared to the mouse, including the timing of zygotic genome activation (ZGA) (Fig. 1), which occurs at the 2-cell (2C) stage in mouse embryos (Flach, Johnson, Braude, Taylor, & Bolton, 1982) and between the 4–8C stage in human embryos (Braude, Bolton, & Moore, 1988). Consistently, fully formed blastocysts are hatched by E4.5 in mice, compared to E7 in humans.

Many mammalian blastocysts are morphologically similar, comprising 3 cell lineages: the (preimplantation) epiblast, primitive endoderm (PrE) and trophectoderm (TE), but some transcriptomic differences are evident (Blakeley et al., 2015; Boroviak et al., 2015, 2018; Petropoulos et al., 2016; Xue et al., 2013). For example, *Klf4, Tfap2c* and *Pou5f1* (encoding OCT4) were detected earlier in mouse embryos (1–2C) compared to their human orthologues (present at 8C), whereas *Gata6* expression was detected at the same stage (8C) in both (Blakeley et al., 2015). Importantly, expression of some genes is species-specific: *Klf2* is present in mouse but not primate embryos, while *KLF17* expression is detected in the inner cell mass (ICM) of primates (Blakeley et al., 2015). Also, while *Esrrb* expression was detected in mouse and marmoset ICM, *ESRRB* was only present in extraembryonic lineages of human blastocysts (Blakeley et al., 2015; Boroviak et al., 2015). Furthermore, while the segregation between $CDX2^{+}$ TE and $OCT4^{+}$ epiblast occurs shortly after the segregation of inside (ICM) and outside cells of the mouse morula (E2.5) (Saiz & Plusa, 2013), spatial restriction of these cell fates does not occur until the blastocyst stage (E6) in humans (Niakan & Eggan, 2013; Petropoulos et al., 2016). Signaling control of lineage segregation may not be conserved across mammals, as WNT signaling may play a role in PrE maintenance in marmoset but not in mouse embryos (Boroviak et al., 2015). These studies show that mammalian developmental events do not have to occur in a precise sequence to establish the basic embryonic structure.

This notion is exemplified later when the epiblast in humans and many other non-rodent mammals undergo epithelialization and develop into a planar epithelium after implantation (E7–8). This flat structure is often

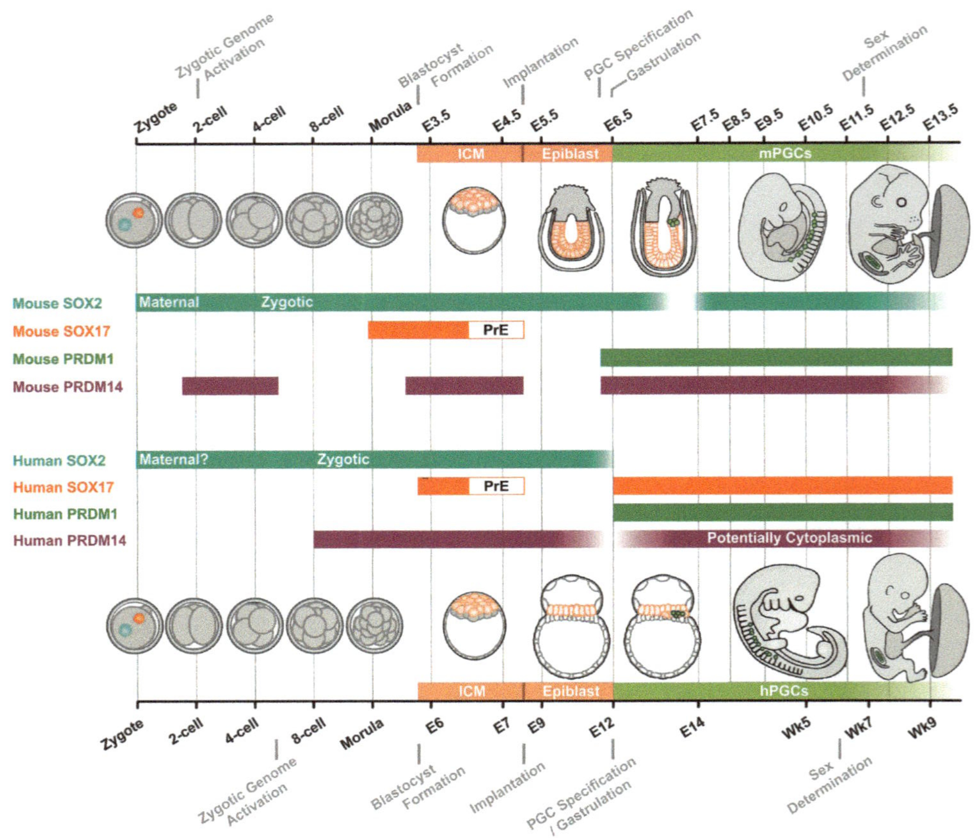

Fig. 1 See legend on opposite page.

referred to as bilaminar disc comprising the (postimplantation) epiblast and the hypoblast (a PrE-derivative equivalent to the VE in mouse). Several other mammals, including rabbit, pig, cow, sheep, cat, dog, primates and even a basal rodent *Lagostomus maximus*, all form a bilaminar disc embryo (Boroviak & Nichols, 2017; Eakin & Behringer, 2004; Kobayashi et al., 2017; Leopardo & Vitullo, 2017; Sasaki et al., 2016; Sheng, 2015). A fundamental difference is that in mouse embryos the TE cells proximal to the epiblast-rosette expand and polarize to form the ExE, but is not the case in non-rodent mammals. Furthermore, the *Lagostomus maxiumus* and *Cavia porcellus* (guinea pig) rodents form ExE, which is located far apart from their epiblast (Chuva de Sousa Lopes & Roelen, 2008; Eakin & Behringer, 2004; Leopardo & Vitullo, 2017) indicating further divergence within rodents.

Primate embryos establish extraembryonic lineages including the amnion during the first 2–3 days after implantation. The amnion is a squamous layer of epithelial cells lining the proximal TE and is directly segregated from the epiblast (Boroviak & Nichols, 2017; Luckett, 1975; Sasaki et al., 2016). Rhesus monkey embryos form the amnion epithelium a day later by upfolding of epiblast margins in contrast to epiblast cavitation in humans (Table 2) (Luckett, 1975). Other mammals such as rabbits and pigs do not form amnion in the same way. The polar TE degenerates concomitantly to epiblast epithelialization, leaving the epiblast exposed like in avian and reptilian embryos (Sheng, 2015). Their amniotic structure is formed

Fig. 1 Developmental timelines of mouse and human early development showing the dynamics of SOX2, SOX17, PRDM1 and PRDM14 expression. Detectable mouse SOX2 protein and *Sox2* transcripts are found from maternal contribution before zygotic expression of SOX2 initiates after ZGA. Mouse SOX2 is transiently downregulated in specifying PGCs before being re-expressed (Yabuta, Kurimoto, Ohinata, Seki, & Saitou, 2006). In humans, SOX2 is downregulated early in hPGCLC specification around the time when SOX17 is upregulated (Irie et al., 2015). Mouse SOX17 was first detected heterogeneously in the mouse morula (solid block) before becoming restricted to the PrE in expanded blastocysts (orange outline) similarly to human embryos except human SOX17 was only first detected in early cavitating blastocysts (Niakan & Eggan, 2013). Human PRDM14 is first detected in 8C embryos (Yan et al., 2013). It is also expressed in gonadal hPGCs, albeit potentially in the cytoplasm (Irie et al., 2015). PRDM14 levels in human peri-implantation embryos and nascent hPGCs are unknown, but it is expressed in their *in vitro* counterparts (conventional and competent hESCs, as well as hPGCLCs). Expression after implantation was only shown in the pluripotent epiblast, therefore early BLIMP1 expression in the visceral endoderm (VE) is not shown. Similarly, expression after gastrulation was only shown in cells adopting PGC fate.

later by an amniochorionic fold extending from the posterior epiblast border at gastrulation stages (Hassoun et al., 2009, 2010) similar to mouse embryos (Table 2) (Pereira et al., 2011). Furthermore, the timing of implantation in rabbits and pigs is uncoupled with the morphogenetic changes following the blastocyst stage. Rabbit embryos undergo gastrulation peri-implantation, whereas pig embryos implant approximately 2 days after the primitive streak is fully formed (Table 2) (Hassoun et al., 2010; Hopf et al., 2011; Irie et al., 2014; Kobayashi et al., 2017; Ramos-Ibeas et al., 2018). These observations emphasize the notion that developmental events are not necessarily interlinked to achieve the same developmental objectives.

Given considerable species-specific diversity during early mammalian development, how does one investigate human development indirectly using a model? While using evolutionary closer organisms is practically feasible (Boroviak et al., 2015, 2018; Sasaki et al., 2016), it will be difficult to rule-out small but significant species-specific differences even among primates (Table 2). Alternatively, one could directly investigate human development *in vitro* (Shao et al., 2017; ten Berge et al., 2008; Warmflash, Sorre, Etoc, Siggia, & Brivanlou, 2014) or ideally, with explicit ethical and legal approvals in place, *ex vivo* (Deglincerti et al., 2016; Shahbazi et al., 2016). Interestingly, the $OCT4^+$ epiblast population of *ex vivo* cultured blastocysts seems to have initiated rosette-organization and lumenogenesis that morphologically resembled a proamniotic cavity as seen in E9 human embryos based on immunofluorescent and nuclei morphological analysis (Deglincerti et al., 2016; Shahbazi et al., 2016). However, further culture resulted in the collapse of the embryo. It is likely that environmental requirements differ between mouse and human embryos. Indeed, marmoset blastocysts exhibit differential response to WNT signaling compared to mouse (Boroviak et al., 2015). Nonetheless transcriptomic analysis of the surviving epiblast population could reveal developmental progression.

3. Modeling peri-implantation development and PGC specification

3.1 PGC-like cell specification from pluripotent cells *in vitro*

Derivation of human embryonic stem cells (hESCs) from the ICM of the blastocyst, or somatic cell reprogramming to induced pluripotent stem cells (hiPSCs), can be used to model early human development (Rossant & Tam, 2017), including PGC specification (Irie et al., 2018). Early attempts to

Table 2 Key developmental events in pigs, rabbits and in selected rodents and primates.

Common name	Species	Order	Expanded blastocyst	Implantation	Polar TE	ExE presence	Epiblast morphology	Amnion formation and timing	Onset of gastrulation	Earliest described stage with germ cells identified	PGC expression of SOX2 or SOX17	References
Mouse	*Mus musculus*	Rodentia	E3.5–4	E4.5–5	Yes	Yes, adjacent	Cup-shape	Amniochorionic fold (E7.5)	E6.5	Gastrula stage/ pregastrula stage (precursors)	SOX2	Kojima, Tam, and Tam (2014), Ohinata et al. (2009), Pereira et al. (2011), and Saiz and Plusa (2013)
Rat	*Rattus rattus*	Rodentia	E4.5–5	E5.5–6	Yes	Yes, adjacent	Cup-shape	Amniochorionic fold (E8.5)	E7.5	Gastrula stage	SOX2	Leitch et al. (2010) and Schneider and Norton (1979)
Guinea pig	*Cavia porcellus*	Rodentia	E5–6	E6–7	Yes	Yes, far away	Cup-shape	Epiblast cavitation/ mesoderm migration	E12–13	Early somite stage		Bookhout (1937), Chuva de Sousa Lopes and Roelen (2008), Eakin and Behringer (2004), and Sansom and Hill (1931)
Plains vizcacha	*Lagostomus maximus*	Rodentia	E18	E18–22	Yes	Yes, far away	Flat disc	Epiblast cavitation (E26–32)	E32–39	Neural plate stage	SOX17	Leopardo and Vitullo (2017) and Roberts and Weir (1973)
Humans	*Homo sapiens*	Primates	E6–7	E7–8	Yes (CT)	No	Flat disc	Epiblast cavitation (E9)	E12–14	Gastrula stage	SOX17	Boroviak and Nichols (2017), Irie, Tang, and Azim Surani (2014), Luckett (1975), and Politzer (1933)
Rhesus macaque	*Macaca mulatta*	Primates	E8	E9	Yes (CT)	No	Flat disc	Epiblast margins upfolding (E11)	E12–13			Boroviak and Nichols (2017) and Luckett (1975, 1978)
Cynomolgus monkey	*Macaca fascicularis*	Primates	E7–9	E9–10	Yes (CT)	No	Flat disc	Epiblast cavitation (E11)	E13–14	Pregastrula stage (in amnion)	SOX17	Nakamura et al. (2016) and Sasaki et al. (2016)

Continued

Table 2 Key developmental events in pigs, rabbits and in selected rodents and primates.—cont'd

Common name	Species	Order	Expanded blastocyst	Implantation	Polar TE	ExE presence	Epiblast morphology	Amnion formation and timing	Onset of gastrulation	Earliest described stage with germ cells identified	PGC expression of SOX2 or SOX17	References
Marmoset	*Callithrix jacchus*	Primates	E7–8	E10–12	Yes (CT)	No	Flat disc	Epiblast cavitation (E12–15)	E20	Early Somite stage (17S, E50)		Aeckerle, Drummer, Debowski, Viebahn, and Behr (2015), Boroviak et al. (2015), Boroviak and Nichols (2017), Enders & Lopata (1999), Hearn and Summers (1986), and Moore, Gems, and Hearn (1985)
Pig	*Sus scrofa domesticus*	Artiodactyla	E8	E14	Degenerated	No	Flat disc (exposed)	Amniochorionic fold (E11–12)	E11.5–12	Gastrula stage/PrePS (precursors)	SOX17	Hassoun, Schwartz, Feistel, Blum, and Viebahn (2009), Hassoun, Schwartz, Rath, Viebahn, and Männer (2010), Irie et al. (2014), Kobayashi et al. (2017), and Ramos-Ibeas, Nichols, and Alberio (2018)
Rabbit	*Oryctolagus cuniculus*	Lagomorpha	E6	E7	Degenerated	No	Flat disc (exposed)	Amniochorionic fold (E7.5–8)	E6.5–7.0	Gastrula stage		Hassoun et al. (2010), Hopf, Viebahn, and Püschel (2011), Irie et al. (2014), and Schäfer-Haas and Viebahn (2000)

The polar TE in primates after implantation becomes the cytotrophoblast (CT). Since this structure is not present in pigs or rabbits, their epiblast morphology is labeled as "exposed." The presence of ExE is indicated along with its relative position to the epiblast, where "adjacent" denotes direct contact of the ExE with the epiblast, whereas "far-away" denotes no contact with the epiblast. Embryonic days (E) in mice and rats denote days after vaginal plug discovery plus 0.5 whereas other species denote days post fertilization. The earliest described stages with identified germ cells are based on identification of cells with germ cell identity. Competence testing or lineage tracing experiments to identify PGC precursors are additionally labeled. Empty fields have not yet been described in literature.

generate hPGCLCs from pluripotent cells were relatively inefficient and not informative (Clark et al., 2004; Gkountela et al., 2013; Hayashi et al., 2012; Kee, Angeles, Flores, Nguyen, & Reijo Pera, 2009). However, several groups have subsequently achieved robust *in vitro* specification of hPGCLCs (Irie et al., 2015; Sasaki et al., 2015; Sugawa et al., 2015).

The initial work was carried out using mouse pluripotent stem cells (mPSCs) (Hayashi et al., 2011). Notably, naïve mESCs cannot directly give rise to mPGCLCs, but following a 48-h exposure to Activin A and fibroblast growth factor 2 (FGF2) they can efficiently form mPGCLCs in response to cytokines, principally BMP4, which recapitulate key characteristics of migratory stage (~E9.5) mPGCs (Hayashi et al., 2011). These germline-competent epiblast-like cells (mEpiLCs) exhibit a transcriptome resembling that of mouse epiblast at pre-gastrulation stage (Hayashi et al., 2011). Longer than 48-h culture in Activin A and FGF2 results in developmentally advanced epiblast stem cells (mEpiSCs) that transcriptionally resemble primitive streak-stage embryos (Brons et al., 2007; Kojima, Kaufman-Francis, et al., 2014; Tesar et al., 2007), with concomitant loss of mPGCLCs induction, consistent with the transient competence window *in vivo* (Hayashi et al., 2011; Hayashi & Surani, 2009; Ohinata et al., 2009).

Although derived from the ICM, hESCs differ from mESCs in their morphology, signaling requirements, epigenetic and transcriptional profile (Lanner & Rossant, 2010; Nichols & Smith, 2009; Rossant, 2015). Unlike mESCs, they rely on FGF2 and Activin/transforming growth factor β (TGFβ) for pluripotency maintenance (Daheron et al., 2004; James, Levine, Besser, & Hemmati-Brivanlou, 2005; Lanner & Rossant, 2010; Sumi, Fujimoto, Nakatsuji, & Suemori, 2004; Thomson et al., 1998). They also feature highly methylated genome and express low levels of naïve pluripotency markers, such as *KLF4*, *TBX3*, and *TFCP2L1* (Hackett & Azim Surani, 2014). Accordingly, hESCs are similar to mEpiSCs (Brons et al., 2007; Tesar et al., 2007) but with notable differences. For example, PRDM14 is not expressed in mEpiSCs, but is critical for pluripotency of hESCs (Chia et al., 2010). Recently derived "ground-state" hESCs share many properties with naïve mESCs, with similar transcriptome and DNA methylome as in human ICM (Pastor et al., 2016; Takashima et al., 2014; Theunissen et al., 2014). However, the extent of functional conservation between pluripotency TFs of naïve hESC and naïve mESC remains to be thoroughly examined.

When the protocol developed for mPGCLC induction is applied to conventional human pluripotent stem cells (hPSCs) such as hESCs, the efficiency of hPGCLC induction is very low (<5%) (Irie et al., 2015). However, when cultured in a so-called 4i medium, hPSCs acquire germline competence (Gafni et al., 2013; Irie et al., 2015) and generate high percentages of hPGCLCs (~50%) upon BMP2 or BMP4 treatment in cell aggregates, called embryoid bodies (EBs) (Irie et al., 2015). 4i-cultured hPSCs maintain key hallmarks of pluripotency (Gafni et al., 2013), but upregulate many primitive-streak related genes and self-renew in a germline-competent state (Irie et al., 2015), which makes them a tractable system to study hPGC competence and specification, especially when combined with gene targeting or pharmacological perturbations (Irie et al., 2018).

Additional methods for hPGCLCs inductions have since been developed and rely on a two-step procedure. First, Activin/Nodal and WNT signaling activation drives conventional hESCs to pre-mesendoderm (preME) (Kobayashi et al., 2017) or incipient mesoderm-like cells (iMeLCs) (Sasaki et al., 2015), which transiently display germline potential; these competent cells are induced to form hPGCLCs in response to cytokines, including BMP2 as described previously. The significance of these systems is their closer recapitulation of the *in vivo* developmental progression in the posterior epiblast.

Both approaches (self-renewing and transient) involve activation of WNT pathway for generation of hPGCLC-competent cells (Irie et al., 2015; Kobayashi et al., 2017; Sasaki et al., 2015). It is therefore likely that similar to mouse, cynomolgus money and porcine embryos (Aramaki et al., 2013; Kobayashi et al., 2017; Ohinata et al., 2009; Sasaki et al., 2016), human pluripotent cells acquire germline potential in response to WNT signaling. Importantly, the resulting hPGCLCs are likely counterparts of pre-migratory hPGCs (Irie et al., 2015; Kobayashi et al., 2017; Sasaki et al., 2015), thus representing an earlier stage compared to mPGCLCs.

The relevance of *in vitro* induced germ cells is underscored by their ability to form mature gametes, at least in mice. mPGCLCs induced by cytokines or by overexpression of key TFs can develop into mature sperm when transplanted to neonatal testes of mice (Hayashi et al., 2011; Ishikura et al., 2016; Nakaki et al., 2013), or into oocytes if aggregated with fetal female gonadal somatic cells (Hikabe et al., 2016). Importantly, the resulting spermatozoa and oocytes can generate fertile offspring. Furthermore, meiosis and haploid germ cell induction from mouse and human PSCs were reported (Easley et al., 2012; Jung et al., 2017; Kee et al., 2009; Zhou et al., 2016). Most recently human oogonia-like cells were obtained by aggregation of hPGCLCs with mouse embryonic ovarian somatic cells for

4 months (Yamashiro et al., 2018). The resulting germ cells resemble their gonadal counterparts, showing similar transcriptome, global DNA demethylation and imprint erasure but only partial X-chromosome reactivation.

The *in vitro* assays to generate PGCLCs from mouse and human pluripotent cells provide evidence for PGC specification from the epiblast. However, the observation of cells with PGC identity in the nascent amnion of cynomolgus monkey embryos (Sasaki et al., 2016) raises the question of the presence of alternative or additional sites of PGC specification in primates (Kobayashi & Surani, 2018; Sasaki et al., 2016), Further studies are necessary to elaborate on this report and verify its applicability to human development.

3.2 Developmental models based on self-organizing properties of pluripotent cells

The self-organizing properties of stem cells have been illustrated by the generation of several tissue organoids *in vitro* including intestinal, liver, lung, cerebral and retinal organoids from multipotent adult stem cells or from PSCs in step-wise approaches to recapitulate the developmental pathways observed *in vivo* (Clevers, 2016; Eiraku et al., 2011; Huch et al., 2013; Lancaster & Knoblich, 2014; Lancaster et al., 2013; Rock et al., 2009; Sato et al., 2009; Spence et al., 2011). Since PSCs represent cells in the early embryo, they should be able to form self-organized patterns and possibly structures observed in early development. This notion is supported by observations where mESC clumps under minimally stimulatory conditions (with transient WNT stimulation) can spontaneously form protrusions with gene expression patterns reflecting the gastrulating embryo, termed "gastruloids" (Beccari et al., 2018; Turner et al., 2017; van den Brink et al., 2014). The requirement for such minimal conditions denotes the presence of endogenous instructive activity in the culture, probably sourced from interactions with other cell types. Indeed, when mESCs were co-cultured with trophoblast stem cells (mTSCs) and extraembryonic endoderm (XEN) cells in a juxtaposed, essentially non-inducing environment (Harrison, Sozen, Christodoulou, Kyprianou, & Zernicka-Goetz, 2017; Sozen et al., 2018), approximately 21% of aggregates formed a fused spheroid between mESCs and mTSCs surrounded by XEN cells, conveniently named "ETX embryoids." These ETX embryoids exhibited T/BRA^{+} cells on one side with $LEFTY1^{+}$ AVE-like identity in XEN cells at the opposing side. Importantly, spatial transcriptomics showed the presence of both anterior and posterior epiblast identities in these spheroids reminiscent of an E7.0 gastrulating mouse embryo (Peng et al., 2016; Sozen et al., 2018). The presence of cells with PGC-like identity co-expressing OCT4, STELLA and T/BRA with an upregulation of *Tfap2c* and *Prdm14* was also reported in ETX embryoids (Harrison et al., 2017). These studies demonstrate the importance of interactions between different cell types for spatiotemporal regulation of developmental events, particularly the endoderm cells, which is well documented in the embryo (Tam & Loebel, 2007).

Modeling human development presents a different challenge due to the planar morphology of the postimplantation embryo. Several models were proposed to "flatten" a mouse epiblast with signaling and morphogenetic regulations giving rise to the spatially confined anteroposterior and germ layer patterns (Beddington & Robertson, 1999; Behringer, Wakamiya, Tsang, & Tam, 2000; Tang et al., 2016). Even though hESCs represent an advanced developmental state compared to mESCs (Nichols & Smith, 2009), their ability to self-organize into developmentally relevant patterns has been demonstrated using spatially constricted hESC colonies in 2D by a method known as "micropatterning" (Warmflash et al., 2014). Under uniformly inductive conditions with BMP4 signaling, these colonies exhibit rings of cellular identities radially around a central pole. With $SOX2^+$ ectoderm in the centre, followed by T/BRA^+ primitive streak-like cells and $SOX17^+$ endoderm in the periphery. It is of interest that some $SOX17^+$ cells are seen scattered in the $SOX2^-$ intermediate layer, warranting further examination of their cellular identity. This remarkable phenomenon is conferred by differential apical-lateral localization of BMP receptors between the periphery and central cells coupled with a reaction-diffusion mechanism involving BMP and its inhibitor NOGGIN (Etoc et al., 2016; Tewary et al., 2017). Furthermore, cellular identities along the anteroposterior primitive streak, including the GSC^+ organizer, can be induced with combinations of WNT and ACTIVIN/NODAL signaling (Martyn, Kanno, Ruzo, Siggia, & Brivanlou, 2018). Evidence of epithelial-to-mesenchymal transition (EMT) was also observed with cells expanding from a monolayer (Martyn et al., 2018; Warmflash et al., 2014) indicating that many of these developmental events can be captured *in vitro* using hESCs.

To challenge the issue of epiblast morphology divergence between mouse and humans, primed pregastrulation mEpiLCs can also recapitulate these radial patterns in spatially constricted colonies under different conditions (Morgani, Metzger, Nichols, Siggia, & Hadjantonakis, 2018) suggesting that the intrinsic self-organizing properties of the epiblast is independent of its shape. In both species, however, anteroposterior axis is not known to be radial, suggesting additional signals are absent in this system.

This may be a mechanosensory mechanism conferred by colony shape (Blin, Picart, Théry, & Puceat, 2017), or alternatively the presence of other cell types like in ETX embryoids (Sozen et al., 2018). The latter will be a challenge, as human trophoblast stem cells (hTSCs) have only been described recently (Okae et al., 2018), but no XEN-equivalent, human hypoblast stem cells have been reported to date. Moreover, the described micropatterns are mostly a monolayer of cells in 2 dimensions (2D), which may be a limiting factor to capturing 3D morphogenetic events *in vitro*.

Symmetry breaking in hESC aggregates can also be achieved in inductive Matrigel-embedded conditions that results in asymmetric distribution of cells with primitive streak identity in a BMP4-induced WNT-dependent manner (ten Berge et al., 2008). However, none of these aforementioned methods report observations of the earliest bifurcation event in human embryos—the formation of the amnion. Nonetheless, hESCs that are epithelial in character can form a lumenized epithelial spheroid when cultured on a differential extracellular-matrix (ECM) gel bed supplemented with media containing a reduced ECM concentration (conventionally labeled as "2.5D" culture). Interestingly, these spheroids are comprised of either columnar epithelial cells with epiblast identity or squamous (flattened) epithelial cells with amnion-like identity, or in some cases both, in a density-dependent manner (Shao, Taniguchi, Gurdziel, et al., 2017; Shao, Taniguchi, Townshend, et al., 2017). The formation of this squamous epithelium is sensitive to BMP4 signaling (Shao, Taniguchi, Townshend, et al., 2017). BMP4 expression was shown in the proximal side of the postimplantation cynomolgus monkey embryos in the polar trophectoderm derivative (cytotrophoblast) and perhaps even in the amnion epithelium itself (Sasaki et al., 2016). These structures could potentially substitute for the absence of ExE in primates, at least for the early inductive roles via BMP signaling. Although hPGCLCs were not observed (Shao, Taniguchi, Townshend, et al., 2017) it remains an open question as to the origin(s) and morphogenetic mechanisms of specifying cells with hPGC-like identity during this early lineage bifurcation event.

4. Gene regulatory networks for mouse and human PGC specification

Loss- and gain of function studies in mouse models, as well as in mouse and human *in vitro* PGCLC specification, have provided critical insights into the signaling and gene regulatory networks of PGC specification and development (Table 3).

Table 3 Gain and loss of function studies showing genes involved in hPGCLC specification.

Gene(s)	LOF/GOF	Markers of hPGCLCs	Cytokines (BMP4, LIF, SCF, EGF)	Phenotype in hPGCLC specification *in vitro*	References
SOX17	Loss	Nanos3 +, TNAP +	+	No hPGCLCs	Irie et al. (2015)
	Loss	Blimp1 +, TFAP2C +, EPCAM +, ITGA6 +	+	No hPGCLCs	Kojima et al. (2017)
	Gain	Nanos3 +, TNAP +	−	Negligible response	Kobayashi et al. (2017)
	Gain	Blimp1 +, TFAP2C +	+	Rescues hPGCLC in EOMES −/−	Kojima et al. (2017)
PRDM1	Loss	Blimp1 +, TFAP2C +	+	Significant reduction in number of hPGCLCs (~8%)	Irie et al. (2015)
	Loss (KD)	VASA +, NANOS3 +, STELLA +	BMP4 + WNT3A	No germ cells	Lin et al. (2014)
	Loss (KD)	TRA-1-81 +, cKIT +	BMP4 + LIF	Reduced number	Sugawa et al. (2015)
	Gain	VASA +, SCP3 +	BMP4 + WNT3A	VASA expression enhancement, suppression of SOX2	Lin et al. (2014)
	Gain	Nanos3 +, TNAP +	−	No hPGCLCs	Kobayashi et al. (2017)
TFAP2C	Loss	Blimp1 +, EPCAM +, ITGA6 +	+	No hPGCLCs by day 6 (late)	Kojima et al. (2017)
	Gain	Nanos3 +, TNAP +	−	No hPGCLCs	Kobayashi et al. (2017)
SOX17 + *PRDM1*	Gain	Nanos3 +, TNAP +	−	Abundance of hPGCLCs in day 1 of EB formation	Kobayashi et al. (2017)

Table 3 Gain and loss of function studies showing genes involved in hPGCLC specification.—cont'd

Gene(s)	LOF/GOF	Markers of hPGCLCs	Cytokines (BMP4, LIF, SCF, EGF)	Phenotype in hPGCLC specification *in vitro*	References
SOX17 + TFAP2C	Gain	Nanos3+, TNAP+	−	No hPGCLCs	Kobayashi et al. (2017)
PRDM1 + TFAP2C	Gain	Nanos3+, TNAP+	−	Negligible response	Kobayashi et al. (2017)
PRDM14	Loss (KD)	TRA-1-81+, cKIT+	BMP4 + LIF	WT number of hPGCLCs	Sugawa et al. (2015)
EOMES	Loss	EPCAM+, ITGA6+	+	Dramatically reduced number	Chen et al. (2017)
	Loss	Blimp1+, TFAP2C+, EPCAM+, ITGA6+	+	No hPGCLCs	Kojima et al. (2017)
	Gain	Blimp1+, TFAP2C+, EPCAM+, ITGA6+	+	hPGCLCs induced only when OE in late iMeLC, larger EB sizes	Kojima et al. (2017)
T/BRA	Loss	Blimp1+, EPCAM+, ITGA6+	+	WT number of hPGCLCs	Kojima et al. (2017)
MIXL1	Loss	Blimp1+, EPCAM+, ITGA6+	+	WT number of hPGCLCs	Kojima et al. (2017)
SOX2	Gain	VASA+, PRDM1+	+	No germ cells, neural differentiation	Lin et al. (2014)
NANOG	Gain	Nanos3+, TNAP+	−	No hPGCLCs	Kobayashi et al. (2017)
TRIM28	Loss	EPCAM+, ITGA6+	+	No hPGCLCs	Tao et al. (2018)

Markers used to identify hPGCLCs in indicated studies are shown. The use of cytokines during hPGCLC specification are marked with (+) in studies which employed BMP4, LIF, SCF and EGF. Studies using a different method with less cytokines are indicated. Overexpression experiments in the absence of cytokines are marked (−). Loss of function studies are divided into knockout and, as indicated, siRNA knockdown (KD) strategies. Combinatorial multi-gene gain of function studies are also shown. WT denotes wildtype.

4.1 The genetic basis for mouse PGC specification

A pivotal study using single-cell cDNA analysis of the founder mPGCs and somatic neighbors at E7.0–7.5 attempted to identify the key regulators of the PGC fate (Saitou et al., 2002). This showed that *Fragilis* (also known as *Ifitm*) and *Stella* are two genes highly specific to nascent mPGCs. Surprisingly, KO of either of these loci led to no detectable effect on mPGC specification (Lange et al., 2008; Payer et al., 2003). Interestingly, mesodermal marker *T (Brachyury)* is expressed in both nascent germ cells and in the surrounding soma, suggesting that, similar to some other organisms, like axolotls, mPGCs are specified from mesoderm-like cells (Saitou et al., 2002). T was later shown to be actively involved in mPGC fate by activating the enhancers of key germ cell genes (Aramaki et al., 2013).

A hallmark of PGC specification is downregulation of the somatic program, for example, homeobox genes such as *Hoxa1, Hoxb1, Lim1* and *Evx1* (Saitou et al., 2002). Further analysis of the single-cell data from (Saitou et al., 2002) identified *Prdm1* (encoding BLIMP1) as a very early (E6.25) and specific mPGC marker, which is a key regulator of mPGC specification, critical for repression of the somatic program (Ohinata et al., 2005). Indeed, *Prdm1* KO led to a significant decrease in number of specified mPGC. These $Prdm1^{-/-}$ cells displayed aberrant proliferation, migration and transcriptional profile (Ohinata et al., 2005).

Subsequent single-cell microarray analysis on nascent $Prdm1^{+}$ mPGCs and surrounding soma from wildtype and *Prdm1*-null mice identified additional regulators of early mPGC development, in particular, *Prdm14*. Loss of function of PRDM14 resulted in normal but infertile mice due to reduced mPGC specification and rapid mPGC loss (Yamaji et al., 2008).

Finally, a third core TF for mPGC fate, TFAP2C (also known as AP2γ), was identified as a direct target of BLIMP1 (Kurimoto, Yabuta, et al., 2008; Magnúsdóttir et al., 2013; Weber et al., 2010). It is expressed in mPGCs from E7.25 until E12.5 (Weber et al., 2010). A conventional KO of *Tfap2c* gene resulted in embryonic lethality due to defects in TE differentiation (Auman et al., 2002; Werling & Schorle, 2002), but a conditional *Tfap2c* KO specifically in mPGCs led to male and female sterility due to mPGC loss around E8.0 (Weber et al., 2010).

Co-overexpression of *Prdm1*, *Prdm14* and *Tfap2c* or of any two of these factors is sufficient to induce mPGCLCs from mEpiLCs even without BMP signaling (Nakaki et al., 2013). Interestingly, forced expression of *Prdm14* alone can also induce mPGCLCs, albeit to a lower efficiency (Nakaki et al., 2013). PRDM14, BLIMP1 and TFAP2C form an interdependent tripartite regulatory network (Magnúsdóttir et al., 2013), where all three factors contribute to the suppression of the somatic program, with a dominant contribution from BLIMP1; BLIMP1 also directly represses proliferation-related genes and, together with PRDM14, promotes epigenetic resetting; PRDM14 and TFAP2C upregulate genes involved in germ cell development and migration (Magnúsdóttir et al., 2013; Nakaki et al., 2013).

4.2 Distinct network for human PGC development

While mouse models have uncovered major genetic and signaling components of mammalian PGC development, they fail to recapitulate some aspects of hPGCs (Kobayashi & Surani, 2018), in particular, the downstream gene regulatory networks differ. Human *in vitro* models combined with studies of gonadal hPGCs showed that, similar to mPGCs, human germ cells express conserved markers of PGCs (BLIMP1, TFAP2C, DND1, NANOS3, UTF1 and KIT) and pluripotency (TNAP, OCT4, NANOG, PRDM14, and LIN28A) (Irie et al., 2015; Kobayashi et al., 2017; Sasaki et al., 2015). It is of interest that early mouse and human PGCs share different sets of pluripotency genes (Tang et al., 2016). Mouse PGCs express *Sox2, Esrrb* and *Klf2* whereas hPGC(LC)s express *KLF4* and *TFCP2L1*. The functional significance of the species-specificity of these and other genes in PGC development remains largely unexplored. Furthermore, functional divergence may not simply be a case of differential expression. Although NANOG is expressed in both mPGCs and hPGCs (Tang et al., 2015), its role in PGC specification seems to differ. *Nanog* loss-of-function specifically in the germline in mice leads to a severe decrease in mPGC numbers at E11.5–E12.5 (Chambers et al., 2007; Yamaguchi et al., 2009), and its overexpression in mEpiLCs drives mPGCLC induction even without BMP, by activating the enhancers of key germ cell genes *Prdm1* and *Prdm14* (Murakami et al., 2016). However, this is not the case for hPGCLC specification (Table 3) (Kobayashi et al., 2017). Mouse and human NANOG show quite extensive amino acid sequence divergence, which may explain functional differences (Hart, Hartley, Ibrahim, & Robb, 2004). It would be informative to perform an inducible loss of NANOG during hPGCLC specification to reveal its roles, if any, in the human germline.

Notably, hPGCs specification depends on SOX17 expression, a TF conventionally known for its role in endoderm specification and dispensable for mPGCs development (Hara et al., 2009; Irie et al., 2015; Kanai-Azuma et al., 2002). There is sequential upregulation of SOX17, BLIMP1 and

TFAP2C during hPGCLC specification (Irie et al., 2015), and *SOX17* KO abrogates hPGCLC induction and precludes upregulation of other germ cell markers, including BLIMP1 (Irie et al., 2015; Kojima et al., 2017). Loss of *PRDM1* (encoding BLIMP1) results in aberrant AP^{+} hPGCLCs that show reduced or absent expression of key germ cell genes and they fail to repress somatic markers, as in *Prdm1*$^{-/-}$ mice (Irie et al., 2015; Kojima et al., 2017; Ohinata et al., 2005). Furthermore, combined overexpression of SOX17 with BLIMP1, but not TFAP2C, is sufficient to drive hPGCLC fate even in the absence of cytokines (Kobayashi et al., 2017). Altogether, this places SOX17 at the top of the hPGC TF hierarchy, upstream of BLIMP1 and TFAP2C. A recent report contradicted some of these findings, suggesting that TFAP2C activation by BMP is independent of SOX17 and that *SOX17*-null cells can transiently upregulate other germ cell markers, such as *NANOS3* and *KLF4* (Kojima et al., 2017). Further, the study suggested that *TFAP2C* deletion had a more pronounced phenotype than *PRDM1*/BLIMP1 KO, at least on day 2 of differentiation (Irie et al., 2015; Kojima et al., 2017).

Notably, a different SOX family member, SOX2, is crucial for mouse, but not human PGCs (Campolo et al., 2013; de Jong et al., 2008; Irie et al., 2015). SOX2 is also a core pluripotency regulator in mouse and human (Wong, Chambers, & Mullin, 2016). It is, however, repressed at the very onset of hPGCLC specification, and is not detected in the gonadal hPGCs (Fig. 1) (de Jong et al., 2008; Irie et al., 2015; Perrett et al., 2008). Notably, the presence of SOX17 and the absence of SOX2 have been demonstrated in PGCs of monkeys and pigs (Table 2), which distinguishes them from mPGCs (Irie et al., 2018).

Furthermore, the role of PRDM14 in hPGC fate remains unclear; although its expression is detectable in hPGCLCs/hPGCs *in vitro* and *in vivo*, its cellular location could be in the cytoplasm (Fig. 1) (Irie et al., 2015). Moreover, a knockdown study suggested that PRDM14 might be dispensable for hPGCLC specification (Sugawa et al., 2015). Notably, in addition to SOX17, hPGCs, but not mPGCs, feature other alternative lineage genes, including *GATA4* (endoderm marker) and *TEAD4* (trophectoderm marker), despite BLIMP1 suppressing most somatic genes (Tang et al., 2015).

Finally, mPGCLCs show modest upregulation of later germ cell markers *Ddx4* and *Dazl* and display substantial epigenetic resetting (Hayashi et al., 2011), which is not the case for hPGCLCs, as they represent a much earlier developmental stage (Irie et al., 2015; Tang et al., 2015).

4.3 Differential roles of SOX factors in mouse and human PGC specification

The *"SRY-box" (SOX)* family of TFs in mouse and human have 20 members each containing a High Mobility Group (HMG) DNA binding domain with at least 50% sequence homology with *SRY*, which are further subdivided based on HMG homology into 8 main groups (*SOXA* to *SOXH*) (Bowles, Schepers, & Koopman, 2000; Pevny & Lovell-Badge, 1997; Schepers, Teasdale, & Koopman, 2002). SOX TFs can be either transcriptional activators (SOXB1, C, E and F) or repressors (SOXB2) and function in a context-dependent manner in many developmental lineages (Kamachi & Kondoh, 2013; Kondoh & Kamachi, 2010; Uchikawa, Kamachi, & Kondoh, 1999).

4.3.1 SOX17 is crucial for human, but not mouse PGC fate

SOX17 in mouse and human belongs to the SOXF group of TFs comprising SOX7, SOX17 and SOX18 (Pevny & Lovell-Badge, 1997; Schepers et al., 2002). SOXF TFs are largely involved in cardiovascular development (Francois, Koopman, & Beltrame, 2010; Lilly, Lacaud, & Kouskoff, 2017). The HMG domain of SOXF proteins binds to the minor groove of the DNA helix in a sequence specific manner to the consensus sequence 5′-($^{A}/_{T}$)($^{A}/_{T}$)CAA($^{A}/_{T}$)G-3′ and causes an 80° bend in DNA (Hosking, Muscat, Koopman, Dowhan, & Dunn, 1995; Palasingam, Jauch, Ng, & Kolatkar, 2009). Similar to SOXB TFs, SOX17 also activates transcription via the C-terminal domain (Hosking et al., 1995; Kanai et al., 1996; Wegner, 2010). SOX17 is primarily an inducer of endoderm cells starting from 32–64C blastocysts in the segregating ICM to the definitive endoderm (DE) of E7.5 embryos (Artus, Piliszek, & Hadjantonakis, 2011; Kanai-Azuma et al., 2002; Niakan et al., 2010).

A temporal series of single-cell mRNA analysis during early mPGC did reveal transient expression of *Sox17* in E7.25 early bud-stage embryos (Yabuta et al., 2006), when specified mPGCs had initiated expression of *Prdm14*, *Sox2/3* and importantly *Stella* (*Dppa3/PGC7*) raising the possibility that *Sox17* may play a role in this early window. However, while *Sox17*-null embryos were devoid of DE and exhibited severe posterior trunk development leading to embryonic lethality at E10.5 (Table 1) (Kanai-Azuma et al., 2002), but the number of AP- and STELLA-positive mPGCs were not affected (Hara et al., 2009). Crucially, the observed migration defect using *Sox17*-null mESCs can be rescued in a *Sox17*$^{+/+}$ chimeric embryo environment establishes the fact that mouse SOX17 does not have a role in mPGC specification but is important for mPGC migration indirectly by regulating hindgut formation (Hara et al., 2009). Interestingly, *Sox17* and a shorter isoform (*t-Sox17*) without DNA binding nor transactivating activity were detected in the mouse testis from early pachytene spermatocyte to the round spermatids stages (Kanai et al., 1996), but the roles of *Sox17* in mouse spermatogenesis have not yet been studied.

Human *SOX17*/SOX17 is first detected in cavitating blastocysts and subsequently restricted to putative PrE cells (Fig. 1) (Blakeley et al., 2015; Niakan & Eggan, 2013). Extensive functional characterization of SOX17 is largely absent and has relied mostly on extrapolation from mouse studies. SOX17 is expressed robustly during hESC differentiation to cells with DE identity (D'Amour et al., 2005; Kobayashi et al., 2017). Constitutive overexpression of *SOX17* in hESCs resulted in the upregulation of DE markers (*CXCR4, CER, GSC, DLX5, ECAD* and *FOXQ1*) (Séguin, Draper, Nagy, & Rossant, 2008). This is in contrast to the activation of extraembryonic endoderm (ExEn) genes (*Gata4, Gata6, LamB1* and *Hnf1β*) following SOX17 overexpression in mESCs (Niakan et al., 2010; Qu, Pan, Zhang, & Huang, 2008). The inability to derive *Sox17*$^{-/-}$ XEN cells (Niakan et al., 2010) and the differential pluripotent states between hESCs and mESCs (Nichols & Smith, 2009) may reflect on an overall inductive role of SOX17 in specifying PrE and DE. Nonetheless, during differentiation of hESCs into several lineages, including DE, SOX17 co-binds genomic sites with other endoderm factors such as GATA4/6 and FOXA1/2 (Tsankov et al., 2015).

SOX17 is expressed in human embryonic testis/ovaries along with OCT4 and NANOG but not SOX2 (Perrett et al., 2008). This is in stark contrast to the presence of SOX2 but not SOX17 in mouse embryonic testis (Perrett et al., 2008). The essential role of SOX17 in hPGCLC specification was established when *SOX17*$^{-/-}$ 4i-cultured hPSCs could not form hPGCLCs in inductive conditions (Table 3) (Irie et al., 2015). Interestingly, *PRDM1*$^{-/-}$ cells (lacking BLIMP1) also failed to induce NANOS3^{+} cells, but exhibited some residual SOX17^{+}/OCT4^{+}/AP^{+} co-expressing cells, indicating that SOX17 is upstream of BLIMP1 (Irie et al., 2015). This differs from mouse, where BLIMP1 and PRDM14 have a more central role in mPGC specification (Ohinata et al., 2005; Yamaji et al., 2008). It is of interest that SOX17 overexpression alone could not efficiently induce hPGCLCs and these residual hPGCLCs showed aberrant co-expression of some endoderm genes, indicating insufficient suppression of somatic gene expression. Indeed, a balanced dosage of SOX17/BLIMP1 co-overexpression was highly effective in inducing hPGCLCs (Kobayashi et al., 2017). This indicates that the somatic-repressive role of BLIMP1 is important to secure hPGCLC identity through repression of the endoderm-inductive activity from SOX17 overexpression (Irie et al., 2015). This hPGCLC-inductive activity does not occur with SOX2 in humans (Lin et al., 2014). Indeed, SOX2 is quickly repressed during hPGCLC induction, and furthermore, SOX2 overexpression is detrimental to the human germline fate (Lin et al., 2014). It is unclear how a transcriptional activator such as SOX17 and a generally repressive BLIMP1 cooperatively activate the germ cell gene-regulatory network on the genomic level. In other words, what factors regulate the context-dependant response to SOX17 in the germline and endoderm lineages? What is the evolutionary significance of the hPGCLC-inductive activity of SOX17 in humans compared to PGC-proliferative role of SOX2 in rodents?

One of the hallmarks of SOX TFs is to partner with different proteins to confer its context-dependent functions (Kondoh & Kamachi, 2010). SOX17 has been proposed to interact with OCT4, a member of POU family of TFs and critical regulator of pluripotency (Niwa, Miyazaki, & Smith, 2000; Stefanovic et al., 2009). OCT4 is localized on the genome at the *SOX17* promoter in differentiating hESCs and this localization is sensitive to the presence of SOX2 (Stefanovic et al., 2009). Importantly, co-immunoprecipitation (coIP) analysis shows the presence of a complex involving OCT4 and SOX17 in differentiating hESCs and in TCam-2, a human seminoma cell line, which shares many features with hPGCs (de Jong et al., 2008; Nettersheim et al., 2016; Stefanovic et al., 2009). These observations were consistent with coIP studies during PrE differentiation in mouse cells (Aksoy et al., 2013) suggesting that conserved mechanism of specific lineage commitment from pluripotent cells involves OCT4 switching partners between SOX2 and SOX17 (Aksoy et al., 2013; Nettersheim et al., 2016; Stefanovic et al., 2009). Given that OCT4 is expressed in both mouse and human PGCs coupled with the observation that OCT4 deletion affected early mouse PGC development (Table 1) (Kehler et al., 2004; Okamura et al., 2008), whether similar mechanisms are in play to establish the hPGC gene regulatory network remains to be examined.

4.3.2 SOX2 is a key pluripotency TF required for mouse PGC development, but absent from human germline

SOX2 in mouse and human belongs to the SOXB1 family of TFs comprising SOX1, SOX2 and SOX3 (Pevny & Lovell-Badge, 1997; Schepers et al., 2002; Uchikawa et al., 1999). SOXB1 TFs are largely involved in the developing neurectoderm (Uchikawa et al., 1999; Wood & Episkopou, 1999). SOX2 is the only SOXB TF expressed in embryos before implantation, where it is initially localized in the cytoplasm in the zygote before becoming restricted to the nucleus in 4C-6C mouse embryos (Avilion et al., 2003; Keramari et al., 2010). Single-cell transcriptomic analysis indicates that *Sox2* expression was associated with the inner cells of the 16C morula that would form the ICM (Fig. 1) (Guo et al., 2010). In human embryos, enrichment of *SOX2* transcripts was detected a little later, in 8C morulae, likely an attribute to their protracted ZGA (Blakeley et al., 2015).

SOX2 is indispensable in maintaining mESC pluripotency and acts downstream of OCT4 (Masui et al., 2007; Niwa, Masui, Chambers, Smith, & Miyazaki, 2002; Wong et al., 2016). However, knockdown studies in hESCs suggested a lineage-regulatory role of pluripotency TFs, where SOX2 inhibited primitive streak-like identity promoted by OCT4 (Wang, Oron, Nelson, Razis, & Ivanova, 2012). Indeed, SOX2 and OCT4 progressively segregate and associate with the neuroectoderm and mesendoderm lineages, respectively, in differentiating mESCs (Thomson et al., 2011), consistent with the primed pluripotent state of hESCs (Nichols & Smith, 2009).

Chromatin immunoprecipitation (ChIP) studies in mESCs revealed that SOX2 colocalizes with OCT4 and NANOG on the genomic DNA in proximity to a cohort of pluripotency-associated genes (Chen et al., 2008; Kim, Chu, Shen, Wang, & Orkin, 2008; Loh et al., 2006; Marson et al., 2008) including genes involved in X-chromosome inactivation such as *Tsix* and *Rnf12* (Navarro, Moffat, Mullin, & Chambers, 2011; Navarro et al., 2010). Motif discovery identified a composite sequence comprising an *Octamer* and *Sox* binding site arranged in a specific orientation (known as *Oct-Sox* motif) within proximity of many pluripotency-associated genes (Chen et al., 2008; Kondoh & Kamachi, 2010; Loh et al., 2006). Many of these pluripotency-related markers are indeed controlled by SOX2 and OCT4 cooperative transcriptional activation (Ambrosetti, Basilico, & Dailey, 1997; Chew et al., 2005; Kuroda et al., 2005; Nakatake et al., 2006; Nishimoto, Fukushima, Okuda, & Muramatsu, 1999; Okumura-Nakanishi, Saito, Niwa, & Ishikawa, 2005; Rodda et al., 2005; Tokuzawa et al., 2003; Tomioka et al., 2002; Yuan, Corbi, Basilico, & Dailey, 1995).

This notion is largely consistent with ChIP studies in hESCs (Boyer et al., 2005) that exhibit more similarity to mEpiSCs than mESCs (Matsuda et al., 2017). These observations indicate that the core regulatory function of SOX2 in mouse and human pluripotency is largely comparable.

The HMG domain of SOX2, like SOX17, binds to the minor groove of DNA with the consensus sequence 5′-($^{A}/_{T}$)($^{A}/_{T}$)CAAAG-3′ (Bowles et al., 2000). The observation that SOXB1 and SOXF TFs bind to rather similar motif sequences indicates that SOX TFs in general bind non-specifically to a rather generic *SOX* motif and their functions are largely conferred by interactions with tissue-specific factors (Kondoh & Kamachi, 2010). The C-terminal domain of SOX2 contains a serine-rich region within a transactivation domain (Ambrosetti, Schöler, Dailey, & Basilico, 2000; Nowling, Johnson, Wiebe, & Rizzino, 2000). This serine-rich region comprises a triple-repeat motif that is crucial for the direct physical interaction with NANOG in mESCs (Gagliardi et al., 2013). The HMG domain itself interacts with the POU-specific (POU_S) domain of OCT4 on DNA where the interaction interface involves five amino acid residues on HMG (Fig. 2) (Ambrosetti et al., 1997; Chambers & Tomlinson, 2009; Reményi et al., 2003; Williams et al., 2004)

The mode of OCT4-SOX2 interaction on DNA was conventionally regarded as stepwise, where SOX2 binding to the *Oct-Sox* motif stabilizes the DNA-bound conformation of OCT4 (Chambers & Tomlinson, 2009). Recent studies monitoring the single-molecule dynamics of SOX2 on chromatin reported a mechanism involving initial genome engaging activity of SOX2 before dwelling on a target motif that seems more prominent in the assembly of the OCT4-SOX2 protein complex (Chen et al., 2014).

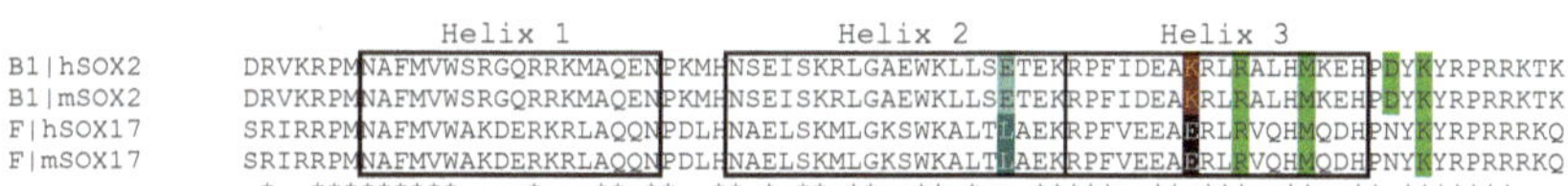

Fig. 2 The OCT4 interaction interface in SOX2 and SOX17 HMG domains. Residues highlighted in red/black are part of the interacting residues with OCT4 predicted from structural studies (Reményi et al., 2003; Williams, Cai, & Clore, 2004) that were shown to switch functions if swapped (Jauch et al., 2011). Residues highlighted in blue were described to change SOX2 binding to compressed *Oct-Sox* motif (Merino et al., 2014; Palasingam et al., 2009). Residues highlighted in green were additionally described in Reményi et al. (2003) and Williams et al. (2004). Asterisks denote identical residues; residue conservation between groups of strongly and weak similarity in chemical properties are labeled with colon (:) and period (.), respectively. Residues comprising the three alpha helices are indicated in boxes.

This observation of independent genome engagement indicated that SOX TFs possess a pioneering activity of establishing a transcriptional complex for target gene regulation (Hou, Srivastava, & Jauch, 2017).

Based on electrophoretic mobility shift assays (EMSA), SOX2 cannot cooperatively bind with OCT4 on "compressed" *Oct-Sox* motifs unlike SOX17, where the distance between the *Octamer* and *Sox* binding site is reduced compared to "canonical" motifs, possibly due to steric hindrance (Jauch et al., 2011). This exclusivity may be important for the partial redistribution of OCT4 between canonical and compressed motifs during lineage commitment (Aksoy et al., 2013). Consistent with this notion, a single Glu122Lys point mutation in the SOX17 HMG domain (SOX17EK), which is part of the interaction interface with OCT4 (Fig. 2), converted the mutant TF to function like SOX2 in supporting the acquisition of induced pluripotency (Jauch et al., 2011; Palasingam et al., 2009; Reményi et al., 2003; Williams et al., 2004). Indeed, SOX17EK showed cooperative binding with OCT4 on the canonical *Oct-Sox* motif (Aksoy et al., 2013; Jauch et al., 2011). Consistently, a reciprocal mutant SOX2KE (Lys59Glu) adopted the endoderm specifying activity of SOX17 when overexpressed in mESCs (Jauch et al., 2011), which upon a further mutation (Glu46Leu) resulted in efficient cooperative binding to the compressed motif with OCT4 (Merino et al., 2014). Given that the HMG domains of SOX2 and SOX17 are essentially identical between mouse and human (Fig. 2), the functional relevance of these mutants in human cells remains to be tested.

SOX2 also has a prominent role in mouse germline development. SOX2 is transiently repressed in BLIMP1$^+$ mPGCs in E7.25 late streak (LS) stage embryos but re-expressed soon after (Campolo et al., 2013; Kurimoto, Yabuta, et al., 2008; Scholer, Dressler, Rohdewohid, & Gruss, 1990; Yabuta et al., 2006). Subsequently, expression level of SOX2 begins to decline in fetal gonadal PGCs from E13.5 to 17.5 (Campolo et al., 2013). SOX2 re-expression is dependent on the presence of *Prdm14*, indicating that activation of *Sox2* is downstream of PRDM14 activity (Yamaji et al., 2008). Using a combination of Cre-expressing mouse lines, deletion of *Sox2* as early as E7.25–7.5 using *Blimp1*-Cre resulted in the reduction of STELLA$^+$ mPGCs in the proximal posterior region of E7.5 bud stage embryos (Campolo et al., 2013). These embryos further showed a complete absence of germ cells in both male and female gonads of E13.5 embryos. However, SOX2 does not have an inductive role in mPGC specification, as forced overexpression of SOX2 abrogates mPGCLC specification even when NANOG is co-overexpressed (Murakami et al., 2016), indicating that the somatic (likely neural) inductive function of SOX2 in mEpiLCs is dominant (Corsinotti et al., 2017; Zhao, Nichols, Smith, & Li, 2004) and further suggests that the transient window of SOX2 repression during mPGC specification may be important to safeguard the PGC fate (Kurimoto, Yabuta, et al., 2008; Yabuta et al., 2006). Nonetheless, deletion of *Sox2* between E9.0 and 10.5 using *TNAP*-Cre resulted in complete depletion of oocyte and pro-spermatogonia in prepuberal ovaries and perinatal testes. No effect in meiotic spermatocytes and oocytes was observed when *Sox2* was deleted using *Spo11*-Cre expressed in meiotic germ cells. These observations indicate that SOX2 is required for mPGC survival across a range of mPGC developmental stages up until meiosis (Campolo et al., 2013). Since OCT4 and NANOG are co-expressed with SOX2 in specified mPGCs (Kurimoto, Yabuta, et al., 2008; Yabuta et al., 2006), it is not clear whether the role of SOX2 in premeiotic mPGCs involves cooperative interactions with OCT4 and NANOG or functions distinctly. In any case, since *SOX2* is not expressed in hPGCs (Irie et al., 2015; Perrett et al., 2008), it is unlikely that any regulatory functions for SOX2 exist in the human germline.

4.4 PRDM family genes in mouse and human PGC specification

PRDM is a metazoan-specific gene family coding for proteins featuring an N-terminal PR domain and, with the exception of PRDM11, a variable-number array of zinc fingers (ZFs) at the C-terminus (Hohenauer & Moore, 2012; Vervoort, Meulemeester, Behague, & Kerner, 2016). There are 17 gene family members in primates (Vervoort et al., 2016). The PR domain is related to the SET domain, characteristic of histone lysine methyltransferases (HMTs), while ZFs are responsible for DNA binding and protein-protein interactions (Hohenauer & Moore, 2012). PRDMs are TFs that modulate gene activity either directly, via their intrinsic HMT activity (which is seen in some of them), or indirectly, by recruiting other TFs and enzymes to specific genomic loci (Hohenauer & Moore, 2012).

PRDM family members can act both as repressors and activators and often display context-dependent roles, mostly defined by histone-modifying enzymes they partner with (Hohenauer & Moore, 2012). PRDMs have been implicated in various developmental and pathological processes, including forelimb development, neurogenesis, blood, vascular and immune system development, as well as leukemia, breast and gastrointestinal tract cancers (Fog, Galli, & Lund, 2012; Morgan et al., 2009). Importantly,

BLIMP1 (PRDM1) and PRDM14 are critical TFs for mouse germ cell specification. In other mammals, the role of BLIMP1 in PGC fate seems to be conserved, while that of PRDM14 is less clear.

4.4.1 BLIMP1 is a conserved PGC regulator in mouse and human

BLIMP1 (encoded by *Prdm1*) is a founding member of the PRDM family. It contains a PR domain, five ZFs, of which ZFs 1–2 preferentially bind to GAAAG DNA motif, and proline-rich domains, that help recruit protein partners (Doody et al., 2010; Su et al., 2009). BLIMP1 was identified as a TF controlling B cell differentiation into plasma cells (Shaffer et al., 2002). However, it is expressed in a variety of tissues during embryonic and postnatal development and has significant roles in PGC, forelimb and placental development, as well as in adult mammary gland formation and skin homeostasis (Ahmed, Elias, Mould, Bikoff, & Robertson, 2016; Magnusdottir et al., 2007; Mould, Morgan, Li, Bikoff, & Robertson, 2012; Ohinata et al., 2005; Robertson et al., 2007; Vincent et al., 2005). BLIMP1 is dispensable for the early axis formation, but *Prdm1* KO embryos die at ~10.5 due to defects in caudal branchial arch and the labyrinthine layer of the placenta (Vincent et al., 2005).

BLIMP1 is an essential factor for mouse germline specification (Ohinata et al., 2005). It is first upregulated in proximal epiblast cells at E6.25 that go on to become mPGCs (Fig. 1), as evidenced by lineage-tracing experiments (Ohinata et al., 2005). *Prdm1* KO led to reduced PGC specification (20 mPGC-like cells instead of ~40 mPGCs in the wildtype); mutant cells failed to repress homeobox genes, showed aberrant proliferation and migration, identifying BLIMP1 as the critical regulator of germ cell fate. Importantly, *Prdm1* KO also derepressed *de novo* DNA methyltransferase *Dnmt3b* and *Uhrf1*, the cofactor of maintenance DNA methyltransferase DNMT1, which resulted in sustained high global DNA methylation compared to normal mPGCs (Magnúsdóttir et al., 2013).

BLIMP1 lacks apparent HMT activity, but exerts its repressive or activating functions by recruiting chromatin modifiers, including HMTs and histone deacetylases. For example, BLIMP1 cooperates with PRTM5, an HMT that symmetrically demethylates arginine 3 on histone H2A and/or H4 (H2A/H4 R3). Interestingly, mPGCs from E8.5 to E11.5 transiently display high levels of H2A/H4 R3 methylation, concomitantly with nuclear co-localization of BLIMP1-PRMT5 complex. The deposition of this mark has functional importance, since PRMT5 loss and concomitant depletion of H2A/H4R3 methylation leads to male and female sterility (Kim et al., 2014).

Many of the aforementioned roles of BLIMP1 seem to be conserved in hPGCs, with a key distinction that in humans it acts downstream of and together with SOX17 (Irie et al., 2015; Kojima et al., 2017). Aggregates of *PRDM1*-null hESCs treated with BMP2/4 can yield a small number of aberrant $SOX17^{+}AP^{+}$ hPGCLCs, which, however, lack *NANOS3*, *KLF4* and *UTF1*, and show significant reduction in the levels of *TFAP2C*, *OCT4* and *NANOG*. After several days in culture, these mutant hPGCLCs are lost, potentially reflecting lack of proliferation (Irie et al., 2015; Tang et al., 2015). Similar to mPGCs, BLIMP1-deficient hPGCLCs derepressed somatic markers, especially endoderm genes, likely induced by SOX17 (Irie et al., 2015). BLIMP1 loss in human seminoma cell line also leads to upregulation of somatic transcripts (Weber et al., 2010), confirming the key role of BLIMP1 in repression of alternative lineages.

A recent study demonstrated that combined overexpression of BLIMP1 with SOX17 in hESCs can induce hPGCLCs even in the absence of relevant signaling cues. This clearly shows that the two factors are critical for hPGC fate. BLIMP1 gain-of-function alone is, however, insufficient to induce hPGCLCs (Kobayashi et al., 2017). Crucially, BLIMP1 is also expressed in germ cells of pigs and primates (Kobayashi et al., 2017; Sasaki et al., 2016), which further suggests its conserved involvement in mammalian PGC development. Interestingly, while frogs and fish have PRDM1 orthologs, their loss does not impair germ cell specification, potentially due to a distinct mode of PGC specification (by germ plasm). Instead, BLIMP1 regulates anterior mesendoderm and prechordal plate development in these species (de Souza et al., 1999; Wilm & Solnica-Krezel, 2005).

4.4.2 PRDM14 is critical for mouse PGC fate with unclear functions in human PGCs

PRDM14 is a critical regulator of mouse PGC specification and pluripotency in both mouse and human (Nakaki & Saitou, 2014; Seki, 2018). It features one PR domain with no reported HMT activity and six zinc fingers responsible for DNA binding (Hohenauer & Moore, 2012). While the PR domain and zinc fingers are conserved among vertebrates, the N-terminal part is more divergent (Hohenauer & Moore, 2012). Nevertheless, human PRDM14 can rescue some defects observed in $Prdm14^{-/-}$ mESCs, suggesting at least partial biochemical conservation (Nady et al., 2015).

In mice, PRDM14 is first expressed from 2- to 8-cell stage embryos (Fig. 1) and higher PRDM14 levels at the 4-cell stage are predictive of the ICM as opposed to the PrE fate (Burton et al., 2013). After transient repression, PRDM14 reappears in the epiblast of the ICM, but not in the trophectoderm (TE) or PrE (Yamaji et al., 2008). In line with this, PRDM14 is expressed in mESCs—the *in vitro* counterparts of the naïve epiblast (Ma, Swigut, Valouev, Rada-Iglesias, & Wysocka, 2011). PRDM14 is then quickly downregulated in the postimplantation embryo followed by upregulation specifically in nascent mPGCs (Yamaji et al., 2008); it persists in mPGCs until E13.5 when female PGCs initiate meiosis and male PGCs undergo mitotic arrest (Nakaki & Saitou, 2014).

In human embryos, PRDM14 is first detected at the 8-cell stage (Fig. 1), it is then continuously expressed in the ICM (Yan et al., 2013). Of note, unlike in the mouse, monkey embryos display sustained PRDM14 expression until after gastrulation (Nakamura et al., 2016). It is possible that the same is true for human embryos, since conventional hESCs, which resemble postimplantation epiblast, also express PRDM14 (Chia et al., 2010; Tsuneyoshi et al., 2008). It is then detected specifically in gonadal hPGCs, however, predominantly in the cytoplasm (Irie et al., 2015).

Reflecting its specific expression in the ICM and mPGCs, PRDM14 has been implicated in pluripotency maintenance and germ cell specification. Interestingly, *Prdm14* KO does not affect mouse embryonic development on a mixed genetic background and *Prdm14*$^{-/-}$ mESCs can be derived in naïve conditions (Yamaji et al., 2008, 2013). Instead, *Prdm14* KO mice are sterile due to complete lack of germ cells in both females and males (Yamaji et al., 2008). PRDM14 is among the earliest known mPGC-specific markers, expressed from E6.5 (Ohinata et al., 2005; Yamaji et al., 2008). It acts together with BLIMP1 and TFAP2C to upregulate germline-specific genes and repress somatic transcripts (Magnúsdóttir et al., 2013; Nakaki et al., 2013). PRDM14 is also key to re-acquisition of potential pluripotency, as *Prdm14*$^{-/-}$ mPGCs fail to fully upregulate *Sox2* and *Stella* and show impaired capacity to generate mouse embryonic germ cells (mEGCs) (Yamaji et al., 2008). The critical role of PRDM14 in mPGC specification is highlighted by its ability to induce mPGC fate *in vitro*. When overexpressed in aggregates of mPGC-competent mEpiLCs, PRDM14 induces mPGCLCs even in the absence of cytokines (Nakaki et al., 2013).

In *Prdm14*$^{-/-}$ mice, low numbers of mPGCs are still specified, but they are progressively lost and fully disappear by E12.5 (Yamaji et al., 2008).

In line with this, *Prdm14*-null mESCs can still generate mPGCLCs *in vitro*, but with aberrant transcriptional and epigenetic profile (Shirane et al., 2016). In particular, mutant mPGCs and mPGCLCs showed delayed and reduced expression of *Prdm1*, *Tfap2c*, *Dnd1* and *Nanos3* and failed to upregulate *Sox2* and *Stella* (Grabole et al., 2013; Shirane et al., 2016). Interestingly, however, some other key pluripotency genes, including *Pou5f1* and *Nanog*, were not affected by PRDM14 loss.

Most prominently, PRDM14 orchestrates global epigenetic reprogramming (Nakaki & Saitou, 2014; Seki, 2018). In both mPGCs and naïve mESCs, PRDM14 has been intimately linked with global DNA demethylation. The latter can occur via two major pathways: (1) progressive replication-associated dilution when *de novo* (DNMT3A, DNMT3B and DNMT3L) and maintenance (DNMT1 and UHRF1) DNA methylation machineries are repressed and (2) active DNA-demethylation through oxidation of the methyl group on 5-methylcytosine (5mC) to 5-hydroxymethylcytosine (5hmC) by the TET (Ten-Eleven Translocation) enzymes. 5hmC cannot serve as a template for DNA methylation maintenance and can be removed by the base excision repair (BER) pathway (Hackett et al., 2013; Hajkova et al., 2010; Hill, Amouroux, & Hajkova, 2014; Tang et al., 2016). PRDM14-deficient cells display defects in both types of DNA demethylation, since they fail to repress *Dnmt3b* and *Uhrf1* in mPGCs or mPGCLC and do not recruit TET enzymes to target loci in mESCs and mPGCs (Okashita et al., 2014, 2015; Shirane et al., 2016). Together this results in considerable genome hypermethylation compared to wildtype (WT) controls.

In addition to defects in DNA demethylation, *Prdm14*$^{-/-}$ mPGCs fail to reduce H3K9me2 and increase H3K27me3 levels, presumably due to derepression of *Ehmt1* and loss of Polycomb repressive complex 2 (PRC2) recruitment, respectively (Yamaji et al., 2008). Furthermore, PRDM14 has been implicated in X-chromosome reactivation. In mESCs it binds to and represses the *Xist* locus, which is known to initiate X inactivation (Ma et al., 2011). Further, *Prdm14* overexpression during conversion of primed mEpiSCs to naïve mESCs promoted X reactivation (Gillich et al., 2012). Conversely, mutant female embryos or iPSCs show delayed or incomplete X-chromosome reactivation (Payer et al., 2013). Overexpression and loss-of-function studies demonstrated that PRDM14 is important for iPSCs generation in both mouse and human (Chia et al., 2010; Payer et al., 2013). In hiPSC it can replace KLF4, which is one of the core reprogramming factors (Chia et al., 2010).

Of note, unlike in the mixed genetic background, C57BL/6 *Prdm14*$^{-/-}$ mice showed reduced survival during postimplantation development, presumably resulting from earlier epigenetic defects before implantation (Payer et al., 2013). Additional evidence for PRDM14 function in early embryogenesis comes from *in vitro* studies. Role of PRDM14 in naïve mESCs is twofold: it promotes DNA hypomethylation (see above) and antagonizes the pro-differentiation FGF receptor (FGFR) signaling (Grabole et al., 2013; Yamaji et al., 2013). PRDM14 KO leads to derepression of *de novo* DNA methyltransferases *Dnmt3a, Dnmt3b,* and *Dnmt3l* and mutant mESCs display a highly methylated genome even in 2i/LIF conditions. Also, PRDM14 represses multiple components of the FGFR signaling, such as *Fgfr1, Fgfr2, Epha4* and *Hs6st2*, reducing cellular response to FGF ligands. Consistently, mESCs cultured in serum are vulnerable to PRDM14 loss and *Prdm14*$^{-/-}$ blastocysts do not yield mESCs in serum/LIF conditions (Yamaji et al., 2013). By contrast, 2i/LIF mESCs, shielded from pro-differentiation FGFR signaling by a MEK inhibitor, can be derived and maintained without PRDM14 (Yamaji et al., 2013). *Prdm14*$^{-/-}$ mESC can even contribute to chimeras, but less efficiently than their WT counterparts. Teratomas derived from *Prdm14*$^{-/-}$ 2i/LIF mESCs contained all three germ layers but were heavily biased toward neuronal lineages (Yamaji et al., 2013). Interestingly, stabilization of the core pluripotency circuitry by GSK3β inhibitor maintains serum-grown *Prdm14*$^{-/-}$ mESCs even without FGFR/MEK inhibition, suggesting compensatory mechanisms (Yamaji et al., 2013). PRDM14 was also shown to activate AKT-mTORC1 signaling via upregulation of *Tcl1* in mESCs, thereby promoting their survival (Grabole et al., 2013; Yamaji et al., 2013). Interestingly, while PRDM14 regulates *Sox2* in mPGCs, it is dispensable for the expression of *Sox2, Nanog* and *Pou5f1* in mESCs (Grabole et al., 2013; Ma et al., 2011; Yamaji et al., 2013).

A knockdown study in serum/LIF conditions showed that PRDM14 prevents PrE differentiation, corroborated by PRDM14 binding to many PrE-associated loci (Ma et al., 2011). A complete KO in more homogenous 2i/LIF mESCs, however, showed that PRDM14 is mainly repressing neuronal differentiation, as evidenced by teratoma assays, microarray and chromatin immunoprecipitation sequencing (ChIP-seq) analysis (Yamaji et al., 2013). Both ChIP-seq studies found that mouse PRDM14 binds predominantly far from promoters, with around 4% (Yamaji et al., 2013) or 10% (Ma et al., 2011) of peaks located at promoter regions, i.e., within 1 kb of the transcription start site (TSS).

Forced *Prdm14* expression prevents upregulation of lineage markers during EB differentiation from both mouse and human ESCs (Okashita et al., 2015; Tsuneyoshi et al., 2008). Also, PRDM14 overexpression in adherent mEpiLC culture reverted the cells back to naïve pluripotency (Okashita et al., 2016) (unlike mPGCLC induction from mEpiLC aggregates). Altogether these findings suggest that PRDM14 is important for pluripotency acquisition and maintenance and its downregulation can promote differentiation.

In hESCs, PRDM14 is a core pluripotency factor that directly regulates *POU5F1* via its proximal enhancer (Chia et al., 2010). PRDM14 knockdown led to reduced levels of OCT4, SOX2 and NANOG and upregulated the expression of lineage markers (Chan et al., 2013; Chia et al., 2010; Tsuneyoshi et al., 2008). Interestingly, ChIP-seq in hESCs showed that the PRDM14 motif is conserved, but the binding profile is drastically different to that in the mouse, with human PRDM14 preferentially binding around gene promoters (Chia et al., 2010). The set of target genes is also poorly conserved. For example, among FGFR signaling genes, only *FGFR1* was found to be a shared PRDM14 target between mouse and human. This is not surprising considering that hESCs require FGFR signaling for pluripotency and survival. Such discrepancy in motif and binding pattern is not unique to PRDM14 and has also been observed for OCT4 and NANOG (Kunarso et al., 2010). This could either reflect broader differences between primed and naïve pluripotent states or the divergence of gene regulatory elements in the two species. Indeed, pluripotency factors often bind to enhancers derived from young transposable elements (TEs) (Kunarso et al., 2010). It would be of interest to test if PRDM14 target sequences have a significant proportion of species-specific TEs. Importantly, PRDM14 is integrated in the core hESC TF network, co-occupying targets with OCT4, SOX2 and NANOG. Its own regulatory regions are co-occupied by these factors, similar to mouse (Chia et al., 2010). Altogether this suggests that while PRDM14 is important for pluripotency in the two species, its functions have diverged significantly.

The function of PRDM14 in hPGC fate has been debated, since it appears to be predominantly cytoplasmic in gonadal hPGCs, showing relatively low but specific expression (Irie et al., 2015). Porcine PGCs also displayed mainly cytoplasmic PRDM14 staining (Kobayashi et al., 2017). Also, in hPGCLCs its expression is lower than in hESCs (Irie et al., 2015). Furthermore, a knockdown study suggested that PRDM14 is dispensable for hPGCLC specification (Sugawa et al., 2015). Importantly, however, the knockdown efficiency was low and the homogeneity and kinetics of PRDM14 depletion were not assessed. Short hairpin RNAs (shRNAs) were delivered to D2 hPGCLC precursors already expressing SOX17 and BLIMP1 (Sugawa et al., 2015), which could further complicate data interpretation.

PRDM14 can both activate and repress genes in mouse and human ESCs, as well as in mPGCs and mPGCLCs. Since PRDM14 does not possess any apparent enzymatic activity (Hohenauer & Moore, 2012), its activating or repressive roles have been attributed to putative binding partners. Some reports have provided evidence for PRDM14 association with PRC2, known to deposit the repressive H3K27me3 mark, as well as PRC2 recruiter JARID2 (Payer et al., 2013; Yamaji et al., 2013). Other studies, however, could not reproduce these results (Nady et al., 2015; Tu et al., 2016). Two groups reported stoichiometric interaction between PRDM14 and corepressor CBFA2T2 (also known as Mtgr1) in mESCs and NCCIT, a human embryonic carcinoma cell line (Nady et al., 2015; Tu et al., 2016). Interestingly, *Cbfa2t2* KO phenocopies many of the effects of *Prdm14* deletion in mESC and early mPGCs (Nady et al., 2015; Tu et al., 2016). CBFA2T2, however, does not have an enzymatic activity either and thus does not resolve the aforementioned question. Gene activation, at least in some contexts, could be related to PRDM14 interaction with ESRRB. Indeed, in mESCs PRDM14-activated targets were enriched in ESSRB motif (Yamaji et al., 2013) and the two proteins can interact physically (Iseki et al., 2016). Importantly, ESSRB has recently been shown to have a key role in mPGC development, capable of rescuing *Nanog* loss (Zhang, Leitch, et al., 2018). The potential cooperation of ESSRB with PRDM14 in this context warrants investigation. Also, as mentioned above, PRDM14 forms a complex with TET proteins thereby contributing to site-specific active DNA-demethylation and subsequent gene activation (Okashita et al., 2014).

5. Perspective

The germline is arguably an "immortal" lineage, which in part can be attributed to its unique epigenetic re-setting. Germ cells generate the totipotent state and transmit genetic and epigenetic information along with cytoplasmic components to subsequent generations (Irie et al., 2018). Whether environmentally induced epigenetic modifications in the mammalian germline persist, possibly transgenerationally, remains to be fully investigated. A deeper understanding of the molecular mechanisms that govern human germ cell development, from PGCs to gametes, is a fundamental biological question with implications for human health.

Recent studies shed light on some unique features of human PGCs compared to the mouse, including the role of SOX17 as the major regulator of the human germline with concomitant repression of SOX2, as well as the upregulation of distinct sets of somatic and naïve pluripotency markers (Tang et al., 2015). Some factors, such as BLIMP1 and TFAP2C, seem to have largely equivalent roles. On the other hand, the roles of PRDM14, if any, in the human germline are yet to be fully investigated, considering its key functions in mPGC development. These and other differences might stem from distinct embryo morphology in rodents and other mammalian species around the time of gastrulation and PGC formation. Indeed, many non-rodent species with a flat bilaminar disc similar to humans show expression of SOX17 in their PGCs (Table 2) (Irie et al., 2018). It would also be informative to test the temporal expression and functions of other PGC- and pluripotency-related TFs in human germ cells and in other non-rodent model species with distinct embryo morphologies to understand the evolutionary significance of developmental diversification.

Aberrant PGC differentiation can lead to infertility and germ cell tumors, which are the most prevalent solid cancer type in young men (Hayes-Lattin & Nichols, 2009; Matzuk & Lamb, 2008). Given the emerging differences between mouse and human PGCs, the ability to model human germ line development *in vitro* is an important step toward elucidating human-specific mechanisms of normal development and disease. Future studies could establish additional experimental models *in vitro*, which could allow obtaining mature sperm or egg from patient-specific iPSCs for potential treatment of infertility. A recent study reported human oogonia-like cells from hPGCLCs *in vitro* (Yamashiro et al., 2018). However, the efficiency of the procedure was low and the resulting cells did not recapitulate some aspects of *bona fide* oogonia. An authentic reproducible model that recapitulates key developmental events *in vivo* is essential for advancing our knowledge of human germline biology.

Acknowledgments

We thank the members of the Surani lab for helpful discussions. We are also grateful to Dr. Christopher Andrew Penfold for critical reading of the manuscript. The Gurdon Institute receives core funding from the Wellcome Trust and Cancer Research UK. M.A.S. is a Wellcome Trust Senior Investigator.

References

Acampora, D., Mazan, S., Lallemand, Y., Avantaggiato, V., Maury, M., Simeone, A., et al. (1995). Forebrain and midbrain regions are deleted in Otx2-/- mutants due to a defective anterior neuroectoderm specification during gastrulation. *Development (Cambridge, England)*, *121*, 3279–3290.

Aeckerle, N., Drummer, C., Debowski, K., Viebahn, C., & Behr, R. (2015). Primordial germ cell development in the marmoset monkey as revealed by pluripotency factor expression: Suggestion of a novel model of embryonic germ cell translocation. *Molecular Human Reproduction*, *21*, 66–80.

Ahmed, M. I., Elias, S., Mould, A. W., Bikoff, E. K., & Robertson, E. J. (2016). The transcriptional repressor Blimp1 is expressed in rare luminal progenitors and is essential for mammary gland development. *Development*, *143*, 1663–1673.

Aksoy, I., Jauch, R., Chen, J., Dyla, M., Divakar, U., Bogu, G. K., et al. (2013). Oct4 switches partnering from Sox2 to Sox17 to reinterpret the enhancer code and specify endoderm. *EMBO Journal*, *32*, 938–953.

Ambrosetti, D. C., Basilico, C., & Dailey, L. (1997). Synergistic activation of the fibroblast growth factor 4 enhancer by Sox2 and Oct-3 depends on protein-protein interactions facilitated by a specific spatial arrangement of factor binding sites. *Molecular and Cellular Biology*, *17*, 6321–6329.

Ambrosetti, D. C., Schöler, H. R., Dailey, L., & Basilico, C. (2000). Modulation of the activity of multiple transcriptional activation domains by the DNA binding domains mediates the synergistic action of Sox2 and Oct-3 on the fibroblast growth factor-4 enhancer. *Journal of Biological Chemistry*, *275*, 23387–23397.

Anderson, R., Copeland, T. K., Scholer, H., Heasman, J., & Wylie, C. (2000). The onset of germ cell migration in the mouse embryo. *Mechanisms of Development*, *91*, 61–68.

Aramaki, S., Hayashi, K., Kurimoto, K., Ohta, H., Yabuta, Y., Iwanari, H., et al. (2013). A mesodermal factor, T, specifies mouse germ cell fate by directly activating germline determinants. *Developmental Cell*, *27*, 516–529.

Artus, J., Piliszek, A., & Hadjantonakis, A. K. (2011). The primitive endoderm lineage of the mouse blastocyst: Sequential transcription factor activation and regulation of differentiation by Sox17. *Developmental Biology*, *350*, 393–404.

Auman, H. J., Nottoli, T., Lakiza, O., Winger, Q., Donaldson, S., & Williams, T. (2002). Transcription factor AP-2gamma is essential in the extra-embryonic lineages for early postimplantation development. *Development*, *129*, 2733–2747.

Avilion, A. A., Nicolis, S. K., Pevny, L. H., Perez, L., Vivian, N., & Lovell-Badge, R. (2003). Multipotent cell lineages in early mouse development depend on SOX2 function. *Genes and Development*, *17*, 126–140.

Beccari, L., Moris, N., Girgin, M., Turner, D. A., Baillie-Johnson, P., Cossy, A. C., et al. (2018). Multi-axial self-organization properties of mouse embryonic stem cells into gastruloids. *Nature*, *562*(7726), 272–276.

Beddington, R. S. (1982). An autoradiographic analysis of tissue potency in different regions of the embryonic ectoderm during gastrulation in the mouse. *Journal of Embryology and Experimental Morphology*, *69*, 265–285.

Beddington, R. S., & Robertson, E. J. (1999). Axis development and early asymmetry in mammals. *Cell*, *96*(2), 195–209. Review.

Behringer, R. R., Wakamiya, M., Tsang, T. E., & Tam, P. P. (2000). A flattened mouse embryo: Leveling the playing field. *Genesis*, *28*(1), 23–30. Review.

Blakeley, P., Fogarty, N. M. E., del Valle, I., Wamaitha, S. E., Hu, T. X., Elder, K., et al. (2015). Defining the three cell lineages of the human blastocyst by single-cell RNA-seq. *Development*, *142*, 3151–3165.

Blin, G., Picart, C., Théry, M., & Puceat, M. (2017). Geometrical confinement guides Brachyury self-patterning in embryonic stem cells. *BioRxiv*, 1–46.

Bookhout, C. G. (1937). The germ cell cycle in the guinea pig. *Cell and Tissue Research*, *25*, 728–748.

Boroviak, T., Loos, R., Lombard, P., Okahara, J., Behr, R., Sasaki, E., et al. (2015). Lineage-specific profiling delineates the emergence and progression of naive pluripotency in mammalian embryogenesis. *Developmental Cell*, *35*, 366–382.

Boroviak, T., & Nichols, J. (2017). Primate embryogenesis predicts the hallmarks of human naïve pluripotency. *Development*, *144*, 175–186.

Boroviak, T., Stirparo, G. G., Dietmann, S., Hernando-Herraez, I., Mohammed, H., Reik, W., et al. (2018). Single cell transcriptome analysis of human, marmoset and mouse embryos reveals common and divergent features of preimplantation development. *Development*, *145* dev167833.

Bowles, J., Schepers, G., & Koopman, P. (2000). Phylogeny of the SOX family of developmental transcription factors based on sequence and structural indicators. *Developmental Biology*, *227*, 239–255.

Boyer, L. A., Tong, I. L., Cole, M. F., Johnstone, S. E., Levine, S. S., Zucker, J. P., et al. (2005). Core transcriptional regulatory circuitry in human embryonic stem cells. *Cell*, *122*, 947–956.

Braude, P., Bolton, V., & Moore, S. (1988). Human gene expression first occurs between the four- and eight-cell stages of preimplantation development. *Nature*, *332*, 459–461.

Brinster, R. L. (2002). Germline stem cell transplantation and transgenesis. *Science*, *296*, 2174–2176.

Brons, I. G. M., Smithers, L. E., Trotter, M. W. B., Rugg-Gunn, P., Sun, B., Chuva De Sousa Lopes, S. M., et al. (2007). Derivation of pluripotent epiblast stem cells from mammalian embryos. *Nature*, *448*, 191–195.

Burton, A., Muller, J., Tu, S., Padilla-Longoria, P., Guccione, E., & Torres-Padilla, M. E. (2013). Single-cell profiling of epigenetic modifiers identifies PRDM14 as an inducer of cell fate in the mammalian embryo. *Cell Reports*, *5*, 687–701.

Campolo, F., Gori, M., Favaro, R., Nicolis, S., Pellegrini, M., Botti, F., et al. (2013). Essential role of Sox2 for the establishment and maintenance of the germ cell line. *Stem Cells*, *31*, 1408–1421.

Chambers, I., Silva, J., Colby, D., Nichols, J., Nijmeijer, B., Robertson, M., et al. (2007). Nanog safeguards pluripotency and mediates germline development. *Nature*, *450*, 1230–1234.

Chambers, I., & Tomlinson, S. R. (2009). The transcriptional foundation of pluripotency. *Development*, *136*, 2311–2322.

Chan, Y. S., Goke, J., Lu, X., Venkatesan, N., Feng, B., Su, I. H., et al. (2013). A PRC2-dependent repressive role of PRDM14 in human embryonic stem cells and induced pluripotent stem cell reprogramming. *Stem Cells*, *31*, 682–692.

Chang, H., Huylebroeck, D., Verschueren, K., Guo, Q., Matzuk, M. M., & Zwijsen, A. (1999). Smad5 knockout mice die at mid-gestation due to multiple embryonic and extraembryonic defects. *Development*, *126*, 1631.

Chang, H., & Matzuk, M. M. (2001). Smad5 is required for mouse primordial germ cell development. *Mechanisms of Development*, *104*, 61–67.

Chatfield, J., O'Reilly, M. A., Bachvarova, R. F., Ferjentsik, Z., Redwood, C., Walmsley, M., et al. (2014). Stochastic specification of primordial germ cells from mesoderm precursors in axolotl embryos. *Development*, *141*, 2429–2440.

Chen, D., Liu, W., Lukianchikov, A., Hancock, G. V., Zimmerman, J., Lowe, M. G., et al. (2017). Germline competency of human embryonic stem cells depends on eomesodermin. *Biology of Reproduction*, *97*, 850–861.

Chen, X., Xu, H., Yuan, P., Fang, F., Huss, M., Vega, V. B., et al. (2008). Integration of external signaling pathways with the core transcriptional network in embryonic stem cells. *Cell*, *133*, 1106–1117.

Chen, J., Zhang, Z., Li, L., Chen, B. C., Revyakin, A., Hajj, B., et al. (2014). Single-molecule dynamics of enhanceosome assembly in embryonic stem cells. *Cell, 156*, 1274–1285.
Chew, J.-l., Tam, W.-L., Yeap, L.-s., Li, P., Ang, Y.-S., Lim, B., et al. (2005). Reciprocal transcriptional regulation of Pou5f1 and Sox2 via the Oct4/Sox2 complex in embryonic stem cells. *Microbiology, 25*, 6031–6046.
Chia, N. Y., Chan, Y. S., Feng, B., Lu, X., Orlov, Y. L., Moreau, D., et al. (2010). A genome-wide RNAi screen reveals determinants of human embryonic stem cell identity. *Nature, 468*, 316–320.
Chiquoine, A. D. (1954). The identification, origin, and migration of the primordial germ cells in the mouse embryo. *The Anatomical Record, 118*, 135–146.
Chuva de Sousa Lopes, S. M., & Roelen, B. A. J. (2008). Primordial germ cell specification: The importance of being 'blimped'. *Histology and Histopathology, 23*, 1553–1561.
Clark, A. T., Bodnar, M. S., Fox, M., Rodriquez, R. T., Abeyta, M. J., Firpo, M. T., et al. (2004). Spontaneous differentiation of germ cells from human embryonic stem cells in vitro. *Human Molecular Genetics, 13*, 727–739.
Clevers, H. (2016). Modeling development and disease with organoids. *Cell, 165*, 1586–1597.
Corsinotti, A., Wong, F. C. K., Tatar, T., Szczerbinska, I., Halbritter, F., Colby, D., et al. (2017). Distinct SoxB1 networks are required for naïve and primed pluripotency. *eLife, 6*, 1–28.
Daheron, L., Opitz, S. L., Zaehres, H., Lensch, M. W., Andrews, P. W., Itskovitz-Eldor, J., et al. (2004). LIF/STAT3 signaling fails to maintain self-renewal of human embryonic stem cells. *Stem Cells, 22*, 770–778.
D'Amour, K. A., Agulnick, A. D., Eliazer, S., Kelly, O. G., Kroon, E., & Baetge, E. E. (2005). Efficient differentiation of human embryonic stem cells to definitive endoderm. *Nature Biotechnology, 23*, 1534–1541.
De Felici, M. (2013). Origin, migration, and proliferation of human primordial germ cells. In G. Coticchio, D. F. Albertini, & L. De Santis (Eds.), *Oogenesis* (pp. 19–37) London, New York: Springer.
de Jong, J., Stoop, H., Gillis, A. J. M., van Gurp, R. J. H. L. M., van de Geijn, G.-J. M., de Boer, M., et al. (2008). Differential expression of SOX17 and SOX2 in germ cells and stem cells has biological and clinical implications. *Journal of Pathology, 215*, 21–30.
de Souza, F. S., Gawantka, V., Gomez, A. P., Delius, H., Ang, S. L., & Niehrs, C. (1999). The zinc finger gene Xblimp1 controls anterior endomesodermal cell fate in Spemann's organizer. *The EMBO Journal, 18*, 6062–6072.
Deglincerti, A., Croft, G. F., Pietila, L. N., Zernicka-Goetz, M., Siggia, E. D., & Brivanlou, A. H. (2016). Self-organization of the in vitro attached human embryo. *Nature, 533*, 251–254.
Doody, G. M., Care, M. A., Burgoyne, N. J., Bradford, J. R., Bota, M., Bonifer, C., et al. (2010). An extended set of PRDM1/BLIMP1 target genes links binding motif type to dynamic repression. *Nucleic Acids Research, 38*, 5336–5350.
Downs, K. M., & Davies, T. (1993). Staging of gastrulating mouse embryos by morphological landmarks in the dissecting microscope. *Development, 118*, 1255–1266.
Eakin, G. S., & Behringer, R. R. (2004). Diversity of germ layer and axis formation among mammals. *Seminars in Cell and Developmental Biology, 15*, 619–629.
Easley, C. A. t., Phillips, B. T., McGuire, M. M., Barringer, J. M., Valli, H., Hermann, B. P., et al. (2012). Direct differentiation of human pluripotent stem cells into haploid spermatogenic cells. *Cell Reports, 2*, 440–446.
Eiraku, M., Takata, N., Ishibashi, H., Kawada, M., Sakakura, E., Okuda, S., et al. (2011). Self-organizing optic-cup morphogenesis in three-dimensional culture. *Nature, 472*, 51–58.

Enders, A. C., & Lopata, A. (1999). Implantation in the marmoset monkey: Expansion of the early implantation site. *Anatomical Record*, *256*, 279–299.

Etoc, F., Metzger, J., Ruzo, A., Kirst, C., Yoney, A., Ozair, M. Z., et al. (2016). A balance between secreted inhibitors and edge sensing controls gastruloid self-organization. *Developmental Cell*, *39*(3), 302–315.

Extavour, C. G., & Akam, M. (2003). Mechanisms of germ cell specification across the metazoans: Epigenesis and preformation. *Development*, *130*, 5869–5884.

Felix, W. (1912). *The development of the urinogenital organs, manual of human embryology II.* Philadelphia: J.B. Lippincott Company, The Washington Square Press.

Flach, G., Johnson, M. H., Braude, P. R., Taylor, R. A., & Bolton, V. N. (1982). The transition from maternal to embryonic control in the 2-cell mouse embryo. *The EMBO Journal*, *1*, 681–686.

Fog, C. K., Galli, G. G., & Lund, A. H. (2012). PRDM proteins: Important players in differentiation and disease. *Bioessays*, *34*, 50–60.

Francois, M., Koopman, P., & Beltrame, M. (2010). SoxF genes: Key players in the development of the cardio-vascular system. *International Journal of Biochemistry and Cell Biology*, *42*, 445–448.

Fuss, A. (1911). Über extraregionare Geschlechtszellen bei einem menschlichen embryo von 4 Wochen. *Anat Am*, *39*, 407–409.

Gafni, O., Weinberger, L., Mansour, A. A., Manor, Y. S., Chomsky, E., Ben-Yosef, D., et al. (2013). Derivation of novel human ground state naive pluripotent stem cells. *Nature*, *504*, 282–286.

Gagliardi, A., Mullin, N. P., Ying Tan, Z., Colby, D., Kousa, A. I., Halbritter, F., et al. (2013). A direct physical interaction between Nanog and Sox2 regulates embryonic stem cell self-renewal. *EMBO Journal*, *32*, 2231–2247.

Gillich, A., Bao, S., Grabole, N., Hayashi, K., Trotter, M. W., Pasque, V., et al. (2012). Epiblast stem cell-based system reveals reprogramming synergy of germline factors. *Cell Stem Cell*, *10*, 425–439.

Ginsburg, M., Snow, M. H. L., & McLaren, A. (1990). Primordial germ cells in the mouse embryo during gastrulation. *Development*, *110*, 521–528.

Gkountela, S., Li, Z., Vincent, J. J., Zhang, K. X., Chen, A., Pellegrini, M., et al. (2013). The ontogeny of cKIT+ human primordial germ cells proves to be a resource for human germ line reprogramming, imprint erasure and in vitro differentiation. *Nature Cell Biology*, *15*, 113–122.

Gkountela, S., Zhang, K. X., Shafiq, T. A., Liao, W. W., Hargan-Calvopina, J., Chen, P. Y., et al. (2015). DNA demethylation dynamics in the human prenatal germline. *Cell*, *161*, 1425–1436.

Grabole, N., Tischler, J., Hackett, J. A., Kim, S., Tang, F., Leitch, H. G., et al. (2013). Prdm14 promotes germline fate and naive pluripotency by repressing FGF signalling and DNA methylation. *EMBO Reports*, *14*, 629–637.

Grigoryan, T., Wend, P., Klaus, A., & Birchmeier, W. (2008). Deciphering the function of canonical Wnt signals in development and disease. Conditional loss- and gain-of-function mutations of β-catenin in mice. *Genes & Development*, *1*, 2308–2341. Grigoryan, Wend, Klaus, Birchmeier.pdf.

Guo, G., Huss, M., Tong, G. Q., Wang, C., Li Sun, L., Clarke, N. D., et al. (2010). Resolution of cell fate decisions revealed by single-cell gene expression analysis from zygote to blastocyst. *Developmental Cell*, *18*, 675–685.

Guo, F., Yan, L., Guo, H., Li, L., Hu, B., Zhao, Y., et al. (2015). The transcriptome and DNA methylome landscapes of human primordial germ cells. *Cell*, *161*, 1437–1452.

Hackett, J. A., & Azim Surani, M. (2014). Regulatory principles of pluripotency: From the ground state up. *Cell Stem Cell*, *15*, 416–430.

Hackett, J. A., Sengupta, R., Zylicz, J. J., Murakami, K., Lee, C., Down, T. A., et al. (2013). Germline DNA demethylation dynamics and imprint erasure through 5-hydroxymethylcytosine. *Science*, *339*, 448–452.
Hajkova, P., Ancelin, K., Waldmann, T., Lacoste, N., Lange, U. C., Cesari, F., et al. (2008). Chromatin dynamics during epigenetic reprogramming in the mouse germ line. *Nature*, *452*, 877–881.
Hajkova, P., Jeffries, S. J., Lee, C., Miller, N., Jackson, S. P., & Surani, M. A. (2010). Genome-wide reprogramming in the mouse germ line entails the base excision repair pathway. *Science*, *329*, 78–82.
Hara, K., Kanai-Azuma, M., Uemura, M., Shitara, H., Taya, C., Yonekawa, H., et al. (2009). Evidence for crucial role of hindgut expansion in directing proper migration of primordial germ cells in mouse early embryogenesis. *Developmental Biology*, *330*, 427–439.
Harrison, S. E., Sozen, B., Christodoulou, N., Kyprianou, C., & Zernicka-Goetz, M. (2017). Assembly of embryonic and extraembryonic stem cells to mimic embryogenesis in vitro. *Science*, *356*(6334).
Hart, A. H., Hartley, L., Ibrahim, M., & Robb, L. (2004). Identification, cloning and expression analysis of the pluripotency promoting nanog genes in mouse and human. *Developmental Dynamics*, *230*, 187–198.
Hassoun, R., Schwartz, P., Feistel, K., Blum, M., & Viebahn, C. (2009). Axial differentiation and early gastrulation stages of the pig embryo. *Differentiation*, *78*, 301–311.
Hassoun, R., Schwartz, P., Rath, D., Viebahn, C., & Männer, J. (2010). Germ layer differentiation during early hindgut and cloaca formation in rabbit and pig embryos. *Journal of Anatomy*, *217*, 665–678.
Hayashi, K., Kobayashi, T., Umino, T., Goitsuka, R., Matsui, Y., & Kitamura, D. (2002). SMAD1 signaling is critical for initial commitment of germ cell lineage from mouse epiblast. *Mechanisms of Development*, *118*, 99–109.
Hayashi, K., Ogushi, S., Kurimoto, K., Shimamoto, S., Ohta, H., & Saitou, M. (2012). Offspring from oocytes derived from in vitro primordial germ cell-like cells in mice. *Science*, *338*, 971–975.
Hayashi, K., Ohta, H., Kurimoto, K., Aramaki, S., & Saitou, M. (2011). Reconstitution of the mouse germ cell specification pathway in culture by pluripotent stem cells. *Cell*, *146*, 519–532.
Hayashi, K., & Surani, M. A. (2009). Self-renewing epiblast stem cells exhibit continual delineation of germ cells with epigenetic reprogramming in vitro. *Development*, *136*, 3549–3556.
Hayes-Lattin, B., & Nichols, C. R. (2009). Testicular cancer: A prototypic tumor of young adults. *Seminars in Oncology*, *36*, 432–438.
Hearn, J. P., & Summers, P. M. (1986). Experimental manipulation of embryo implantation in the marmoset monkey and exotic equids. *Theriogenology*, *25*, 3–11.
Hikabe, O., Hamazaki, N., Nagamatsu, G., Obata, Y., Hirao, Y., Hamada, N., et al. (2016). Reconstitution in vitro of the entire cycle of the mouse female germ line. *Nature*, *539*, 299–303.
Hill, M. A. (2018). *Embryonic development.* [Online]. Retrieved from https://embryology.med.unsw.edu.au/embryology/index.php/Main_Page.
Hill, P. W., Amouroux, R., & Hajkova, P. (2014). DNA demethylation, Tet proteins and 5-hydroxymethylcytosine in epigenetic reprogramming: An emerging complex story. *Genomics*, *104*, 324–333.
Hill, J. P., & Florian, J. (1931). A young human embryo (Embryo Dobbin) with head-process and prochordal plate. *Philosophical Transactions of the Royal Society of London Series B*, *219*, 443–486.
Hohenauer, T., & Moore, A. W. (2012). The Prdm family: Expanding roles in stem cells and development. *Development*, *139*, 2267–2282.

Hopf, C., Viebahn, C., & Püschel, B. (2011). BMP signals and the transcriptional repressor BLIMP1 during germline segregation in the mammalian embryo. *Development Genes and Evolution*, *221*, 209–223.

Hosking, B. M., Muscat, G. E. O., Koopman, P. A., Dowhan, D. H., & Dunn, T. L. (1995). Trans-activation and DNA-binding properties of the transcription factor, Sox-18. *Nucleic Acids Research*, *23*, 2626–2628.

Hou, L., Srivastava, Y., & Jauch, R. (2017). Molecular basis for the genome engagement by Sox proteins. *Seminars in Cell and Developmental Biology*, *63*, 2–12.

Huch, M., Dorrell, C., Boj, S. F., van Es, J. H., Li, V. S. W., van de Wetering, M., et al. (2013). In vitro expansion of single Lgr5 + liver stem cells induced by Wnt-driven regeneration. *Nature*, *494*, 247.

Irie, N., Sybirna, A., & Surani, M. A. (2018). What can stem cell models tell us about human germ cell biology? *Current Topics in Developmental Biology*, *129*, 25–65.

Irie, N., Tang, W. W. C., & Azim Surani, M. (2014). Germ cell specification and pluripotency in mammals: A perspective from early embryogenesis. *Reproductive Medicine and Biology*, *13*, 203–215.

Irie, N., Weinberger, L., Tang, W. W., Kobayashi, T., Viukov, S., Manor, Y. S., et al. (2015). SOX17 is a critical specifier of human primordial germ cell fate. *Cell*, *160*, 253–268.

Iseki, H., Nakachi, Y., Hishida, T., Yamashita-Sugahara, Y., Hirasaki, M., Ueda, A., et al. (2016). Combined overexpression of JARID2, PRDM14, ESRRB, and SALL4A dramatically improves efficiency and kinetics of reprogramming to induced pluripotent stem cells. *Stem Cells*, *34*, 322–333.

Ishikura, Y., Yabuta, Y., Ohta, H., Hayashi, K., Nakamura, T., Okamoto, I., et al. (2016). In vitro derivation and propagation of spermatogonial stem cell activity from mouse pluripotent stem cells. *Cell Reports*, *17*, 2789–2804.

James, D., Levine, A. J., Besser, D., & Hemmati-Brivanlou, A. (2005). TGFbeta/activin/nodal signaling is necessary for the maintenance of pluripotency in human embryonic stem cells. *Development*, *132*, 1273–1282.

Jauch, R., Aksoy, I., Hutchins, A. P., Ng, C. K. L., Tian, X. F., Chen, J., et al. (2011). Conversion of Sox17 into a pluripotency reprogramming factor by reengineering its association with Oct4 on DNA. *Stem Cells*, *29*, 940–951.

Johnson, A. D., & Alberio, R. (2015). Primordial germ cells: The first cell lineage or the last cells standing? *Development*, *142*, 2730–2739.

Juliano, C. E., Swartz, S. Z., & Wessel, G. M. (2010). A conserved germline multipotency program. *Development*, *137*, 4113–4126.

Jung, D., Xiong, J., Ye, M., Qin, X., Li, L., Cheng, S., et al. (2017). In vitro differentiation of human embryonic stem cells into ovarian follicle-like cells. *Nature Communications*, *8*, 15680.

Kamachi, Y., & Kondoh, H. (2013). Sox proteins: Regulators of cell fate specification and differentiation. *Development*, *140*, 4129.

Kanai, Y., Kanai-Azuma, M., Noce, T., Saido, T. C., Shiroishi, T., Hayashi, Y., et al. (1996). Identification of two Sox17 messenger RNA isoforms, with and without the high mobility group box region, and their differential expression in mouse spermatogenesis. *Journal of Cell Biology*, *133*, 667–681.

Kanai-Azuma, M., Kanai, Y., Gad, J. M., Tajima, Y., Taya, C., Kurohmaru, M., et al. (2002). Depletion of definitive gut endoderm in Sox17-null mutant mice. *Development*, *129*, 2367–2379.

Kee, K., Angeles, V. T., Flores, M., Nguyen, H. N., & Reijo Pera, R. A. (2009). Human DAZL, DAZ and BOULE genes modulate primordial germ-cell and haploid gamete formation. *Nature*, *462*, 222–225.

Kehler, J., Tolkunova, E., Koschorz, B., Pesce, M., Gentile, L., Boiani, M., et al. (2004). Oct4 is required for primordial germ cell survival. *EMBO Reports*, *5*, 1078–1083.

Keramari, M., Razavi, J., Ingman, K. A., Patsch, C., Edenhofer, F., Ward, C. M., et al. (2010). Sox2 is essential for formation of trophectoderm in the preimplantation embryo. *PLoS One*, *5*, e13952.
Kim, Y., & Capel, B. (2006). Balancing the bipotential gonad between alternative organ fates: A new perspective on an old problem. *Developmental Dynamics*, *235*, 2292–2300.
Kim, J., Chu, J., Shen, X., Wang, J., & Orkin, S. H. (2008). An extended transcriptional network for pluripotency of embryonic stem cells. *Cell*, *132*, 1049–1061.
Kim, S., Gunesdogan, U., Zylicz, J. J., Hackett, J. A., Cougot, D., Bao, S., et al. (2014). PRMT5 protects genomic integrity during global DNA demethylation in primordial germ cells and preimplantation embryos. *Molecular Cell*, *56*, 564–579.
Kobayashi, T., & Surani, M. A. (2018). On the origin of the human germline. *Development*, *145* dev150433.
Kobayashi, T., Zhang, H., Tang, W. W. C., Irie, N., Withey, S., Klisch, D., et al. (2017). Principles of early human development and germ cell program from conserved model systems. *Nature*, *546*, 416–420.
Kojima, Y., Kaufman-Francis, K., Studdert, J. B., Steiner, K. A., Power, M. D., Loebel, D. A. F., et al. (2014). The transcriptional and functional properties of mouse epiblast stem cells resemble the anterior primitive streak. *Cell Stem Cell*, *14*, 107–120.
Kojima, Y., Sasaki, K., Yokobayashi, S., Sakai, Y., Nakamura, T., Yabuta, Y., et al. (2017). Evolutionarily distinctive transcriptional and signaling programs drive human germ cell lineage specification from pluripotent stem cells. *Cell Stem Cell*, *21*517–532.e515.
Kojima, Y., Tam, O. H., & Tam, P. P. L. (2014). Timing of developmental events in the early mouse embryo. *Seminars in Cell and Developmental Biology*, *34*, 65–75.
Kondoh, H., & Kamachi, Y. (2010). SOX-partner code for cell specification: Regulatory target selection and underlying molecular mechanisms. *International Journal of Biochemistry and Cell Biology*, *42*, 391–399.
Kunarso, G., Chia, N. Y., Jeyakani, J., Hwang, C., Lu, X., Chan, Y. S., et al. (2010). Transposable elements have rewired the core regulatory network of human embryonic stem cells. *Nature Genetics*, *42*, 631–634.
Kurimoto, K., Yabuta, Y., Ohinata, Y., Shigeta, M., Yamanaka, K., & Saitou, M. (2008). Complex genome-wide transcription lineage in mice for the specification of the germ cell dynamics orchestrated by Blimp1. *Genes & Development*, *22*, 1617–1635.
Kurimoto, K., Yamaji, M., Seki, Y., & Saitou, M. (2008). Specification of the germ cell lineage in mice: A process orchestrated by the PR-domain proteins, Blimp1 and Prdm14. *Cell Cycle*, *7*, 3514–3518.
Kuroda, T., Tada, M., Kubota, H., Kimura, H., Hatano, S.-y., Suemori, H., et al. (2005). Octamer and Sox elements are required for transcriptional cis regulation of nanog gene expression. *Molecular and Cellular Biology*, *25*, 2475–2485.
Lancaster, M. A., & Knoblich, J. A. (2014). Organogenesis in a dish: Modeling development and disease using organoid technologies. *Science*, *345* 1247125.
Lancaster, M. A., Renner, M., Martin, C. A., Wenzel, D., Bicknell, L. S., Hurles, M. E., et al. (2013). Cerebral organoids model human brain development and microcephaly. *Nature*, *501*, 373–379.
Lange, U. C., Adams, D. J., Lee, C., Barton, S., Schneider, R., Bradley, A., et al. (2008). Normal germ line establishment in mice carrying a deletion of the Ifitm/Fragilis gene family cluster. *Molecular and Cellular Biology*, *28*, 4688–4696.
Lanner, F., & Rossant, J. (2010). The role of FGF/Erk signaling in pluripotent cells. *Development*, *137*, 3351–3360.
Lawson, K. A. (1999). Fate mapping the mouse embryo. *International Journal of Developmental Biology*, *775*, 773–775.
Lawson, K. A., Dunn, N. R., Roelen, B. A., Zeinstra, L. M., Davis, A. M., Wright, C. V., et al. (1999). Bmp4 is required for the generation of primordial germ cells in the mouse embryo. *Genes & Development*, *13*, 424–436.

Lawson, K. A., & Hage, W. J. (1994). Clonal analysis of the origin of primordial germ cells in the mouse. *Ciba Foundation Symposium*, *182*, 68–84. discussion 84–91. Review.

Lawson, K., Meneses, J., & Pedersen, R. (1991). Clonal analysis of epiblast fate during germ layer formation in the mouse embryo. *Development (Cambridge, England)*, *113*, 891–911.

Leitch, H. G., Blair, K., Mansfield, W., Ayetey, H., Humphreys, P., Nichols, J., et al. (2010). Embryonic germ cells from mice and rats exhibit properties consistent with a generic pluripotent ground state. *Development*, *137*, 2279–2287.

Leitch, H. G., Mcewen, K. R., Turp, A., Encheva, V., Carroll, T., Grabole, N., et al. (2013). Naive pluripotency is associated with global DNA hypomethylation. *Nature Structural and Molecular Biology*, *20*, 311–316.

Leopardo, N. P., & Vitullo, A. D. (2017). Early embryonic development and spatiotemporal localization of mammalian primordial germ cell-associated proteins in the basal rodent Lagostomus maximus. *Scientific Reports*, *7*, 1–11.

Li, L., Dong, J., Yan, L., Yong, J., Liu, X., Hu, Y., et al. (2017). Single-cell RNA-Seq analysis maps development of human germline cells and gonadal niche interactions. *Cell Stem Cell*, *20*, 858–873. e854.

Lilly, A. J., Lacaud, G., & Kouskoff, V. (2017). SOXF transcription factors in cardiovascular development. *Seminars in Cell and Developmental Biology*, *63*, 50–57.

Lin, I. Y., Chiu, F. L., Yeang, C. H., Chen, H. F., Chuang, C. Y., Yang, S. Y., et al. (2014). Suppression of the SOX2 neural effector gene by PRDM1 promotes human germ cell fate in embryonic stem cells. *Stem Cell Reports*, *2*, 189–204.

Liu, P., Wakamiya, M., Shea, M. J., Albrecht, U., Behringer, R. R., & Bradley, A. (1999). Requirement for Wnt3 in vertebrate axis formation. *Nature Genetics*, *22*, 361–365.

Loh, Y. H., Wu, Q., Chew, J. L., Vega, V. B., Zhang, W., Chen, X., et al. (2006). The Oct4 and Nanog transcription network regulates pluripotency in mouse embryonic stem cells. *Nature Genetics*, *38*, 431–440.

Luckett, P. (1975). The development of primordial and difinitive amniotic cavities in early rhesus monkey and human embryos. *American Journal of Anatomy*, *144*, 149–167.

Luckett, P. (1978). Origin and differentiation of the yolk sac and extraembryonic mesoderm in presomite humans and rhesus monkey embryos. *American Journal of Anatomy*, *152*, 59–97.

Ma, Z., Swigut, T., Valouev, A., Rada-Iglesias, A., & Wysocka, J. (2011). Sequence-specific regulator Prdm14 safeguards mouse ESCs from entering extraembryonic endoderm fates. *Nature Structural & Molecular Biology*, *18*, 120–127.

Magnúsdóttir, E., Dietmann, S., Murakami, K., Günesdogan, U., Tang, F., Bao, S., et al. (2013). A tripartite transcription factor network regulates primordial germ cell specification in mice. *Nature Cell Biology*, *15*, 905–915.

Magnusdottir, E., Kalachikov, S., Mizukoshi, K., Savitsky, D., Ishida-Yamamoto, A., Panteleyev, A. A., et al. (2007). Epidermal terminal differentiation depends on B lymphocyte-induced maturation protein-1. *Proceedings of the National Academy of Sciences of the United States of America*, *104*, 14988–14993.

Magnúsdóttir, E., & Surani, M. A. (2014). How to make a primordial germ cell. *Development*, *141*, 245–252.

Marson, A., Foreman, R., Chevalier, B., Bilodeau, S., Kahn, M., Young, R. A., et al. (2008). Wnt signaling promotes reprogramming of somatic cells to pluripotency. *Cell Stem Cell*, *3*, 132–135.

Martyn, I., Kanno, T. Y., Ruzo, A., Siggia, E. D., & Brivanlou, A. H. (2018). Self-organization of a human organizer by combined Wnt and Nodal signalling. *Nature*, *558*(7708), 132–135.

Masui, S., Nakatake, Y., Toyooka, Y., Shimosato, D., Yagi, R., Takahashi, K., et al. (2007). Pluripotency governed by Sox2 via regulation of Oct3/4 expression in mouse embryonic stem cells. *Nature Cell Biology*, *9*, 625–635.

Matsuda, K., Mikami, T., Oki, S., Iida, H., Andrabi, M., Boss, J. M., et al. (2017). ChIP-seq analysis of genomic binding regions of five major transcription factors highlights a central role for ZIC2 in the mouse epiblast stem cell gene regulatory network. *Development, 144*, 1948–1958.

Matzuk, M. M., & Lamb, D. J. (2008). The biology of infertility: Research advances and clinical challenges. *Nature Medicine, 14*, 1197.

Mehlmann, L. M. (2005). Stops and starts in mammalian oocytes: Recent advances in understanding the regulation of meiotic arrest and oocyte maturation. *Reproduction, 130*, 791–799.

Merino, F., Ng, C. K. L., Veerapandian, V., Schöler, H. R., Jauch, R., & Cojocaru, V. (2014). Structural basis for the SOX-dependent genomic redistribution of OCT4 in stem cell differentiation. *Structure, 22*, 1274–1286.

Moore, H. D. M., Gems, S., & Hearn, J. P. (1985). Early implantation stages in the Marmoset monkey. *The American Journal of Anatomy, 172*, 265–278.

Morgan, M. A., Magnusdottir, E., Kuo, T. C., Tunyaplin, C., Harper, J., Arnold, S. J., et al. (2009). Blimp-1/Prdm1 alternative promoter usage during mouse development and plasma cell differentiation. *Molecular and Cellular Biology, 29*, 5813–5827.

Morgani, S. M., Metzger, J. J., Nichols, J., Siggia, E. D., & Hadjantonakis, AK. (2018). Micropattern differentiation of mouse pluripotent stem cells recapitulates embryo regionalized cell fate patterning. *eLife*, 7pii: e32839.

Mould, A., Morgan, M. A., Li, L., Bikoff, E. K., & Robertson, E. J. (2012). Blimp1/Prdm1 governs terminal differentiation of endovascular trophoblast giant cells and defines multipotent progenitors in the developing placenta. *Genes & Development, 26*, 2063–2074.

Muller, W. E. (2006). The stem cell concept in sponges (Porifera): Metazoan traits. *Seminars in Cell & Developmental Biology, 17*, 481–491.

Murakami, K., Günesdogan, U., Zylicz, J. J., Tang, W. W. C., Sengupta, R., Kobayashi, T., et al. (2016). NANOG alone induces germ cells in primed epiblast in vitro by activation of enhancers. *Nature, 529*, 403–407.

Nady, N., Gupta, A., Ma, Z., Swigut, T., Koide, A., Koide, S., et al. (2015). ETO family protein Mtgr1 mediates Prdm14 functions in stem cell maintenance and primordial germ cell formation. *eLife, 4*, e10150.

Nakaki, F., Hayashi, K., Ohta, H., Kurimoto, K., Yabuta, Y., & Saitou, M. (2013). Induction of mouse germ-cell fate by transcription factors in vitro. *Nature, 501*, 222–226.

Nakaki, F., & Saitou, M. (2014). PRDM14: A unique regulator for pluripotency and epigenetic reprogramming. *Trends in Biochemical Sciences, 39*, 289–298.

Nakamura, T., Okamoto, I., Sasaki, K., Yabuta, Y., Iwatani, C., Tsuchiya, H., et al. (2016). A developmental coordinate of pluripotency among mice, monkeys and humans. *Nature, 537*, 57–62.

Nakatake, Y., Fukui, N., Iwamatsu, Y., Masui, S., Takahashi, K., Yagi, R., et al. (2006). Klf4 cooperates with Oct3/4 and Sox2 to activate the Lefty1 core promoter in embryonic stem cells. *Molecular and Cellular Biology, 26*, 7772–7782.

Navarro, P., Moffat, M., Mullin, N. P., & Chambers, I. (2011). The X-inactivation trans-activator Rnf12 is negatively regulated by pluripotency factors in embryonic stem cells. *Human Genetics, 130*, 255–264.

Navarro, P., Oldfield, A., Legoupi, J., Festuccia, N., Dubois, A. S., Attia, M., et al. (2010). Molecular coupling of Tsix regulation and pluripotency. *Nature, 468*, 457–460.

Nettersheim, D., Heimsoeth, A., Jostes, S., Schneider, S., Fellermeyer, M., Hofmann, A., et al. (2016). SOX2 is essential for in vivo reprogramming of seminoma-like TCam-2 cells to an embryonal carcinoma-like fate. *Oncotarget, 7*, 47095–47110.

Niakan, K. K., & Eggan, K. (2013). Analysis of human embryos from zygote to blastocyst reveals distinct gene expression patterns relative to the mouse. *Developmental Biology, 375*, 54–64.

Niakan, K. K., Ji, H., Maehr, R., Vokes, S. A., Rodolfa, K. T., Sherwood, R. I., et al. (2010). Sox17 promotes differentiation in mouse embryonic stem cells by directly regulating extraembryonic gene expression and indirectly antagonizing self-renewal. *Genes and Development*, *24*, 312–326.

Nichols, J., & Smith, A. (2009). Naive and primed pluripotent states. *Cell Stem Cell*, *4*, 487–492.

Nishimoto, M., Fukushima, A., Okuda, A., & Muramatsu, M. (1999). The gene for the embryonic stem cell coactivator UTF1 carries a regulatory element which selectively interacts with a complex composed of Oct-3/4 and Sox-2. *Molecular and Cellular Biology*, *19*, 5453–5465.

Niwa, H., Masui, S., Chambers, I., Smith, A. G., & Miyazaki, J.-i. (2002). Phenotypic complementation establishes requirements for specific POU domain and generic transactivation function of Oct-3/4 in embryonic stem cells. *Molecular and Cellular Biology*, *22*, 1526–1536.

Niwa, H., Miyazaki, J.-i., & Smith, A. G. (2000). Quantitative expression of Oct-3/4 defines differentiation, dedifferentiation or self-renewal of ES cells. *Nature Genetics*, *24*, 2–6.

Nomura, M., & Li, E. (1998). Smad2 role in mesoderm formation, left-right patterning and craniofacial development. *Nature*, *393*, 786–790.

Nowling, T. K., Johnson, L. R., Wiebe, M. S., & Rizzino, A. (2000). Identification of the transactivation domain of the transcription factor Sox-2 and an associated co-activator. *Journal of Biological Chemistry*, *275*, 3810–3818.

Ohinata, Y., Ohta, H., Shigeta, M., Yamanaka, K., Wakayama, T., & Saitou, M. (2009). A signaling principle for the specification of the germ cell lineage in mice. *Cell*, *137*, 571–584.

Ohinata, Y., Payer, B., O'Carroll, D., Ancelin, K., Ono, Y., Sano, M., et al. (2005). Blimp1 is a critical determinant of the germ cell lineage in mice. *Nature*, *436*, 207–213.

Okae, H., Toh, H., Sato, T., Hiura, H., Takahashi, S., Shirane, K., et al. (2018). Derivation of human trophoblast stem cells. *Cell Stem Cell*, *22*(1)50.e6–63.e6.

Okamura, D., Tokitake, Y., Niwa, H., & Matsui, Y. (2008). Requirement of Oct3/4 function for germ cell specification. *Developmental Biology*, *317*, 576–584.

Okashita, N., Kumaki, Y., Ebi, K., Nishi, M., Okamoto, Y., Nakayama, M., et al. (2014). PRDM14 promotes active DNA demethylation through the ten-eleven translocation (TET)-mediated base excision repair pathway in embryonic stem cells. *Development*, *141*, 269–280.

Okashita, N., Sakashita, N., Ito, K., Mitsuya, A., Suwa, Y., & Seki, Y. (2015). PRDM14 maintains pluripotency of embryonic stem cells through TET-mediated active DNA demethylation. *Biochemical and Biophysical Research Communications*, *466*, 138–145.

Okashita, N., Suwa, Y., Nishimura, O., Sakashita, N., Kadota, M., Nagamatsu, G., et al. (2016). PRDM14 drives OCT3/4 recruitment via active demethylation in the transition from primed to naive pluripotency. *Stem Cell Reports*, 7, 1072–1086.

Okumura-Nakanishi, S., Saito, M., Niwa, H., & Ishikawa, F. (2005). Oct-3/4 and Sox2 regulate Oct-3/4 gene in embryonic stem cells. *Journal of Biological Chemistry*, *280*, 5307–5317.

Palasingam, P., Jauch, R., Ng, C. K. L., & Kolatkar, P. R. (2009). The structure of Sox17 bound to DNA reveals a conserved bending topology but selective protein interaction platforms. *Journal of Molecular Biology*, *388*, 619–630.

Pastor, W. A., Chen, D., Liu, W., Kim, R., Sahakyan, A., Lukianchikov, A., et al. (2016). Naive human pluripotent cells feature a methylation landscape devoid of blastocyst or germline memory. *Cell Stem Cell*, *18*, 323–329.

Payer, B., Rosenberg, M., Yamaji, M., Yabuta, Y., Koyanagi-Aoi, M., Hayashi, K., et al. (2013). Tsix RNA and the germline factor, PRDM14, link X reactivation and stem cell reprogramming. *Molecular Cell*, *52*, 805–818.

Payer, B., Saitou, M., Barton, S. C., Thresher, R., Dixon, J. P. C., Zahn, D., et al. (2003). Stella is a maternal effect gene required for normal early development in mice. *Current Biology*, *13*, 2110–2117.

Peng, G., Suo, S., Chen, J., Chen, W., Liu, C., Yu, F., et al. (2016). Spatial transcriptome for the molecular annotation of lineage fates and cell identity in mid-gastrula mouse embryo. *Developmental Cell*, *36*(6), 681–697.

Pereira, P. N., Dobreva, M. P., Graham, L., Huylebroeck, D., Lawson, K. A., & Zwijsen, A. (2011). Amnion formation in the mouse embryo: The single amniochorionic fold model. *BMC Developmental Biology*, *11*, 48.

Perrett, R. M., Turnpenny, L., Eckert, J. J., O'Shea, M., Sonne, S. B., Cameron, I. T., et al. (2008). The early human germ cell lineage does not express SOX2 during in vivo development or upon in vitro culture. *Biology of Reproduction*, *78*, 852–858.

Petropoulos, S., Edsgärd, D., Reinius, B., Deng, Q., Panula, S. P., Codeluppi, S., et al. (2016). Single-cell RNA-Seq reveals lineage and X chromosome dynamics in human preimplantation embryos. *Cell*, *167*, 285.

Pevny, L. H., & Lovell-Badge, R. (1997). Sox genes find their feet. *Current Opinion in Genetics and Development*, *7*, 338–344.

Politzer, G. (1930). Uber einen menschlichen Embryo mit 7 Urwirbelpaaren. 3. Das Entoderm. *Z Anat Entw Gesch*, *93*, 386–428.

Politzer, G. (1933). The germ line of the human being. *Journal of Anatomy and History of Development*, *100*, 331–360.

Qu, X. B., Pan, J., Zhang, C., & Huang, S. Y. (2008). Sox17 facilitates the differentiation of mouse embryonic stem cells into primitive and definitive endoderm in vitro. *Development, Growth & Differentiation*, *50*, 585–593.

Ramos-Ibeas, P., Nichols, J., & Alberio, R. (2018). States and origins of mammalian embryonic pluripotency in vivo and in a dish. *Current Topics in Developmental Biology*, *128*, 151–179.

Reményi, A., Lins, K., Nissen, L. J., Reinbold, R., Schöler, H. R., & Wilmanns, M. (2003). Crystal structure of a POU/HMG/DNA ternary complex suggests differential assembly of Oct4 and Sox2 on two enhancers. *Genes and Development*, *17*, 2048–2059.

Rivera-Pérez, J. A., & Magnuson, T. (2005). Primitive streak formation in mice is preceded by localized activation of Brachyury and Wnt3. *Developmental Biology*, *288*, 363–371.

Roberts, C. M., & Weir, B. J. (1973). Implantation in the plains viscacha, Lagostomux maximus. *Journal of Reproduction and Fertility*, *33*, 299–307.

Robertson, E. J., Charatsi, I., Joyner, C. J., Koonce, C. H., Morgan, M., Islam, A., et al. (2007). Blimp1 regulates development of the posterior forelimb, caudal pharyngeal arches, heart and sensory vibrissae in mice. *Development*, *134*, 4335–4345.

Rock, J. R., Onaitis, M. W., Rawlins, E. L., Lu, Y., Clark, C. P., Xue, Y., et al. (2009). Basal cells as stem cells of the mouse trachea and human airway epithelium. *Proceedings of the National Academy of Sciences of the United States of America*, *106*, 12771.

Rodda, D. J., Chew, J. L., Lim, L. H., Loh, Y. H., Wang, B., Ng, H. H., et al. (2005). Transcriptional regulation of Nanog by OCT4 and SOX2. *Journal of Biological Chemistry*, *280*, 24731–24737.

Rossant, J. (2015). Mouse and human blastocyst-derived stem cells: Vive les differences. *Development*, *142*, 9–12.

Rossant, J., & Tam, P. P. L. (2017). New insights into early human development: Lessons for stem cell derivation and differentiation. *Cell Stem Cell*, *20*, 18–28.

Saitou, M., Barton, S. C., & Surani, M. A. (2002). A molecular programme for the specification of germ cell fate in mice. *Nature*, *418*, 293–300.

Saiz, N., & Plusa, B. (2013). Early cell fate decisions in the mouse embryo. *Reproduction*, *145*, R65–R80.

Sakamoto, Y., Hara, K., Kanai-Azuma, M., Matsui, T., Miura, Y., Tsunekawa, N., et al. (2007). Redundant roles of Sox17 and Sox18 in early cardiovascular development of mouse embryos. *Biochemical and Biophysical Research Communications, 360*, 539–544.

Sansom, G. S., & Hill, J. P. (1931). Observations on the structure and mode of implantation of the blastocyst of Cavia. *The Transactions of the Zoological Society of London, 21*, 295–354.

Sasaki, K., Nakamura, T., Okamoto, I., Yabuta, Y., Iwatani, C., Tsuchiya, H., et al. (2016). The germ cell fate of cynomolgus monkeys is specified in the nascent amnion. *Developmental Cell, 39*, 169–185.

Sasaki, K., Yokobayashi, S., Nakamura, T., Okamoto, I., Yabuta, Y., Kurimoto, K., et al. (2015). Robust in vitro induction of human germ cell fate from pluripotent stem cells. *Cell Stem Cell, 17*, 178–194.

Sato, T., Vries, R. G., Snippert, H. J., Van De Wetering, M., Barker, N., Stange, D. E., et al. (2009). Single Lgr5 stem cells build crypt-villus structures in vitro without a mesenchymal niche. *Nature, 459*, 262–265.

Schäfer-Haas, A., & Viebahn, C. (2000). The term cell epitope PG-2 is expressed in primordial germ cells and in hypoblast cells of the gastrulating rabbit embryo. *Anatomy and Embryology, 202*, 13–23.

Schepers, G. E., Teasdale, R. D., & Koopman, P. (2002). Twenty pairs of Sox: Extent, homology, and nomenclature of the mouse and human Sox transcription factor gene families. *Developmental Cell, 3*, 167–170.

Schneider, B. F., & Norton, S. (1979). Equivalent ages in rat, mouse and chick embryos. *Teratology, 19*, 273–278.

Scholer, H. R., Dressler, G. R., Rohdewohid, H., & Gruss, P. (1990). Oct-4: A germline-specific transcription factor mapping to the mouse t-complex. *Trends in Genetics, 6*, 314.

Séguin, C. A., Draper, J. S., Nagy, A., & Rossant, J. (2008). Establishment of endoderm progenitors by SOX transcription factor expression in human embryonic stem cells. *Cell Stem Cell, 3*, 182–195.

Seki, Y. (2018). PRDM14 is a unique epigenetic regulator stabilizing transcriptional networks for pluripotency. *Frontiers in Cell and Development Biology, 6*, 12.

Seki, Y., Hayashi, K., Itoh, K., Mizugaki, M., Saitou, M., & Matsui, Y. (2005). Extensive and orderly reprogramming of genome-wide chromatin modifications associated with specification and early development of germ cells in mice. *Developmental Biology, 278*, 440–458.

Seki, Y., Yamaji, M., Yabuta, Y., Sano, M., Shigeta, M., Matsui, Y., et al. (2007). Cellular dynamics associated with the genome-wide epigenetic reprogramming in migrating primordial germ cells in mice. *Development, 134*, 2627–2638.

Shaffer, A. L., Lin, K. I., Kuo, T. C., Yu, X., Hurt, E. M., Rosenwald, A., et al. (2002). Blimp-1 orchestrates plasma cell differentiation by extinguishing the mature B cell gene expression program. *Immunity, 17*, 51–62.

Shahbazi, M. N., Jedrusik, A., Vuoristo, S., Recher, G., Hupalowska, A., Bolton, V., et al. (2016). Self-organization of the human embryo in the absence of maternal tissues. *Nature Cell Biology, 18*, 700–708.

Shao, Y., Taniguchi, K., Gurdziel, K., Townshend, R. F., Xue, X., Yong, K. M. A., et al. (2017). Self-organized amniogenesis by human pluripotent stem cells in a biomimetic implantation-like niche. *Nature Materials, 16*, 419–427.

Shao, Y., Taniguchi, K., Townshend, R. F., Miki, T., Gumucio, D. L., & Fu, J. (2017). A pluripotent stem cell-based model for post-implantation human amniotic sac development. *Nature Communications, 8*, 208.

Sheng, G. (2015). Epiblast morphogenesis before gastrulation. *Developmental Biology, 401*(1), 17–24.

Shirane, K., Kurimoto, K., Yabuta, Y., Yamaji, M., Satoh, J., Ito, S., et al. (2016). Global landscape and regulatory principles of DNA methylation reprogramming for germ cell specification by mouse pluripotent stem cells. *Developmental Cell*, *39*, 87–103.

Sozen, B., Amadei, G., Cox, A., Wang, R., Na, E., Czukiewska, S., et al. (2018). Self-assembly of embryonic and two extra-embryonic stem cell types into gastrulating embryo-like structures. *Nature Cell Biology*, *20*(8), 979–989.

Spence, J. R., Mayhew, C. N., Rankin, S. A., Kuhar, M. F., Vallance, J. E., Tolle, K., et al. (2011). Directed differentiation of human pluripotent stem cells into intestinal tissue in vitro. *Nature*, *470*, 105–110.

Stefanovic, S., Abboud, N., Désilets, S., Nury, D., Cowan, C., & Pucéat, M. (2009). Interplay of Oct4 with Sox2 and Sox17: A molecular switch from stem cell pluripotency to specifying a cardiac fate. *Journal of Cell Biology*, *186*, 665–673.

Strome, S., & Updike, D. (2015). Specifying and protecting germ cell fate. *Nature Reviews. Molecular Cell Biology*, *16*, 406–416.

Su, S. T., Ying, H. Y., Chiu, Y. K., Lin, F. R., Chen, M. Y., & Lin, K. I. (2009). Involvement of histone demethylase LSD1 in Blimp-1-mediated gene repression during plasma cell differentiation. *Molecular and Cellular Biology*, *29*, 1421–1431.

Sugawa, F., Arauzo-Bravo, M. J., Yoon, J., Kim, K.-P., Aramaki, S., Wu, G., et al. (2015). Human primordial germ cell commitment in vitro associates with a unique PRDM14 expression profile. *The EMBO Journal*, *34*, 1009–1024.

Sumi, T., Fujimoto, Y., Nakatsuji, N., & Suemori, H. (2004). STAT3 is dispensable for maintenance of self-renewal in nonhuman primate embryonic stem cells. *Stem Cells*, *22*, 861–872.

Takashima, Y., Guo, G., Loos, R., Nichols, J., Ficz, G., Krueger, F., et al. (2014). Resetting transcription factor control circuitry toward ground-state pluripotency in human. *Cell*, *158*, 1254–1269.

Tam, P. P. L., & Behringer, R. R. (1997). Mouse gastrulation: The formation of a mammalian body plan. *Mechanisms of Development*, *68*, 3–25.

Tam, P. P., & Loebel, D. A. (2007). Gene function in mouse embryogenesis: Get set for gastrulation. *Nature Reviews Genetics*, *8*(5), 368–381.

Tam, P. P., & Zhou, S. X. (1996). The allocation of epiblast cells to ectodermal and germ-line lineages is influenced by the position of the cells in the gastrulating mouse embryo. *Developmental Biology*, *178*(1), 124–132.

Tang, W. W., Dietmann, S., Irie, N., Leitch, H. G., Floros, V. I., Bradshaw, C. R., et al. (2015). A unique gene regulatory network resets the human germline epigenome for development. *Cell*, *161*, 1453–1467.

Tang, W. W., Kobayashi, T., Irie, N., Dietmann, S., & Surani, M. A. (2016). Specification and epigenetic programming of the human germ line. *Nature Reviews. Genetics*, *17*, 585–600.

Tao, Y., Yen, M. R., Chitiashvili, T., Nakano, H., Kim, R., Hosohama, L., et al. (2018). TRIM28-regulated transposon repression is required for human germline competency and not primed or naive human pluripotency. *Stem Cell Reports*, *10*, 243–256.

ten Berge, D., Koole, W., Fuerer, C., Fish, M., Eroglu, E., & Nusse, R. (2008). Wnt signaling mediates self-organization and axis formation in embryoid bodies. *Cell Stem Cell*, *3*(5), 508–518.

Tesar, P. J., Chenoweth, J. G., Brook, F. A., Davies, T. J., Evans, E. P., Mack, D. L., et al. (2007). New cell lines from mouse epiblast share defining features with human embryonic stem cells. *Nature*, *448*, 196–199.

Tewary, M., Ostblom, J., Prochazka, L., Zulueta-Coarasa, T., Shakiba, N., Fernandez-Gonzalez, R., et al. (2017). A stepwise model of reaction-diffusion and positional information governs self-organized human peri-gastrulation-like patterning. *Development*, *144*(23), 4298–4312.

Theunissen, T. W., Powell, B. E., Wang, H., Mitalipova, M., Faddah, D. A., Reddy, J., et al. (2014). Systematic identification of culture conditions for induction and maintenance of naive human pluripotency. *Cell Stem Cell, 15*, 471–487.

Thomson, J. A., Itskovitz-Eldor, J., Shapiro, S. S., Waknitz, M. A., Swiergiel, J. J., Marshall, V. S., et al. (1998). Embryonic stem cell lines derived from human blastocysts. *Science, 282*, 1145–1147.

Thomson, M., Liu, S. J., Zou, L. N., Smith, Z., Meissner, A., & Ramanathan, S. (2011). Pluripotency factors in embryonic stem cells regulate differentiation into germ layers. *Cell, 145*, 875–889.

Tokuzawa, Y., Kaiho, E., Maruyama, M., Takahashi, K., Mitsui, K., Maeda, M., et al. (2003). Fbx15 is a novel target of Oct3/4 but is dispensable for embryonic stem cell self-renewal and mouse development. *Molecular and Cellular Biology, 23*, 2699–2708.

Tomioka, M., Nishimoto, M., Miyagi, S., Katayanagi, T., Fukui, N., Niwa, H., et al. (2002). Identification of Sox-2 regulatory region which is under the control of Oct-3/4–Sox-2 complex. *Nucleic Acids Research, 30*, 3202–3213.

Tsankov, A. M., Gu, H., Akopian, V., Ziller, M. J., Donaghey, J., Amit, I., et al. (2015). Transcription factor binding dynamics during human ES cell differentiation. *Nature, 518*, 344–349.

Tsuneyoshi, N., Sumi, T., Onda, H., Nojima, H., Nakatsuji, N., & Suemori, H. (2008). PRDM14 suppresses expression of differentiation marker genes in human embryonic stem cells. *Biochemical and Biophysical Research Communications, 367*, 899–905.

Tu, S., Narendra, V., Yamaji, M., Vidal, S. E., Rojas, L. A., Wang, X., et al. (2016). Co-repressor CBFA2T2 regulates pluripotency and germline development. *Nature, 534*, 387–390.

Turner, D. A., Girgin, M., Alonso-Crisostomo, L., Trivedi, V., Baillie-Johnson, P., Glodowski, C. R., et al. (2017). Anteroposterior polarity and elongation in the absence of extra-embryonic tissues and of spatially localised signalling in gastruloids: Mammalian embryonic organoids. *Development, 144*(21), 3894–3906.

Uchikawa, M., Kamachi, Y., & Kondoh, H. (1999). Two distinct subgroups of Group B Sox genes for transcriptional activators and repressors: Their expression during embryonic organogenesis of the chicken. *Mechanisms of Development, 84*, 103–120.

van den Brink, S. C., Baillie-Johnson, P., Balayo, T., Hadjantonakis, A. K., Nowotschin, S., Turner, D. A., et al. (2014). Symmetry breaking, germ layer specification and axial organisation in aggregates of mouse embryonic stem cells. *Development, 141*(22), 4231–4242.

Vervoort, M., Meulemeester, D., Behague, J., & Kerner, P. (2016). Evolution of Prdm genes in animals: Insights from comparative genomics. *Molecular Biology and Evolution, 33*, 679–696.

Vincent, S. D., Dunn, N. R., Sciammas, R., Shapiro-Shalef, M., Davis, M. M., Calame, K., et al. (2005). The zinc finger transcriptional repressor Blimp1/Prdm1 is dispensable for early axis formation but is required for specification of primordial germ cells in the mouse. *Development, 132*, 1315–1325.

Wang, Z., Oron, E., Nelson, B., Razis, S., & Ivanova, N. (2012). Distinct lineage specification roles for NANOG, OCT4, and SOX2 in human embryonic stem cells. *Cell Stem Cell, 10*, 440–454.

Warmflash, A., Sorre, B., Etoc, F., Siggia, E. D., & Brivanlou, A. H. (2014). A method to recapitulate early embryonic spatial patterning in human embryonic stem cells. *Nature Methods, 11*(8), 847–854. https://doi.org/10.1038/nmeth.3016.

Watanabe, H., Hoang, V. T., Mattner, R., & Holstein, T. W. (2009). Immortality and the base of multicellular life: Lessons from cnidarian stem cells. *Seminars in Cell & Developmental Biology, 20*, 1114–1125.

Weber, S., Eckert, D., Nettersheim, D., Gillis, A. J., Schafer, S., Kuckenberg, P., et al. (2010). Critical function of AP-2 gamma/TCFAP2C in mouse embryonic germ cell maintenance. *Biology of Reproduction, 82*, 214–223.

Wegner, M. (2010). All purpose Sox: The many roles of Sox proteins in gene expression. *International Journal of Biochemistry and Cell Biology*, *42*, 381–390.
Werling, U., & Schorle, H. (2002). Transcription factor gene AP-2 gamma essential for early murine development. *Molecular and Cellular Biology*, *22*, 3149–3156.
Western, P. S., Miles, D. C., van den Bergen, J. A., Burton, M., & Sinclair, A. H. (2008). Dynamic regulation of mitotic arrest in fetal male germ cells. *Stem Cells*, *26*, 339–347.
Williams, D. C., Cai, M., & Clore, G. M. (2004). Molecular basis for synergistic transcriptional activation by Oct1 and Sox2 revealed from the solution structure of the 42-kDa Oct1·Sox2·Hoxb1-DNA ternary transcription factor complex. *Journal of Biological Chemistry*, *279*, 1449–1457.
Wilm, T. P., & Solnica-Krezel, L. (2005). Essential roles of a zebrafish prdm1/blimp1 homolog in embryo patterning and organogenesis. *Development*, *132*, 393–404.
Winnier, G., Blessing, M., Labosky, P. A., & Hogan, B. L. M. (1995). Bone morphogenetic protein-4 is required for mesoderm formation and patterning in the mouse. *Genes and Development*, *9*, 2105–2116.
Witschi, E. (1948). Migration of germ cells of human embryos from the yolk sac to the primitive gonadal folds. In *Contributions to embryology*. *209* (67–80). Carnegie Institution of Washington publication.
Wong, F. C. K., Chambers, I., & Mullin, N. P. (2016). *SOX2-dependent regulation of pluripotent stem cells*. Academic Press.
Wood, H. B., & Episkopou, V. (1999). Comparative expression of the mouse Sox1, Sox2 and Sox3 genes from pre-gastrulation to early somite stages. *Mechanisms of Development*, *86*, 197–201.
Xue, Z., Huang, K., Cai, C., Cai, L., Jiang, C. Y., Feng, Y., et al. (2013). Genetic programs in human and mouse early embryos revealed by single-cell RNA sequencing. *Nature*, *500*, 593–597.
Yabuta, Y., Kurimoto, K., Ohinata, Y., Seki, Y., & Saitou, M. (2006). Gene expression dynamics during germline specification in mice identified by quantitative single-cell gene expression profiling. *Biology of Reproduction*, *75*, 705–716.
Yamaguchi, S., Kurimoto, K., Yabuta, Y., Sasaki, H., Nakatsuji, N., Saitou, M., et al. (2009). Conditional knockdown of Nanog induces apoptotic cell death in mouse migrating primordial germ cells. *Development*, *136*, 4011–4020.
Yamaji, M., Seki, Y., Kurimoto, K., Yabuta, Y., Yuasa, M., Shigeta, M., et al. (2008). Critical function of Prdm14 for the establishment of the germ cell lineage in mice. *Nature Genetics*, *40*, 1016–1022.
Yamaji, M., Ueda, J., Hayashi, K., Ohta, H., Yabuta, Y., Kurimoto, K., et al. (2013). PRDM14 ensures naive pluripotency through dual regulation of signaling and epigenetic pathways in mouse embryonic stem cells. *Cell Stem Cell*, *12*, 368–382.
Yamamoto, M., Meno, C., Sakai, Y., Shiratori, H., Mochida, K., Ikawa, Y., et al. (2001). The transcription factor FoxH1 (FAST) formation in the mouse anterior-posterior patterning and node mediates nodal signaling during. *Genes & Development*, *15*, 1242–1256.
Yamashiro, C., Sasaki, K., Yabuta, Y., Kojima, Y., Nakamura, T., Okamoto, I., et al. (2018). Generation of human oogonia from induced pluripotent stem cells in vitro. *Science*, *362*, 356–360.
Yan, L., Yang, M., Guo, H., Yang, L., Wu, J., Li, R., et al. (2013). Single-cell RNA-Seq profiling of human preimplantation embryos and embryonic stem cells. *Nature Structural and Molecular Biology*, *20*, 1131–1139.
Ying, Y., Liu, X. M., Marble, A., Lawson, K., & Zhao, G. (2000). Requirement of Bmp8b for the generation of primordial germ cells in the mouse. *Molecular Endocrinology*, *14*, 1053–1063.

Ying, Y., Qi, X., & Zhao, G. Q. (2001). Induction of primordial germ cells from murine epiblasts by synergistic action of BMP4 and BMP8B signaling pathways. *Proceedings of the National Academy of Sciences of the United States of America*, *98*, 7858–7862.
Ying, Y., & Zhao, G. Q. (2001). Cooperation of endoderm-derived BMP2 and extraembryonic ectoderm-derived BMP4 in primordial germ cell generation in the mouse. *Developmental Biology*, *232*, 484–492.
Yuan, H., Corbi, N., Basilico, C., & Dailey, L. (1995). Developmental-specific activity of the FGF-4 enhancer requires the synergistic action of Sox2 and Oct-3. *Genes and Development*, *9*, 2635–2645.
Zhang, M., Leitch, H. G., Tang, W. W. C., Festuccia, N., Hall-Ponsele, E., Nichols, J., et al. (2018). Esrrb complementation rescues development of nanog-null germ cells. *Cell Reports*, *22*, 332–339.
Zhang, J., Zhang, M., Acampora, D., Vojtek, M., Yuan, D., Simeone, A., et al. (2018). OTX2 restricts entry to the mouse germline. *Nature*, *562*, 595–599.
Zhao, G. Q., Deng, K., Labosky, P. A., Liaw, L., & Hogan, B. L. M. (1996). The gene encoding bone morphogenetic protein 8B is required for the initiation and maintenance of spermatogenesis in the mouse. *Genes and Development*, *10*, 1657–1669.
Zhao, S., Nichols, J., Smith, A. G., & Li, M. (2004). SoxB transcription factors specify neuroectodermal lineage choice in ES cells. *Molecular and Cellular Neuroscience*, *27*, 332–342.
Zhou, Q., Wang, M., Yuan, Y., Wang, X., Fu, R., Wan, H., et al. (2016). Complete meiosis from embryonic stem cell-derived germ cells in vitro. *Cell Stem Cell*, *18*, 330–340.

CHAPTER THREE

Germ cell reprogramming

Kazuki Kurimoto[a,*], Mitinori Saitou[b,c,d]
[a]Department of Embryology, Nara Medical University, Nara, Japan
[b]Institute for the Advanced Study of Human Biology (ASHBi), Kyoto University, Kyoto, Japan
[c]Department of Anatomy and Cell Biology, Graduate School of Medicine, Kyoto University, Kyoto, Japan
[d]Center for iPS Cell Research and Application, Kyoto University, Kyoto, Japan
*Corresponding author: e-mail address: kurimoto@naramed-u.ac.jp

Contents

1. Introduction 92
2. Mouse germ cell reprogramming 93
   2.1 Specification and development of mouse primordial germ cells 93
   2.2 *In vitro* reconstitution of mouse PGC specification from pluripotent stem cells 96
   2.3 Epigenome reprogramming during mouse PGC specification *in vitro* 96
   2.4 Epigenome reprogramming in migrating mPGCs and gonadal germ cells before sex differentiation 100
   2.5 Epigenome programming in sexually differentiating germ cells in embryonic gonads 104
   2.6 *In vitro* proliferation of migrating mPGCs on plane culture 105
   2.7 Induction of the female germ cell pathway and initiation of meiosis *in vitro* 105
   2.8 Derivation of the male germ cell pathway *in vitro* 106
   2.9 Epigenome programming in mouse oogenesis 108
   2.10 Epigenome programming during mouse spermatogenesis 110
3. Human germ cell reprogramming 111
   3.1 Specification and development of human PGCs 111
   3.2 Development of human gonadal germ cells 112
   3.3 Epigenome reprogramming of human gonadal germ cells 113
   3.4 Heterogeneous transcriptome dynamics in human germ cell development 114
   3.5 *In vitro* reconstitution of human PGCs from pluripotent stem cells 114
   3.6 Epigenome reprogramming during *in vitro* reconstitution of human PGCs and their differentiation into oogonia 116
4. Conclusion and perspectives 117
Acknowledgment 117
Funding 117
References 117

Current Topics in Developmental Biology, Volume 135
ISSN 0070-2153
https://doi.org/10.1016/bs.ctdb.2019.04.005

Abstract

Germ cells undergo epigenome reprogramming for proper development of the next generation. The achievement of *in vitro* germ cell derivation from human and mouse pluripotent stem cells and further differentiation in a plane culture and in aggregation with gonadal somatic cells offers unprecedented opportunities for investigation of the germ cell development. Moreover, advances in low-input/single-cell genomics have enabled detailed investigation of epigenome dynamics during germ cell development. These technologies have advanced our knowledge of epigenome reprogramming during the specification and development of primordial germ cells, their sex differentiation, and gametogenesis. Key findings include details of chromatin remodeling and transcriptional regulation, progressive and comprehensive DNA demethylation, and tight links between DNA demethylation and histone marks during the development of primordial germ cells, acquisition of unique totipotent epigenome during oogenesis (e.g., broad H3K4me3 domains and low-level three-dimensional genomic organization), and unexpected organization of the sperm genome. Moreover, these studies suggest the importance of epigenome analyses for in-depth evaluations of *in vitro* gametogenesis.

1. Introduction

Germ cells are the origin of individual organisms and the carriers of hereditary information and genetic diversity. Since mammalian germ cells are differentiated from the same developmental origin as somatic lineages, they reset their genome to acquire totipotency, the ability to form the next generation in a comprehensive manner. This involves epigenome reprogramming in primordial germ cells (PGCs) and programming of the male and female epigenome in more differentiated germ cells. Erasure and re-establishment of the genomic imprint from parental origins and reactivation of X chromosomes in females are an important part of the epigenome reprogramming, as well as the genome-wide DNA demethylation and chromatin remodeling.

Recently, several groups have achieved efficient reconstitution of physiologically functional mouse primordial germ cells (mPGCs) from pluripotent stem cells (mPGC-like cells; mPGCLCs) (Hayashi et al., 2012; Hayashi, Ohta, Kurimoto, Aramaki, & Saitou, 2011), reconstruction of mouse gametes using mPGCLCs (Hikabe et al., 2016; Ishikura et al., 2016; Zhou et al., 2016), expansion of mPGCLCs on feeders in a plane culture (Ohta et al., 2017), entry of the female germ cell pathway and meiosis from cultured mPGCLCs (Miyauchi et al., 2017), induction of PGCLCs from human pluripotent stem cells (human PGCLCs; hPGCLCs) (Irie et al., 2015; Sasaki et al., 2015; Sugawa et al., 2015), and differentiation of oogonia-like cells from hPGCLCs in xenogeneic reconstituted ovaries (Yamashiro et al., 2018). These studies highlight the possibility of *in vitro* reconstitution of human gametes, which would contribute to human biology and reproductive medicine, and provide unprecedented opportunities for understanding the mechanisms of mammalian germ cell development, including transcriptional and epigenome regulations. In addition, epigenome investigations of *in vitro*-derived germ cells will provide key information for appropriate evaluation of their cellular states, by comparison with their counterparts *in vivo*.

Moreover, recent advances in low-input/single-cell genomics, including RNA-sequencing (RNA-seq), whole-genome-bisulfite sequencing (WGBS), and chromatin immunoprecipitation combined with sequencing (ChIP-seq), have enabled in-depth analyses of transcriptome and epigenome dynamics in PGC development, gametogenesis, and early embryogenesis in humans and mice (Kelsey, Stegle, & Reik, 2017). In this review, we will discuss current knowledge on epigenome dynamics and its regulation in human and mouse germ cell development, and in *in vitro* germ cell reconstitution systems.

2. Mouse germ cell reprogramming

The mouse is the most widely used animal model for the experimental investigation of mammalian development. In this section, we will discuss epigenome programming and reprogramming during the development of mouse germ cells and during their *in vitro* reconstitution along the developmental time course.

2.1 Specification and development of mouse primordial germ cells

The mouse germ cell lineage is originated from the epiblast, a simple pluripotent epithelium in the postimplantation embryo (Fig. 1). In mice, at the onset of gastrulation around embryonic day (E) 6.0, the most posterior part of the epiblast is specified as PGCs, the origin of all germ cell lineages, in response to the stimuli from bone morphogenetic protein 4 (BMP4) (Lawson et al., 1999; Saitou, Barton, & Surani, 2002), which is secreted from the extraembryonic ectoderm (ExE).

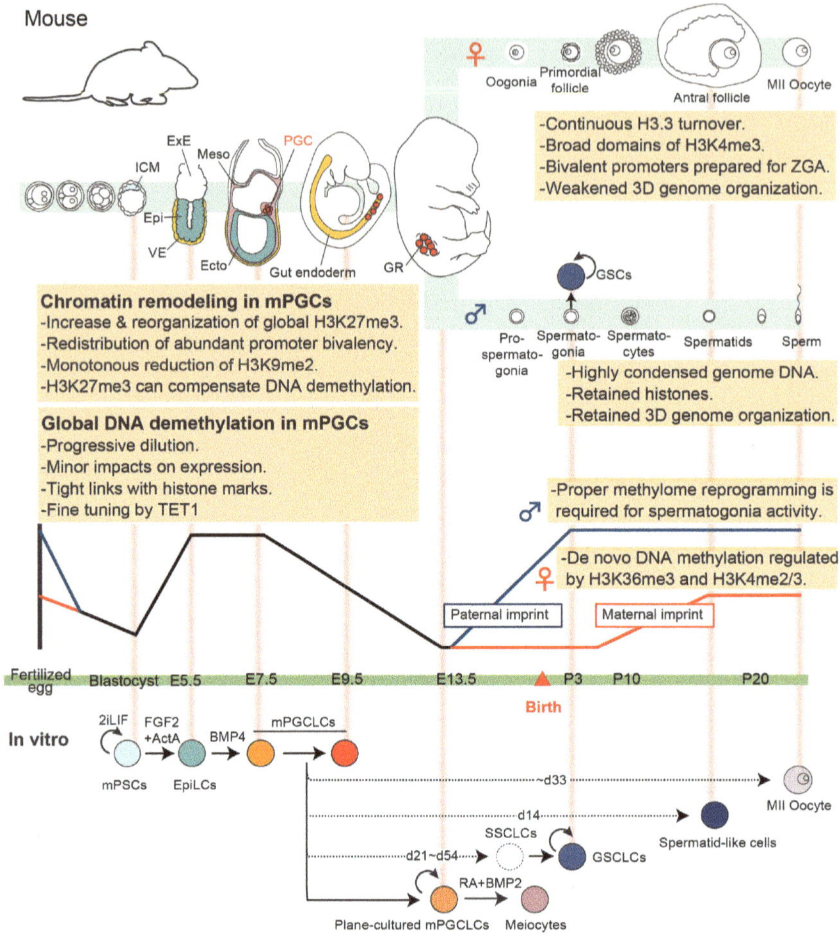

Fig. 1 Schematic representation of mouse germ cell development *in vivo* and *in vitro*. The developmental process of mouse germ cells is schematically represented. Key events in epigenome reprogramming and/or programming are listed. DNA methylation levels are represented with a graph. *In vitro* reconstituted cells are represented with colored circles. *Abbreviations*: *Ect*, ectoderm; *Epi*, epiblast; *GR*, genital ridge; *ICM*, inner cell mass; *PGC*, primordial germ cells; *Meso*, mesoderm; *VE*, visceral endoderm.

The signaling mechanism of mPGC specification involves a combination of BMP4, BMP4 antagonists, and WNT signaling: (1) antagonists of BMP4 (e.g., Cerberus 1 [CER1]) are secreted from the anterior visceral endoderm (AVE), and inhibit anterior epiblast cells from becoming germ cells; and (2) WNT3, secreted from the epiblast itself partly in response to BMP4, provides the epiblast with the competence for the mPGC fate (Ohinata et al., 2009).

The specified mPGCs form a cluster of 30–40 cells in the extraembryonic mesoderm at around E7.0 (Kurimoto & Saitou, 2015). Two transcription factors, B-lymphocyte maturation protein 1 (BLIMP1, also known as PR domain containing 1 [PRDM1]) and PRDM14, play key roles in the mPGC specification (Ohinata et al., 2005; Yamaji et al., 2008). Interestingly, PRDM14 also ensures the naïve state of pluripotent stem cells (Ma, Swigut, Valouev, Rada-Iglesias, & Wysocka, 2011; Yamaji et al., 2013) in complex with co-repressor CBFA2T2 (Nady et al., 2015; Tu et al., 2016). Moreover, it has recently been reported that down-regulation of a transcription factor, OTX2, in the posterior epiblast in response to BMP4 is critical for the mPGC specification (Zhang, Zhang, et al., 2018).

Because germ cell specification occurs under the strong influence of mesoderm development, mPGCs undergo drastic changes in their gene expression profile, which have been analyzed by means of single-cell gene expression profiling (Kurimoto et al., 2006, 2008; Yabuta, Kurimoto, Ohinata, Seki, & Saitou, 2006). mPGCs transiently have a gene expression profile nearly indistinguishable from that of the surrounding somatic mesoderm, and this mesoderm program is swiftly repressed by BLIMP1. Concomitantly with repression of the somatic program, the mPGCs reacquire genes for naïve pluripotency, and show down-regulation of *de novo* DNA methyltransferases (DNMT3A, DNMT3B, DNMT3L), a maintenance factor for DNA methylation (UHRF1), and histone H3K9 methyltransferase GLP.

These expression changes are soon followed by genome-wide reprogramming of epigenetic marks detectable by immunofluorescence; reduction of genomic DNA methylation; increase of histone H3 lysine 27 tri-methylation (H3K27me3) (catalyzed by polycomb repressive complex 2 [PRC2]); and decrease of H3K9 dimethylation (H3K9me2) (catalyzed by the G9A/GLP complex) (Seki et al., 2005). In female embryos, both of the X chromosomes are reactivated in blastocysts and one of them is randomly inactivated upon implantation. During the development of female mPGCs, both of the X chromosomes are again reactivated until the germ cells begin sex differentiation (Avner & Heard, 2001).

The established mPGCs then migrate into the hindgut endoderm, and through the dorsal mesentery, enter and colonize the developing gonads at around E10.5–E11.5. Concomitantly with the morphological change of male and female gonads, the sex differentiation of germ cells become evident in gene expression patterns at around E12.5–E13.5. Then, male germ cells are differentiated into prospermatogonia and enter mitotic arrest. After birth, prospermatogonia resume mitotic proliferation, providing abundant germ cells for spermatogenesis. Female germ cells directly enter and arrest at the prophase of the first meiosis (meiosis I). These cells interact with the surrounding gonadal somatic cells and form primordial follicles. Upon puberty, the primordial follicles start oogenesis.

2.2 *In vitro* reconstitution of mouse PGC specification from pluripotent stem cells

Based on the signaling principle underlying mPGC specification, mouse pluripotent stem cells (mPSCs) cultured under the 2iLIF condition (inhibitors of MAPK and GSK3 [2i], and LIF) (Ying et al., 2008) are induced into pre-gastrulating mouse epiblast-like cells (EpiLCs) by stimulation with activin and b-FGF (Hayashi et al., 2011). Floating aggregates of the EpiLCs are further induced into mPGC-like cells (mPGCLCs) (Hayashi et al., 2011) *via* stimulation by a cytokine cocktail sufficient for induction of the *ex vivo* epiblast into mPGCs (Ohinata et al., 2009). Transcriptome and epigenome features of the mPGCLC induction precisely recapitulate the mPGC specification pathway, and, at day 6 of induction, the mPGCLCs exhibit cellular properties equivalent to those of the E9.5 migrating mPGCs (Hayashi et al., 2011).

The physiological function of male mPGCLCs is demonstrated by the robust capacities for the production of offspring through the transplantation to neonatal testes (Hayashi et al., 2011). When female mPGCLCs are aggregated with somatic cells in E12.5 mouse embryonic ovaries, these cells form reconstituted ovaries that contain follicle-like structures, and transplantation of the reconstituted ovaries to the mouse ovarian bursa results in fertile offspring (Hayashi et al., 2012). More recently, by combining these advances with a culture system for the fully *in vitro* growth and maturation of ovarian follicles (Morohaku et al., 2016), the entire process of oogenesis from mPSCs has been reconstituted *in vitro* (Hikabe et al., 2016).

This system is not only a breakthrough toward *in vitro* gametogenesis, but also a foundation for the experimental investigation of germ cell development, particularly that of early mPGCs, which are relatively small in number in embryos.

2.3 Epigenome reprogramming during mouse PGC specification *in vitro*

mPGC specification is immediately followed by epigenome reprogramming, and thus constitutes the onset of germ cell reprogramming. Transcriptional control, chromatin remodeling, and DNA methylation reprogramming during the induction of mPGCLCs have been investigated with ChIP-seq, WGBS, and inducible overexpression of key transcription factors (TFs).

2.3.1 Transcriptional regulation

The cytokine signaling for mPGC specification from the epiblast can be largely replaced by inducible overexpression of the three key TFs for mPGC specification, BLIMP1, PRDM14, and TFAP2C, in floating aggregates of EpiLCs (Nakaki et al., 2013). mPGCLCs induced by these TFs have a transcriptome profile very similar to that of the cytokine-induced mPGCLCs, and the physiological activity of germ cells, as shown by transplantation into the seminiferous tubules of neonatal mice (Nakaki et al., 2013). Thus, the transcriptional network for mPGC specification can be driven and maintained by these three TFs. Interestingly, genes for early mesoderm development, which are initially up-regulated upon the mPGCLC induction by cytokines, are not detected after overexpression of these TFs. Thus, the transient activation of the mesoderm program itself is essentially dispensable for the mPGC specification.

However, the mesoderm program does drive the core transcriptional network of mPGC specification, as shown by the knockout and inducible overexpression of T (BRACHYURY), a key TF for mesoderm development; T directly activates expression of BLIMP1 and PRDM14, through binding to their enhancers (Aramaki et al., 2013).

Interestingly, overexpression of these key TFs induces mPGC markers in embryonic carcinoma cells (ECCs) (Magnusdottir et al., 2013). However, the BLIMP1-binding patterns are markedly different between the mPGCLCs and the BLIMP1-overexpressed ECCs, showing the importance of the cellular contexts (Kurimoto et al., 2015; Magnusdottir et al., 2013). The impact of the cellular contexts on TF-binding profiles is also highlighted in a systematic comparison of the BLIMP1-binding patterns across developmental lineages (Mitani et al., 2017).

Moreover, inducible overexpression of NANOG in EpiLC aggregates also substantially up-regulates mPGC genes (Murakami et al., 2016), although the NANOG protein is not detected in the epiblast. The overexpressed NANOG binds to the enhancers of *Blimp1* and *Prdm14* in EpiLCs but not in mESCs, probably due to the difference of epigenetic contexts. Thus, although *Blimp1* and *Prdm14* are activated by T under physiological conditions, they can also be activated by overexpression of NANOG. Recently, it has been reported that *Nanog* can be replaced with *Esrrb*, a direct downstream target of *Nanog*, in mouse germ cell development (Zhang, Leitch, et al., 2018), and it would be interesting to examine whether an inducible overexpression of *Esrrb* can drive the mPGC program. Thus, overexpression and/or replacement of key TFs perturb the transcriptional networks for mPGC development, thereby revealing detailed genetic interactions among those TFs.

2.3.2 Dynamics of chromatin remodeling

In the transition from the naïve pluripotent state of mESCs to the primed states of EpiLCs and EpiSCs, drastic changes of active enhancers (H3K27 acetylation [H3K27ac]) occur (Buecker et al., 2014; Factor et al., 2014; Kurimoto et al., 2015). Moreover, the induction of EpiLCs from mESCs accompanies widespread relocation of the OCT4-binding sites mediated by the EpiLC-specific bindings of OTX2 (Buecker et al., 2014). Thus, the down-regulation of OTX2 during the mPGC specification (Zhang, Zhang, et al., 2018) might rewire the OCT4-binding pattern.

H3K27me3, which marks repressive states by PRC2, is broadly distributed and present at high levels in mESCs and EpiLCs (Kurimoto et al., 2015; Marks et al., 2012). During the specification of mPGCLCs, the global level of H3K27me3 is transiently down-regulated upon the activation of the early mesoderm program, and is reacquired during the establishment of the germ cell program. Bivalency of H3K4me3 (active promoters) and H3K27me3, which marks poised promoters, shows a similar kinetics; the bivalent promoters transiently decrease in number upon activation of the early mesoderm program, and are reacquired in mPGCLCs. On the other hand, H3K9me2 is monotonously reduced throughout mPGCLC induction.

T activates the mesoderm program by recruiting H3K27ac, and, subsequently, BLIMP1 represses a broad range of developmental regulators by spreading H3K27me3, as revealed by ChIP-seq analyses (Kurimoto et al., 2015). Thus, mPGCLCs acquire a strong, H3K27me3-dependent repressive state, which is enriched with promoter bivalency and is regulated at least in part by BLIMP1. BLIMP1 is also indispensable for the late mPGC development, while the BLIMP1-binding pattern is prepared upon specification (Kurimoto et al., 2015; Mitani et al., 2017; Yamashiro et al., 2016). These properties underlie the developmental totipotency of germ cells.

2.3.3 Dynamics of DNA methylation

DNA methylation undergoes at least two waves of genome-wide erasure in the life cycle. One occurs in the preimplantation embryos, and then DNA methylation is reacquired in the epiblast upon implantation (Eckersley-Maslin, Alda-Catalinas, & Reik, 2018). In this process, imprints of parental origin are maintained. The other DNA methylation erasure occurs during mPGC development until E13.5, at which time germ cells enter the sex differentiation stage in the male and female gonads. Importantly, mPGCs erase imprints.

The DNA methylation reprogramming during mPGCLC specification has been quantitatively analyzed with WGBS (Miyoshi et al., 2016; Shirane et al., 2016; von Meyenn et al., 2016). Differentiation of EpiLCs from mESCs, which recapitulates the transition from the inner cell mass (ICM) of the blastocyst to the epiblast upon implantation, accompanies not only drastic elevation of the overall DNA methylation level, but also genome-wide re-organization of the methylation pattern. In contrast, during the mPGCLC specification from EpiLCs, the overall DNA methylation progressively decreases, while the methylation pattern is generally preserved. These facts support the argument that the DNA demethylation occurs through a replication-coupled dilution (Arand et al., 2015; Kagiwada, Kurimoto, Hirota, Yamaji, & Saitou, 2013; Seisenberger et al., 2012).

On the other hand, the level of hydroxymethylated DNA, a product of the oxidation of methylated DNA by the ten-eleven translocation (TET) enzymes, is much lower than that of methylated DNA (1%–3%) throughout the mPGCLC specification. This suggests that the hydroxymethylated DNA plays only a minor role in driving the genome-wide demethylation (Shirane et al., 2016).

Importantly, the DNA methylation pattern of mPGCLCs is quite similar to that of gonadal germ cells at later developmental stages (E10.5–E13.5), while the overall methylation level is substantially lower in the gonadal germ cells. This suggests that the gonadal germ cells may maintain the DNA methylation pattern of the epiblast, and mPGCLCs will recapitulate the methylome of gonadal mPGCs if demethylation continues in a replication-coupled manner (see below).

Female mPGCLCs become more rapidly demethylated than their male counterparts (Shirane et al., 2016). Because female mESCs have two active X chromosomes and are extremely demethylated (Habibi et al., 2013), the rapid demethylation in the female mPGCLCs may reflect the initial methylation state in mESCs. In female mESCs, an X-linked MAPK phosphatase, DUSP9, is responsible for the hypomethylation (Choi et al., 2017). Similar mechanisms might drive rapid demethylation during female mPGCLC specification.

During mPGCLC specification, DNA methylation dynamics in subsets of genomic regions are tightly linked with H3K27ac and H3K27me3 histone modifications (Shirane et al., 2016; von Meyenn, Berrens, et al., 2016). There are domains of several kilobases that are methylated at very low levels in mESCs, covering ~2.1% of the genome and enriched in pluripotency genes [e.g., *Pou5f1, Prdm14, Klf4*] and the active H3K27ac mark. In contrast, another group of domains of similar size is very rapidly demethylated during mPGCLC induction, covering ~0.05% of the genome and enriched for developmental regulators [e.g., *Hoxa-Hoxd* clusters] and the polycomb repressive mark, H3K27me3 (Shirane et al., 2016).

Polycomb-bound promoters are protected from *de novo* DNA methylation by KDM2B (Boulard, Edwards, & Bestor, 2015). This mechanism may block DNA methylation by DNMT3A and DNMT3B that may remain in mPGCs at low levels, and may contribute to the rapid demethylation associated with H3K27me3. Thus, the reprogramming of DNA methylation is linked with histone marks, and such associations are different between mESCs and mPGCLCs. This differential regulation shows that the mechanisms of epigenome reprogramming during preimplantation and germ cell development are distinct, while the decrease in overall DNA methylation levels appears to be held in common between these two events.

2.4 Epigenome reprogramming in migrating mPGCs and gonadal germ cells before sex differentiation

During migration, mPGCs are mitotically inactive (Kagiwada et al., 2013; Seki et al., 2007), but around the stage when they enter and colonize the gonads (E10.5–E13.5), mPGCs restart mitotic proliferation and increase in number.

2.4.1 Continuous DNA demethylation

Genome-wide DNA methylation of migrating mPGCs at E9.5 onward and gonadal germ cells has been investigated with WGBS and reduced representation bisulfite sequencing (RRBS) (Kobayashi et al., 2013; Popp et al., 2010; Seisenberger et al., 2012). The genome-wide DNA methylation levels continuously decrease until E13.5, at which time overall DNA methylation reaches its lowest level (~6% CpG methylation).

Specific genomic loci, including imprint regions, promoters of genes expressed in late germ cell development ("germline genes") (Borgel et al., 2010; Guibert, Forne, & Weber, 2012; Weber et al., 2007), and retrotransposons, have slower demethylation kinetics (Kobayashi et al., 2013; Popp et al., 2010; Seisenberger et al., 2012). Imprint regions and promoters of germline genes are eventually demethylated until E13.5, while some retrotransposons escape the comprehensive demethylation (Kobayashi et al., 2013; Ohta et al., 2017; Seisenberger et al., 2012). Remarkably, the erasure of imprint DNA methylation occurs in a manner strictly parallel to the replication and proliferation of germ cells (Kagiwada et al., 2013), suggesting that DNA demethylation with different kinetics (genome-wide and imprint) are mainly driven by the same mechanism, i.e., replication-coupled dilution.

The replication-coupled, passive dilution of DNA methylation during germ cell development is at least partly explained by the down-regulation of the mRNA level of *Uhrf1* (Kagiwada et al., 2013; Kurimoto et al., 2008), which encodes a protein that recruits the maintenance DNA methyltransferase DNMT1 to the replication foci (Bostick et al., 2007; Sharif et al., 2007). UHRF1 has also been reported to be localized in the cytoplasm of gonadal germ cells (Seisenberger et al., 2012). Passive DNA demethylation also occurs in preimplantation embryos and mESCs under the 2iLIF condition (Arand et al., 2015; von Meyenn et al., 2016), and includes degradation of UHRF1 in the proteasome mediated by PRAMEL7 (Graf et al., 2017). These post-translational regulations of UHRF1 may also contribute to the DNA demethylation during the germ cell development.

On the other hand, UHRF1 directly binds to H3K9me2/me3 (Nady et al., 2011; Rottach et al., 2010). The G9A/GLP complex, which catalyzes H3K9 methylation, is required for DNA methylation in mESCs (Dong et al., 2008; Tachibana, Matsumura, Fukuda, Kimura, & Shinkai, 2008). This enzyme complex also catalyzes methylation of a histone H3K9-like mimic of DNA ligase 1, and this methylation recruits UHRF1 to replicating DNA (Ferry et al., 2017). These mechanisms may contribute to the link between the reduction of H3K9me2 and DNA demethylation in germ cells.

Similarly to the case of the mPGCLC specification, TET proteins seem to play only limited roles, if any, in the genome-wide DNA demethylation during late germ cell development (Hill et al., 2018; Yamaguchi et al., 2013), and the phenotypes of deletion mutants are subtle and/or variable (Dawlaty et al., 2011; SanMiguel, Abramowitz, & Bartolomei, 2018). Interestingly, however, TET1 plays a role in maintaining low-level DNA methylation in the gonadal germ cells (Hill et al., 2018; Yamaguchi, Shen, Liu, Sendler, & Zhang, 2013). A similar mechanism is also found in genome-wide oscillations of DNA methylation that are balanced by active demethylation and *de novo* methylation during the transition from the naïve to the primed state of pluripotency (Rulands et al., 2018). Thus, DNA demethylation during germ cell development is driven by a replication-coupled mechanism, and seems to be fine-tuned by TET1.

2.4.2 Histone modification

The distribution of the histone modifications has been also investigated using low-input ChIP-seq methods. During gonadal germ cell development, numerous stage-specific active enhancers marked by H3K27ac have been identified (Ng et al., 2013). Similarly to the case for mPGCLC specification (Kurimoto et al., 2015), abundant H3K4me3-H3K27me3 bivalent promoters are detected at developmental genes in gonadal germ cells at E11.5, and many of these are retained until E13.5 (Sachs et al., 2013). Thus, enrichment of bivalent promoters in developmental regulators seems to be a continuous feature of mPGC development from specification to the beginning of sex differentiation (Kurimoto et al., 2015; Sachs et al., 2013).

It has been reported that a set of developmentally important genes are bivalent in germ cell development throughout sex differentiation and meiosis in both sexes (Lesch, Dokshin, Young, McCarrey, & Page, 2013). The continuous existence of bivalent promoters during germ cell development lead to the idea that such bivalency carries epigenetic information on poised transcriptional states. An interspecies comparison of the bivalent genes suggests that bivalency of developmentally important regulators in germ cells is evolutionarily conserved among amniotes (Lesch, Silber, McCarrey, & Page, 2016).

Importantly, polycomb regulates maintenance of the mPGC state and the appropriate timing of further germ cell differentiation (Yokobayashi et al., 2013). Two central components of polycomb repressive complex 1 (PRC1), RING1 and RNF1, are essential for the mPGC development between E10.5 and E11.5 in a dosage-dependent manner. Subsequently, RNF1 is required for female gonadal germ cells to maintain the pluripotency program, and to prevent premature induction of meiotic gene expression and premature entry into the meiotic prophase (Yokobayashi et al., 2013).

Interplay between histone modifications and DNA methylation reprogramming is also detected in gonadal germ cell development. In contrast to the vast majority of the genome, retrotransposons resist the erasure of methylation during reprogramming in mPGCs and gonadal germ cells

(Kobayashi et al., 2013; Seisenberger et al., 2012; Shirane et al., 2013). Repressive histone marks are at least partly involved in this demethylation resistance (Liu et al., 2014). In E13.5 gonadal germ cells, H3K9me3 and H3K27me3 are distributed predominantly in distinct genomic regions, whereas these repressive marks overlap in retrotransposons. The germ-cell-specific deletion of SETDB1, an enzyme required for H3K9me3, by the *TNAP*-Cre system leads to reduction of both of H3K9me3 and H3K27me3, and DNA methylation in the H3K9me3-enriched regions. This mutation results in derepression of many retrotransposons, highlighting critical roles of H3K9me3 and SETDB1 in silencing of retrotransposons in germ cell development.

The SETDB1-mediated deposition of H3K9me3 interacts with the UHRF1-mediated maintenance of DNA methylation in mESCs (Sharif et al., 2016). If DNMT1, the maintenance DNA methyltransferase, is inducibly deleted in mESCs that have the wild-type UHRF1, some retrotransposons are transiently derepressed for 8–10 days, and subsequently become repressed again. Conversely, if UHRF1 is inducibly deleted in mESCs that have the wild-type DNMT1 or is deleted in combination with DNMT1, such retrotransposons remain silenced. This is because UHRF1 binds to hemimethylated DNA at the replication foci in a prolonged manner in the absence of DNMT1, and such prolonged binding disrupts the SETDB1-mediated H3K9me3 deposition. Thus, there is an antagonistic biochemical interaction between UHRF1 and SETDB1 (Sharif et al., 2016). This mechanism may explain why UHRF1, but not DNMT1, is down-regulated during the germ cell reprogramming.

On the other hand, conditional deletion of DNMT1 in mPGCs at specification by *Blimp1*-Cre (without deletion of UHRF1) results in a significant loss of mPGCs (Hargan-Calvopina et al., 2016). Derepression of retrotransposons is not detected in late germ cells depleted of DNMT1. Intriguingly, a relatively small number of surviving mPGCs seem to have premature gene expression for oocytes and spermatogonia markers (Hargan-Calvopina et al., 2016). Earlier phenotypes of mPGC-specific deletion of DNMT1 and/or UHRF1 may contribute to an understanding of the mechanism that regulates DNA methylation and transposon repression by DNMT1, UHRF1 and SETDB1.

Thus, during the mPGC development, polycomb (H3K27me3) and SETDB1 (H3K9me3) become the predominant repressive mechanisms, concomitantly with the reduction of H3K9me2 and DNA methylation.

2.5 Epigenome programming in sexually differentiating germ cells in embryonic gonads

At E13.5, DNA methylation levels and patterns are similar between male and female germ cells (Kobayashi et al., 2013; Seisenberger et al., 2012). Thus, while the onset of sex differentiation causes rapid transcriptome changes in gonadal germ cells at E12.5 onward, it has little effect on the DNA methylome (Jameson et al., 2012; Seisenberger et al., 2012).

In contrast, between E13.5 and E16.5, the DNA methylome of the gonadal germ cells shows an obvious sex difference. Female germ cells, which enter meiotic prophase in embryonic ovaries, retain the lowest level of genome-wide DNA methylation, including imprint regions. *De novo* DNA methylation in the female germ cells starts after birth.

In embryonic testes, germ cells differentiate into prospermatogonia and enter mitotic arrest. In these cells, the overall DNA methylation level is drastically elevated (Kobayashi et al., 2013; Seisenberger et al., 2012), and it continues to increase during the development to spermatogonia (Hammoud et al., 2014; Ishikura et al., 2016; Kubo et al., 2015). While global DNA methylation is up-regulated, promoters of genes that function in spermatogenesis, including genes involved in piRNA-mediated repression of transposable elements and/or meiosis (Juliano, Wang, & Lin, 2011), remain hypomethylated during the development of prospermatogonia. These genes are not necessarily expressed in spermatogonia, but may be "prepared" for future expression during spermatogenesis.

Interestingly, retrotransposons resist rapid *de novo* DNA methylation upon the onset of prospermatogonia differentiation, and the methylation levels of such elements remain very similar to those in E13.5 germ cells (Kobayashi et al., 2013; Molaro et al., 2014). This reflects a difference in the mechanisms of genome-wide DNA methylation and retrotransposon methylation, the latter of which is mediated by the piRNA pathway (Molaro et al., 2014).

These studies have elucidated many aspects of the epigenomic regulations of germ cell development. On the other hand, gonadal germ cells undergo extensive epigenome reprogramming, and simultaneously receive signals from the surrounding somatic cells that promote further differentiation. This makes it difficult to examine how reprogramming and gonadal signaling influence germ cell differentiation; further investigation of the mechanisms of gonadal germ cell development requires an *in vitro* culture system under defined conditions as described below.

2.6 *In vitro* proliferation of migrating mPGCs on plane culture

Recently, a method was developed to expand mPGCs and mPGCLCs in a plane culture (Miyauchi, Ohta, & Saitou, 2018; Ohta et al., 2017). This method utilizes m220 cells—which express a membrane-bound form of stem cell factor (SCF), a cytokine essential for the mPGC survival—as feeder cells (Dolci et al., 1991; Majumdar, Feng, Medlock, Toksoz, & Williams, 1994). In the plane culture on this feeder system, mPGCLCs are efficiently propagated by forskolin and rolipram, up-regulators of cyclic AMP. Cultured mPGCLCs expand 20-fold by day 7, corresponding to 4–5 doublings. Cultured mPGCLCs contribute to spermatogenesis and to fertile offspring, when transplanted to the neonatal testes of infertile mice, and thus these cells are physiologically functional germ cells (Ohta et al., 2017).

Interestingly, the cultured mPGCLCs have a gene expression profile that remains similar to that of pre-expansion mPGCLCs and E9.5–E11.5 migrating and gonadal mPGCs. A small number of germline genes, which are highly methylated in the epiblast and resistant to demethylation during mPGC development, are slightly up-regulated, but at much lower levels than in sexually differentiating germ cells at E13.5 in the male and female gonads. Consistent with this finding, during mPGCLC expansion, only a limited change is detected in the genome-wide distribution of H3K27ac, an active enhancer mark, and thus, this plane culture does not cause a large change in the transcriptional state of mPGCLCs.

On the other hand, genome-wide DNA methylation levels of the expanding mPGCLCs continue to decrease, and, at day 7, their methylation levels and patterns are remarkably similar to those of the E13.5 germ cells. Thus, the DNA methylation reprogramming does not drive further germ cell differentiation on its own. Interestingly, the promoter H3K27me3 levels are up-regulated during mPGCLC expansion in a manner that compensates for extensive DNA demethylation, and this may contribute to the maintenance of the early-mPGC-like transcriptome. This reprogrammed epigenetic state, established under a condition free from the extrinsic differentiation signals, represents an epigenetic "blank slate" (Ohta et al., 2017).

2.7 Induction of the female germ cell pathway and initiation of meiosis *in vitro*

Female germ cells enter meiosis in the embryonic ovaries after E13.5, but the signaling pathway sufficient to drive this event has long been elusive.

Using the plane-culture expansion system, signaling molecules have been screened for their potential to induce mPGCLCs along the female pathway (Miyauchi et al., 2017). While retinoic acid (RA) has been identified as a critical signal for meiotic entry and female sex differentiation of gonadal germ cells (Bowles et al., 2006; Koubova et al., 2006), RA alone is not sufficient to induce these events in cultured mPGCLCs. In contrast, a combination of RA and BMP signaling drives female germ cell fate and initiates meiotic prophase in cultured mPGCLCs, faithfully recapitulating the cytological and transcriptome progression of the gonadal female germ cells during E12.5–E14.5 (Miyauchi et al., 2017).

Importantly, mPGCLCs before the plane culture do not properly respond to RA and BMP (Miyauchi et al., 2017). Promoters of genes for late germ cells and/or meiosis (e.g., *Ddx4, Dazl, Scp1*) initially have substantial methylation levels but undergo drastic demethylation during expansion (Ohta et al., 2017; Shirane et al., 2016). Thus, promoter demethylation may provide mPGCLCs with epigenetic competence for the female fate and meiosis, by putting germline genes in permissive states. This may explain why BMP signaling plays different roles between specification and female sex differentiation of mouse germ cells (Miyauchi et al., 2017).

2.8 Derivation of the male germ cell pathway *in vitro*

The male germ cell pathway forms a spermatogonial stem cell system that continuously generates abundant sperm, and may be more complex than the female pathway.

Functional spermatid-like cells have been derived from mPGCLCs through a 2-week co-culture with neonatal testicular somatic cells (Zhou et al., 2016). Considering that mPGCLCs are equivalent to the E9.5 migrating mPGCs and that the first spermatogenesis wave begins at around postnatal day 10 (P10), *in vitro* male germ cell development progresses faster than the *in vivo* counterpart by more than 2 weeks. Importantly, during these 2 weeks, extensive epigenome reprogramming and programming occur as described above, and thus the transcriptome and epigenome dynamics in the culture system should be examined systematically.

Another approach for reconstituting male germ cell development utilizes aggregation of mPGCLCs with testicular somatic cells in E12.5 embryos for up to 54 days (Ishikura et al., 2016). In this aggregation culture system, the aggregates form seminiferous tubule-like structures, and, at day 21 of the culture, germ cells contained in the seminiferous tubules become positive for PLZF, a marker of spermatogonia. From these reconstituted testes, several lines of germline stem cells (GSCs) (Kanatsu-Shinohara et al., 2003) can be established (GSC-like cells), and a limited fraction (20%) of these GSC-like cells exhibit the physiological activity of spermatogonial stem cells, as demonstrated by their implantation into seminiferous tubules in adult mice (Ishikura et al., 2016). The cause of this heterogeneous impairment of spermatogonial stem cell activity in the GSC-like cells may be that the mPGCLCs are not fully reprogrammed and retain genomic DNA methylation at a substantial level (Shirane et al., 2016).

Consistent with the finding that GSC-like cells more-or-less exhibit spermatogonial stem cell activity, these cells have an overall transcriptome and methylome similar to those of GSCs, which show the robust activity of spermatogonial stem cells. However, GSC-like cells bear hyper-methylations in a subset of promoters (3%–5% of all genes), and, interestingly, these hyper-methylations are linked with aberrant repression of genes for meiosis and transposon silencing. These aberrations may explain the impairments in spermatogonial stem cell activity of the GSC-like cells. The importance of proper epigenome regulation during the gametogenesis is also highlighted here.

The mechanisms of germ cell specification, development, and sex differentiation have been investigated in detail using *in vitro* reconstitution systems of mouse germ cells. It would be difficult to reveal such mechanisms merely through experiments using germ cells in mouse embryos. Thus, further advances in germ cell reconstitution, including induction of the male pathway under defined conditions, will contribute to our understanding of mouse germ cell development.

Collectively, these studies on the epigenomic regulations of mPGCs and gonadal germ cells *in vivo* and *in vitro* highlight a specific principle of germ cell reprogramming during embryogenesis. Upon their specification, mPGCs undergo drastic chromatin remodeling and begin genome-wide DNA demethylation, including imprint erasure. The resultant repressive epigenome state is highly dependent on H3K27me3 (polycomb) and H3K9me3 (SETDB1), and is enriched for H3K4me3-H3K27me3 bivalency. Accordingly, DNA methylation erasure has only a limited impact on gene expression in mPGCs, and instead confers a permissive state for genes that are expressed later during germ cell differentiation. DNA demethylation may also be compensated by repressive histone modifications, H3K27me3 and H3K9me3. This epigenome state is generally retained until the germ cell sex differentiation begins. Then, in response to extrinsic signaling, male and female germ cells start to acquire their unique epigenomes in a sex-specific manner.

2.9 Epigenome programming in mouse oogenesis

After birth, the epigenome of female germ cells is gradually acquired during oogenesis. Recently, the establishment of *de novo* DNA methylation, histone modification, open chromatin, and higher-order genome organization during oogenesis have been analyzed, using low-input epigenome techniques.

2.9.1 DNA methylation and chromatin remodeling

de novo DNA methylation occurs in concordance with oocyte growth, beginning in mice at around P10 and largely finishing at the germinal vesicle (GV) stage. GV oocytes are transcriptionally inactive and form a uniquely condensed heterochromatin around the nucleolus. *de novo* DNA methylation in oocytes occurs in transcribed gene bodies, and transcription is required for the methylation of many maternal imprint regions (Stewart, Veselovska, & Kelsey, 2016). The interplay between transcription and DNA methylation is apparent by the pattern of histone modification associated with active transcription. During mouse oogenesis, H3K4me2 and H3K4me3 (markers for active promoters) are reduced at maternal imprint regions in primary and growing oocytes before DNA methylation occurs, and H3K36me3 (a marker of transcriptional elongation) is specifically increased at these regions in growing oocytes (Stewart et al., 2015). Indeed, metaphase II (MII) oocytes deficient in H3K4 demethylases, KDM1A and KDM1B, exhibit defects in the DNA methylation of the imprint regions and CpG islands (Ciccone et al., 2009; Stewart et al., 2015). Thus, dynamic changes of histone modifications that are positively associated with transcription precede DNA methylation pattern in oocytes, especially maternal imprints.

Recently, low-input ChIP-seq methods have identified a striking feature of the histone modification profile during oogenesis. In mature oocytes, there are unique, broad domains of H3K4me3 that cover about 22% of the genome and are not associated with gene promoters and gene expression (Dahl et al., 2016; Hanna et al., 2018; Liu et al., 2016; Zhang et al., 2016). These domains encompass intergenic regions, putative enhancers and silent promoters marked by H3K27me3. In MII oocytes, H3K27me3 is highly pervasive in regions depleted of transcription and DNA methylation (Zhang et al., 2016).

In oogenesis, H3K4me3 is initially restricted to active promoters, then gradually accumulates in the broad domains during the oocyte growth in a transcription-independent manner (Dahl et al., 2016; Hanna et al., 2018; Liu et al., 2016). MLL2, a H3K4 methyltransferase, is essential for the transcription-independent accumulation of H3K4me3 (Hanna et al., 2018). The broad H3K4me3 domains are distributed in negative correlation with DNA methylation, and in fact, DNA methylation protects genomic regions from acquiring H3K4me3, as shown by deletion of *Dnmt3a* and *Dnmt3b* (Dahl et al., 2016; Hanna et al., 2018; Zhang et al., 2016). Together with the finding that H3K4me2 and H3K4me3 block *de novo* DNA methylation in oocytes (Stewart et al., 2015), these results suggest a mutual exclusion between DNA methylation and H3K4me3 during oogenesis. In addition to the broad domains, H3K4me3 is also found in the promoters of many genes—including those activated during zygotic gene activation (ZGA)—in mature oocytes, thereby marking permissive promoters for early preimplantation development (Dahl et al., 2016).

After fertilization, in the late two-cell stage, the broad H3K4me3 domains disappear (Dahl et al., 2016; Liu et al., 2016; Zhang et al., 2016), in concordance with the beginning of the major ZGA. Active removal of H3K4me3 from the broad domains by KDM5A and KDM5B is required for normal ZGA and is essential for early embryonic development (Dahl et al., 2016). Then, similarly to other cell types, H3K4me3 becomes confined to the promoters of transcriptionally active genes.

2.9.2 Histone replacement

Because oocytes are arrested at prophase of meiosis I, histone turnover independent from DNA replication also plays important roles in the establishment of a unique epigenome during oogenesis (Nashun et al., 2015). Conditional knockout of HIRA, the histone H3.3 chaperone, in growing oocytes by *Gdf9*-Cre and *Zp3*-Cre alters the chromatin structure, and results in increased DNase I sensitivity, accumulation of DNA damage, and eventually severe fertility defects.

2.9.3 Three-dimensional organization of the genome

The genome is organized in a three-dimensional structure (Bonev & Cavalli, 2016; Dekker, Marti-Renom, & Mirny, 2013). Topologically associating domains (TADs) (400–500 kb in length) are defined by the frequency of long-range interactions between genomic regions. TADs are further organized in the A and B compartments (~5 Mb in length), which correspond to the transcriptionally active and inactive regions, respectively.

A single-cell Hi-C-seq analysis has reported that the strength of such three-dimensional organization becomes weaker during the maturation of GV oocytes (Flyamer et al., 2017). The broad H3K4me3 domains have no link with the A and B compartments, consistent with the notion that these H3K4me3 domains are not correlated with gene expression (Ke et al., 2017). After fertilization, such structures gradually become evident during the early preimplantation development (Ke et al., 2017).

The above findings demonstrate that unique, totipotent epigenomic architectures, including DNA methylation, histone modification, and three-dimensional organization, are formed during oogenesis, and that extensive reprogramming during preimplantation development may signify the transition from a totipotent to a pluripotent epigenome.

2.10 Epigenome programming during mouse spermatogenesis

In sperm, histones are widely replaced with protamines, and thereby genomic DNA is highly condensed. Thus, sperm have long been thought to make little or no epigenetic contribution to embryonic development. In the last decade, however, micrococcal nuclease digestion methods combined with sequencing have revealed that nucleosomes are retained in the genomes of human and mouse sperm (Brykczynska et al., 2010; Carone et al., 2014; Erkek et al., 2013; Hammoud et al., 2009; Samans et al., 2014), and such retained nucleosomes are distributed around developmental regulators with active and repressive histone marks (Brykczynska et al., 2010; Erkek et al., 2013; Hammoud et al., 2009) or in gene deserts (Carone et al., 2014; Samans et al., 2014). The differences in the distribution of the retained nucleosomes might derive from the different methods used for the analyses (Saitou & Kurimoto, 2014).

Moreover, in the sperm genome, long-range interactions form TADs similar to those of the E3.5 blastocyst and mESCs, as reported in the recent Hi-C-seq analyses (Jung et al., 2017; Ke et al., 2017). Extra-long-range interactions (>4 Mb) and interchromosomal interactions are also frequently detected in the sperm genome (Ke et al., 2017). These findings might suggest that sperm transmit extensive epigenetic information to the next generations, thereby mediating the transgenerational epigenetic inheritance (Rando, 2016).

3. Human germ cell reprogramming

Mechanisms of human germ cell development, especially those for the specification of human PGCs (hPGCs), have long been elusive, since the limited accessibility of human embryos at post-implantation stages has made it difficult to investigate hPGCs experimentally. Moreover, the morphology and chronology of human embryogenesis are significantly different from those of mice. Recent progress in the transcriptome and epigenome analyses of human embryos and the *in vitro* reconstitution of hPGCs (hPGCLCs) followed by further differentiation have advanced our understanding of human germ cell biology.

3.1 Specification and development of human PGCs

The early postimplantation development of human embryos is substantially different from that of mice, in which the epiblast is cup-shaped, the amnion derives from the epiblast and the extraembryonic mesoderm during gastrulation (Fig. 1).

At around week (Wk) 2, the post-implantation human embryo forms a bilaminar disc composed of the epiblast and hypoblast (Fig. 2). The polar trophoectoderm above the epiblast differentiates into the syncytiotrophoblast, which invades the endometrium, and the cytotrophoblast, which surrounds the epiblast. The amnioblast derives directly from the epiblast. The newly formed amnion contacts the cytotrophoblast and forms the amniotic cavity. Then, at the beginning of Wk3, gastrulation begins (see also chapter "Genetic basis for primordial germ cells specification in mouse and human" by Sybirna et al.).

hPGCs are first identified in the dorsal wall of the yolk sac endoderm around Wk3 (De Felici, 2013). Then, the hPGCs migrate into the midgut and hindgut endoderm from around Wk4, penetrate the dorsal mesentery surrounding the gut epithelium at the beginning of Wk5, and enter and colonize the genital ridges at Wk5–6. Finally, the hPGCs undergo sex differentiation in the embryonic testes and ovaries.

The origin of hPGCs is currently unknown. In cynomolgus monkeys, PGCs differentiate from the nascent amnion at around E11, prior to gastrulation (Sasaki et al., 2016). BMP4 and WNT3A are expressed in the amnion and cytotrophoblast, respectively. On the other hand, porcine PGCs are first

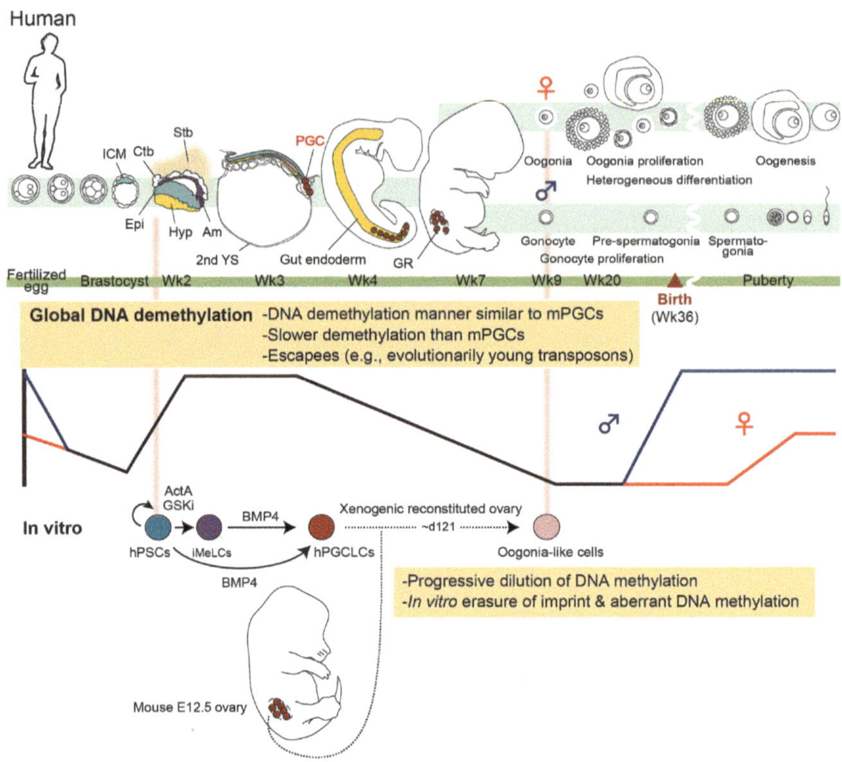

Fig. 2 Schematic representation of human germ cell development *in vivo* and *in vitro*. The developmental process of human germ cells is schematically represented. Key events in epigenome reprogramming and/or programming are listed. DNA methylation levels are represented with a graph. *In vitro* reconstituted cells are represented with colored circles. *Abbreviations*: *2nd YS*, secondary yolk sac; *Am*, amnion; *Ctb*, cytotrophoblast; *Hyp*, hypoblast; *Stb*, syncytiotrophoblast.

detected in the disc-shaped epiblast at the early primitive streak stage (at E11.5) (Kobayashi et al., 2017). Considering the morphological similarity between the gastrulae of humans and cynomolgus monkeys, hPGCs most likely originate from the nascent amnion.

3.2 Development of human gonadal germ cells

In the male, embryonic testes begin to differentiate around Wk5–6 (De Felici, 2013). From Wk7, gonocytes (the male PGCs after reaching the testes) and Sertoli cells form sex cords. During the first trimester, the gonocytes are mitotically active, form a homogeneous cell population with round morphology, and express pluripotency genes and hPGC markers.

During the second trimester, most gonocytes progressively lose their mitotic activity, and differentiate into fetal spermatogonia (prespermatogonia).

In the female, embryonic ovaries begin to differentiate at around Wk6–8. During Wk9, the female hPGCs differentiate into oogonia, which show a high mitotic activity and tend to form clusters of dividing cells. The oogonia enter meiosis and differentiate into primary oocytes at Wk10–11, while their mitotic proliferation continues at least until Wk20. Thus, until Wk20, mitotic oogonia and primary oocytes coexist in the ovaries. Folliculogenesis progresses in the embryonic ovaries, and antral follicles occasionally appear before birth. At around birth, the vast majority of the oogonia and all of the embryonic primary oocytes are eliminated by extensive germ cell death, and a limited population of oogonia remain to contribute to the next generation.

Human germ cells proliferate during migration and in the genital ridges; the estimated numbers of hPGCs are from ~40 at Wk3 to ~150,000 in males and ~300,000 in females at Wk9, and at Wk20, the numbers of prespermatogonia and oogonia have been estimated to be as high as ~4,000,000 and ~7,000,000, respectively. Thus, the developmental process of human germ cells is generally similar to that of mice, whereas human germ cells develop over a much longer time frame and have greater heterogeneity than mouse germ cells (De Felici, 2013).

3.3 Epigenome reprogramming of human gonadal germ cells

Recent studies have reported the transcriptome dynamics and DNA methylation reprogramming of the gonadal germ cells in human embryos. DNA methylation reprogramming of human gonadal germ cells (Wk5.5–20) proceeds in a manner generally similar to that in mice (Eguizabal et al., 2016; Gkountela et al., 2015; Guo et al., 2016, 2015; Tang et al., 2015). Namely, when hPGCs enter the male and female gonads, they are already substantially demethylated (20%–40% CpG at around Wk5–7). DNA methylation of gonadal germ cells reaches its lowest level (~5%) at around Wk9–11 (Gkountela et al., 2015; Guo et al., 2015; Tang et al., 2015). In contrast to the global demethylation, evolutionarily young transposable elements are resistant to the erasure of DNA methylation. These features of the demethylation dynamics are comparable to those of mouse germ cells at E10.5–13.5 (Kobayashi et al., 2013; Seisenberger et al., 2012).

On the other hand, the mitotic behavior of the reprogrammed human germ cells is markedly different from that of mouse germ cells, which become mitotically inactive after being comprehensively demethylated

(~E13.5). As described above, from Wk9 to Wk20 (80 days), human gonocytes and oogonia undergo active mitosis and expand by 20–30 times; this corresponds to at least 4–5 mitotic divisions, and the estimated doubling time is 16–20 days. Thus, human germ cells seem to retain a minimum of DNA methylation during this proliferative period.

3.4 Heterogeneous transcriptome dynamics in human germ cell development

Human germ cells at heterogeneous developmental stages are simultaneously contained in individual embryonic gonads from Wk11 onward, as shown by the single-cell transcriptome analyses.

In particular, female gonadal germ cells undergo four distinct sequential phases, i.e., mitosis, RA signaling, meiotic prophase, and oogenesis (Li et al., 2017). Interestingly, among these, the RA-responsive phase shows an intermediate character between the mitotic oogonia and primary oocytes at the meiotic prophase. Germ cells at this phase express *STRA8*, the key target of the RA signaling, at a higher expression level than those at the meiotic prophase. These cells are positive for late germ cell markers (e.g., *DDX4*, *DAZL*), negative for the early hPGC markers and pluripotency genes, and only weakly positive for meiotic genes (e.g., *SYCP3*, *SPO11*) (Li et al., 2017). Genome-wide, heterogeneous reactivation of the inactive X chromosome in female gonadal germ cells has also been detected (Vertesy et al., 2018).

On the other hand, male germ cells develop through the stages of migration, mitosis, and cell-cycle arrest (Li et al., 2017). These findings may highlight the developmental heterogeneity of human germ cells, and it may be important to consider such heterogeneity when investigating cellular properties in human germ cell populations.

3.5 *In vitro* reconstitution of human PGCs from pluripotent stem cells

Consistent with the idea that the morphology and developmental time frame of the early postimplantation human embryos differ from those of mice, human PSCs (hPSCs) (i.e., human ESCs [hESCs] and human induced PSCs [hiPSCs]) differ substantially from mPSCs in many respects. In the early embryos of cynomolgus monkeys, which show similar morphology to human embryos, the post-implantation epiblast has characteristics similar to those of PSCs from human and cynomolgus monkeys with the primed pluripotent states, as shown by a single-cell transcriptome analysis (Nakamura et al., 2016, 2015, 2017).

Based on the paradigm established for mPGCLCs, hPGCLCs are induced by BMP4 from hPSCs directly and/or *via* incipient mesoderm/primitive streak-like cells (iMeLCs) (Irie et al., 2015; Sasaki et al., 2015; Sugawa et al., 2015). Interestingly, in the nascent amnion of cynomolgus monkeys, PGCs and a few non-PGC cells express genes for the mesoderm and/or primitive streak (*T* and *MSX2*) at low levels (Sasaki et al., 2016), and this observation supports the biological relevance of iMeLCs.

hPGCLCs are positive for the early PGC genes (e.g., *BLIMP1*, *TFAP2C*) but negative for the late PGC markers (e.g., *DAZL*, *DDX4*), and show expression profiles similar to those of early PGCs in cynomolgus monkeys (E13–20) (Sasaki et al., 2016), suggesting that hPGCLCs are equivalent to the early hPGCs. Importantly, SOX17 plays a critical role for the induction of hPGCLCs as an upstream factor for BLIMP1 (Irie et al., 2015; Kojima et al., 2017). The role of SOX17 in the hPGCLC induction is in stark contrast to that in mPGC specification, in which *Sox17* is expressed but is not required (Hara et al., 2009; Kurimoto et al., 2008).

Despite the species differences of the genetic pathways of PGC specification, hPGCLCs have histone mark properties similar to those of mPGCs (high H3K27me3 and low H3K9me2), as shown by immunofluorescence, suggesting that similar chromatin remodeling occurs in human and mouse PGC specification (Sasaki et al., 2015; Tang et al., 2015). On the other hand, DNA demethylation is very limited, if any, in the hPGCLCs (Sasaki et al., 2015; Tang et al., 2015), perhaps reflecting the longer time frame of human germ cell development.

Like mPGCLC induction and mPGC specification, hPGCLC induction requires WNT signaling (Chen et al., 2017; Kobayashi et al., 2017; Kojima et al., 2017). Unlike in the case in mice, transient activation of EOMES, but not T, induces the expression of SOX17 in hPGCLCs (Chen et al., 2017; Kojima et al., 2017). BMP signaling and SOX17 drive BLIMP1 and TFAP2C, which further activate the germ cell program (Irie et al., 2015; Kojima et al., 2017; Sasaki et al., 2015). The somatic program is gradually repressed in hPGCLC induction, in contrast to the swift repression by BLIMP1 in mPGC specification.

Importantly, the induction efficiency of hPGCLCs varies substantially among hPSC clones, as reflected by the gene expression patterns in the iMeLCs (e.g., up-regulations of *EOMES* and *T* as positive indicators) (Chen et al., 2017; Yokobayashi et al., 2017). This clonal variation might be associated with genetic and/or epigenetic variation. Interestingly, *OTX2* expression in iMELCs is a positive indicator of hPGCLC induction

(Yokobayashi et al., 2017), apparently in contrast to the inhibitory role of OTX2 in mPGC specification (Zhang, Zhang, et al., 2018). These findings highlight the evolutionary conservation and divergence of the regulatory mechanisms for human and mouse PGC fates.

3.6 Epigenome reprogramming during *in vitro* reconstitution of human PGCs and their differentiation into oogonia

Recently, it has been shown that hPGCLCs differentiate progressively into oogonia-like cells during long-term culture in xenogeneic reconstituted ovaries with ovarian somatic cells in mouse E12.5 embryos (Yamashiro et al., 2018). hPGCLCs induced from female hiPSCs survive in the xenogeneic reconstituted ovaries for ~4 months (120 days). The transcriptome of the hPGCLC-derived cells changes according to the culture period, and becomes very similar to that of Wk9-Wk11 gonadal germ cells (Tang et al., 2015). At day 120 of aggregation, the hPGCLC-derived cells have a gene expression profile similar to gonadal germ cells at the RA-responsive phase in human embryonic ovaries (Li et al., 2017).

The DNA methylation dynamics during hPGCLC specification from hiPSCs and differentiation into oogonia-like cells has been analyzed by WGBS (Yamashiro et al., 2018). The genome-wide levels of DNA methylation in hiPSCs and iMeLCs are ~80% CpG, and they decrease slightly in hPGCLCs before the xenogeneic aggregation. The methylation levels progressively decrease in hPGCLC-derived cells in the xenogeneic reconstituted ovaries, reaching ~13% CpG at day 120 of aggregation, a level comparable to that in gonadal germ cells at Wk7–Wk10 (Guo et al., 2015; Tang et al., 2015). The methylation patterns of the hPGCLC-derived cells are also very similar to those of gonadal germ cells, but not to those of the blastocyst and naïve hESCs. Thus, hPGCLC-derived cells undergo DNA methylation reprogramming in a manner similar to human gonadal germ cells, but not to early embryos.

Moreover, during the differentiation in the xenogeneic reconstituted ovaries, the hPGCLC-derived cells erase parental imprints and aberrant hyper-methylation in hiPSCs. The inactive X chromosomes are progressively demethylated and partially reactivated. Thus, this study recapitulates the epigenome reprogramming of human germ cells *in vitro*, and may provide a critical step toward human *in vitro* gametogenesis (Yamashiro et al., 2018).

Collectively, the above findings show that research into the mechanisms of human germ cell development is still relatively primitive. To advance this field, more investigations of human germ cells *in vivo* will be needed, as well as further developments in human *in vitro* gametogenesis.

4. Conclusion and perspectives

Recent studies have revealed details of the genetic programs, chromatin remodeling, TF binding patterns, and DNA methylation reprogramming of the germ cell development in mice. *In vitro* reconstitution of germ cell lineages and low-input genomics have contributed significantly to advances in our knowledge of germ cell development. The genetic, epigenetic, and signaling studies have also revealed principles underlying the key mechanisms of human and mouse PGC fates and gametogenesis, as well as the diversity of these mechanisms. Nonetheless, the regulatory mechanisms of human germ cell development, including epigenome reprogramming, are still less understood than those of mice. Further discoveries in the field of human *in vitro* gametogenesis derived from hPGCLCs will greatly contribute to the establishment of a framework of human germ cell biology.

Acknowledgment

We thank the members of our laboratory for their helpful input.

Funding

This work was supported in part by Japan Society for the Promotion (JSPS) KAKENHI grants (JP18H05553, JP16H04720, and JP18K19295 to K.K, and JP17H06098 to M.S.), by funds from the Uehara Memorial Foundation (to K.K), and by a Japan Science and Technology Agency-Exploratory Research for Advanced Technology (JST-ERATO) grant (JPMJER1104 to M.S.).

References

Aramaki, S., Hayashi, K., Kurimoto, K., Ohta, H., Yabuta, Y., Iwanari, H., et al. (2013). A mesodermal factor, T, specifies mouse germ cell fate by directly activating germline determinants. *Developmental Cell*, *27*, 516–529.

Arand, J., Wossidlo, M., Lepikhov, K., Peat, J. R., Reik, W., & Walter, J. (2015). Selective impairment of methylation maintenance is the major cause of DNA methylation reprogramming in the early embryo. *Epigenetics & Chromatin*, *8*, 1.

Avner, P., & Heard, E. (2001). X-chromosome inactivation: Counting, choice and initiation. *Nature Reviews. Genetics*, *2*, 59–67.

Bonev, B., & Cavalli, G. (2016). Organization and function of the 3D genome. *Nature Reviews. Genetics*, *17*, 661–678.

Borgel, J., Guibert, S., Li, Y., Chiba, H., Schubeler, D., Sasaki, H., et al. (2010). Targets and dynamics of promoter DNA methylation during early mouse development. *Nature Genetics*, *42*, 1093–1100.

Bostick, M., Kim, J. K., Esteve, P. O., Clark, A., Pradhan, S., & Jacobsen, S. E. (2007). UHRF1 plays a role in maintaining DNA methylation in mammalian cells. *Science*, *317*, 1760–1764.

Boulard, M., Edwards, J. R., & Bestor, T. H. (2015). FBXL10 protects polycomb-bound genes from hypermethylation. *Nature Genetics*, *47*, 479–485.

Bowles, J., Knight, D., Smith, C., Wilhelm, D., Richman, J., Mamiya, S., et al. (2006). Retinoid signaling determines germ cell fate in mice. *Science*, *312*, 596–600.

Brykczynska, U., Hisano, M., Erkek, S., Ramos, L., Oakeley, E. J., Roloff, T. C., et al. (2010). Repressive and active histone methylation mark distinct promoters in human and mouse spermatozoa. *Nature Structural & Molecular Biology*, *17*, 679–687.

Buecker, C., Srinivasan, R., Wu, Z., Calo, E., Acampora, D., Faial, T., et al. (2014). Reorganization of enhancer patterns in transition from naive to primed pluripotency. *Cell Stem Cell*, *14*, 838–853.

Carone, B. R., Hung, J. H., Hainer, S. J., Chou, M. T., Carone, D. M., Weng, Z., et al. (2014). High-resolution mapping of chromatin packaging in mouse embryonic stem cells and sperm. *Developmental Cell*, *30*, 11–22.

Chen, D., Liu, W., Lukianchikov, A., Hancock, G., Zimmerman, J., Lowe, M., et al. (2017). Germline competency of human embryonic stem cells depends on eomesodermin. *Biology of Reproduction*, *97*, 850–861.

Choi, J., Clement, K., Huebner, A. J., Webster, J., Rose, C. M., Brumbaugh, J., et al. (2017). DUSP9 modulates DNA hypomethylation in female mouse pluripotent stem cells. *Cell Stem Cell*, *20*. 706-719.e7.

Ciccone, D. N., Su, H., Hevi, S., Gay, F., Lei, H., Bajko, J., et al. (2009). KDM1B is a histone H3K4 demethylase required to establish maternal genomic imprints. *Nature*, *461*, 415–418.

Dahl, J. A., Jung, I., Aanes, H., Greggains, G. D., Manaf, A., Lerdrup, M., et al. (2016). Broad histone H3K4me3 domains in mouse oocytes modulate maternal-to-zygotic transition. *Nature*, *537*, 548–552.

Dawlaty, M. M., Ganz, K., Powell, B. E., Hu, Y. C., Markoulaki, S., Cheng, A. W., et al. (2011). Tet1 is dispensable for maintaining pluripotency and its loss is compatible with embryonic and postnatal development. *Cell Stem Cell*, *9*, 166–175.

De Felici, M. (2013). *Origin, migration, and proliferation of human primordial germ cells*. London: Springer-Verlag.

Dekker, J., Marti-Renom, M. A., & Mirny, L. A. (2013). Exploring the three-dimensional organization of genomes: Interpreting chromatin interaction data. *Nature Reviews. Genetics*, *14*, 390–403.

Dolci, S., Williams, D. E., Ernst, M. K., Resnick, J. L., Brannan, C. I., Lock, L. F., et al. (1991). Requirement for mast cell growth factor for primordial germ cell survival in culture. *Nature*, *352*, 809–811.

Dong, K. B., Maksakova, I. A., Mohn, F., Leung, D., Appanah, R., Lee, S., et al. (2008). DNA methylation in ES cells requires the lysine methyltransferase G9a but not its catalytic activity. *The EMBO Journal*, *27*, 2691–2701.

Eckersley-Maslin, M. A., Alda-Catalinas, C., & Reik, W. (2018). Dynamics of the epigenetic landscape during the maternal-to-zygotic transition. *Nature Reviews. Molecular Cell Biology*, *19*, 436–450.

Eguizabal, C., Herrera, L., de Onate, L., Montserrat, N., Hajkova, P., & Izpisua Belmonte, J. C. (2016). Characterisation of the epigenetic changes during human gonadal primordial germ cells reprogramming. *Stem Cells*, *34*, 2418–2428.

Erkek, S., Hisano, M., Liang, C. Y., Gill, M., Murr, R., Dieker, J., et al. (2013). Molecular determinants of nucleosome retention at CpG-rich sequences in mouse spermatozoa. *Nature Structural & Molecular Biology*, *20*, 868–875.

Factor, D. C., Corradin, O., Zentner, G. E., Saiakhova, A., Song, L., Chenoweth, J. G., et al. (2014). Epigenomic comparison reveals activation of "seed" enhancers during transition from naive to primed pluripotency. *Cell Stem Cell*, *14*, 854–863.

Ferry, L., Fournier, A., Tsusaka, T., Adelmant, G., Shimazu, T., Matano, S., et al. (2017). Methylation of DNA ligase 1 by G9a/GLP recruits UHRF1 to replicating DNA and regulates DNA methylation. *Molecular Cell*, *67*. 550-565.e555.

Flyamer, I. M., Gassler, J., Imakaev, M., Brandao, H. B., Ulianov, S. V., Abdennur, N., et al. (2017). Single-nucleus Hi-C reveals unique chromatin reorganization at oocyte-to-zygote transition. *Nature*, *544*, 110–114.

Gkountela, S., Zhang, K. X., Shafiq, T. A., Liao, W. W., Hargan-Calvopina, J., Chen, P. Y., et al. (2015). DNA demethylation dynamics in the human prenatal germline. *Cell*, *161*, 1425–1436.

Graf, U., Casanova, E. A., Wyck, S., Dalcher, D., Gatti, M., Vollenweider, E., et al. (2017). Pramel7 mediates ground-state pluripotency through proteasomal-epigenetic combined pathways. *Nature Cell Biology*, *19*, 763–773.

Guibert, S., Forne, T., & Weber, M. (2012). Global profiling of DNA methylation erasure in mouse primordial germ cells. *Genome Research*, *22*, 633–641.

Guo, H., Hu, B., Yan, L., Yong, J., Wu, Y., Gao, Y., et al. (2016). DNA methylation and chromatin accessibility profiling of mouse and human fetal germ cells. *Cell Research*, *27*, 165–183.

Guo, F., Yan, L., Guo, H., Li, L., Hu, B., Zhao, Y., et al. (2015). The transcriptome and DNA methylome landscapes of human primordial germ cells. *Cell*, *161*, 1437–1452.

Habibi, E., Brinkman, A. B., Arand, J., Kroeze, L. I., Kerstens, H. H., Matarese, F., et al. (2013). Whole-genome bisulfite sequencing of two distinct interconvertible DNA methylomes of mouse embryonic stem cells. *Cell Stem Cell*, *13*, 360–369.

Hammoud, S. S., Low, D. H., Yi, C., Carrell, D. T., Guccione, E., & Cairns, B. R. (2014). Chromatin and transcription transitions of mammalian adult germline stem cells and spermatogenesis. *Cell Stem Cell*, *15*, 239–253.

Hammoud, S. S., Nix, D. A., Zhang, H., Purwar, J., Carrell, D. T., & Cairns, B. R. (2009). Distinctive chromatin in human sperm packages genes for embryo development. *Nature*, *460*, 473–478.

Hanna, C. W., Taudt, A., Huang, J., Gahurova, L., Kranz, A., Andrews, S., et al. (2018). MLL2 conveys transcription-independent H3K4 trimethylation in oocytes. *Nature Structural & Molecular Biology*, *25*, 73–82.

Hara, K., Kanai-Azuma, M., Uemura, M., Shitara, H., Taya, C., Yonekawa, H., et al. (2009). Evidence for crucial role of hindgut expansion in directing proper migration of primordial germ cells in mouse early embryogenesis. *Developmental Biology*, *330*, 427–439.

Hargan-Calvopina, J., Taylor, S., Cook, H., Hu, Z., Lee, S. A., Yen, M. R., et al. (2016). Stage-specific demethylation in primordial germ cells safeguards against precocious differentiation. *Developmental Cell*, *39*, 75–86.

Hayashi, K., Ogushi, S., Kurimoto, K., Shimamoto, S., Ohta, H., & Saitou, M. (2012). Offspring from oocytes derived from in vitro primordial germ cell-like cells in mice. *Science*, *338*, 971–975.

Hayashi, K., Ohta, H., Kurimoto, K., Aramaki, S., & Saitou, M. (2011). Reconstitution of the mouse germ cell specification pathway in culture by pluripotent stem cells. *Cell*, *146*, 519–532.

Hikabe, O., Hamazaki, N., Nagamatsu, G., Obata, Y., Hirao, Y., Hamada, N., et al. (2016). Reconstitution in vitro of the entire cycle of the mouse female germ line. *Nature*, *539*, 299–303.

Hill, P. W. S., Leitch, H. G., Requena, C. E., Sun, Z., Amouroux, R., Roman-Trufero, M., et al. (2018). Epigenetic reprogramming enables the transition from primordial germ cell to gonocyte. *Nature*, *555*, 392–396.

Irie, N., Weinberger, L., Tang, W. W., Kobayashi, T., Viukov, S., Manor, Y. S., et al. (2015). SOX17 is a critical specifier of human primordial germ cell fate. *Cell*, *160*, 253–268.

Ishikura, Y., Yabuta, Y., Ohta, H., Hayashi, K., Nakamura, T., Okamoto, I., et al. (2016). In vitro derivation and propagation of spermatogonial stem cell activity from mouse pluripotent stem cells. *Cell Reports*, *17*, 2780–2804.

Jameson, S. A., Natarajan, A., Cool, J., DeFalco, T., Maatouk, D. M., Mork, L., et al. (2012). Temporal transcriptional profiling of somatic and germ cells reveals biased lineage priming of sexual fate in the fetal mouse gonad. *PLoS Genetics*, *8*, e1002575.

Juliano, C., Wang, J., & Lin, H. (2011). Uniting germline and stem cells: The function of Piwi proteins and the piRNA pathway in diverse organisms. *Annual Review of Genetics*, *45*, 447–469.

Jung, Y. H., Sauria, M. E. G., Lyu, X., Cheema, M. S., Ausio, J., Taylor, J., et al. (2017). Chromatin states in mouse sperm correlate with embryonic and adult regulatory landscapes. *Cell Reports*, *18*, 1366–1382.

Kagiwada, S., Kurimoto, K., Hirota, T., Yamaji, M., & Saitou, M. (2013). Replication-coupled passive DNA demethylation for the erasure of genome imprints in mice. *The EMBO Journal*, *32*, 340–353.

Kanatsu-Shinohara, M., Ogonuki, N., Inoue, K., Miki, H., Ogura, A., Toyokuni, S., et al. (2003). Long-term proliferation in culture and germline transmission of mouse male germline stem cells. *Biology of Reproduction*, *69*, 612–616.

Ke, Y., Xu, Y., Chen, X., Feng, S., Liu, Z., Sun, Y., et al. (2017). 3D chromatin structures of mature gametes and structural reprogramming during mammalian embryogenesis. *Cell*, *170*. 367-381.e320.

Kelsey, G., Stegle, O., & Reik, W. (2017). Single-cell epigenomics: Recording the past and predicting the future. *Science*, *358*, 69–75.

Kobayashi, H., Sakurai, T., Miura, F., Imai, M., Mochiduki, K., Yanagisawa, E., et al. (2013). High-resolution DNA methylome analysis of primordial germ cells identifies gender-specific reprogramming in mice. *Genome Research*, *23*, 616–627.

Kobayashi, T., Zhang, H., Tang, W. W. C., Irie, N., Withey, S., Klisch, D., et al. (2017). Principles of early human development and germ cell program from conserved model systems. *Nature*, *546*, 416–420.

Kojima, Y., Sasaki, K., Yokobayashi, S., Sakai, Y., Nakamura, T., Yabuta, Y., et al. (2017). Evolutionarily distinctive transcriptional and signaling programs drive human germ cell lineage specification from pluripotent stem cells. *Cell Stem Cell*, *21*. 517-532.e515.

Koubova, J., Menke, D. B., Zhou, Q., Capel, B., Griswold, M. D., & Page, D. C. (2006). Retinoic acid regulates sex-specific timing of meiotic initiation in mice. *Proceedings of the National Academy of Sciences of the United States of America*, *103*, 2474–2479.

Kubo, N., Toh, H., Shirane, K., Shirakawa, T., Kobayashi, H., Sato, T., et al. (2015). DNA methylation and gene expression dynamics during spermatogonial stem cell differentiation in the early postnatal mouse testis. *BMC Genomics*, *16*, 624.

Kurimoto, K., & Saitou, M. (2015). Mechanism and reconstitution in vitro of germ cell development in mammals. *Cold Spring Harbor Symposia on Quantitative Biology*, *80*, 147–154.

Kurimoto, K., Yabuta, Y., Hayashi, K., Ohta, H., Kiyonari, H., Mitani, T., et al. (2015). Quantitative dynamics of chromatin remodeling during germ cell specification from mouse embryonic stem cells. *Cell Stem Cell*, *16*, 517–532.

Kurimoto, K., Yabuta, Y., Ohinata, Y., Ono, Y., Uno, K. D., Yamada, R. G., et al. (2006). An improved single-cell cDNA amplification method for efficient high-density oligonucleotide microarray analysis. *Nucleic Acids Research*, *34*, e42.

Kurimoto, K., Yabuta, Y., Ohinata, Y., Shigeta, M., Yamanaka, K., & Saitou, M. (2008). Complex genome-wide transcription dynamics orchestrated by Blimp1 for the specification of the germ cell lineage in mice. *Genes & Development*, *22*, 1617–1635.

Lawson, K. A., Dunn, N. R., Roelen, B. A., Zeinstra, L. M., Davis, A. M., Wright, C. V., et al. (1999). Bmp4 is required for the generation of primordial germ cells in the mouse embryo. *Genes & Development*, *13*, 424–436.

Lesch, B. J., Dokshin, G. A., Young, R. A., McCarrey, J. R., & Page, D. C. (2013). A set of genes critical to development is epigenetically poised in mouse germ cells from fetal stages through completion of meiosis. *Proceedings of the National Academy of Sciences of the United States of America*, *110*, 16061–16066.

Lesch, B. J., Silber, S. J., McCarrey, J. R., & Page, D. C. (2016). Parallel evolution of male germline epigenetic poising and somatic development in animals. *Nature Genetics*, *48*(8), 888–894.

Li, L., Dong, J., Yan, L., Yong, J., Liu, X., Hu, Y., et al. (2017). Single-cell RNA-Seq analysis maps development of human germline cells and gonadal niche interactions. *Cell Stem Cell*, *20*(6), 891–892.

Liu, S., Brind'Amour, J., Karimi, M. M., Shirane, K., Bogutz, A., Lefebvre, L., et al. (2014). Setdb1 is required for germline development and silencing of H3K9me3-marked endogenous retroviruses in primordial germ cells. *Genes & Development*, *28*, 2041–2055.

Liu, X., Wang, C., Liu, W., Li, J., Li, C., Kou, X., et al. (2016). Distinct features of H3K4me3 and H3K27me3 chromatin domains in pre-implantation embryos. *Nature*, *537*, 558–562.

Ma, Z., Swigut, T., Valouev, A., Rada-Iglesias, A., & Wysocka, J. (2011). Sequence-specific regulator Prdm14 safeguards mouse ESCs from entering extraembryonic endoderm fates. *Nature Structural & Molecular Biology*, *18*, 120–127.

Magnusdottir, E., Dietmann, S., Murakami, K., Gunesdogan, U., Tang, F., Bao, S., et al. (2013). A tripartite transcription factor network regulates primordial germ cell specification in mice. *Nature Cell Biology*, *15*, 905–915.

Majumdar, M. K., Feng, L., Medlock, E., Toksoz, D., & Williams, D. A. (1994). Identification and mutation of primary and secondary proteolytic cleavage sites in murine stem cell factor cDNA yields biologically active, cell-associated protein. *The Journal of Biological Chemistry*, *269*, 1237–1242.

Marks, H., Kalkan, T., Menafra, R., Denissov, S., Jones, K., Hofemeister, H., et al. (2012). The transcriptional and epigenomic foundations of ground state pluripotency. *Cell*, *149*, 590–604.

Mitani, T., Yabuta, Y., Ohta, H., Nakamura, T., Yamashiro, C., Yamamoto, T., et al. (2017). Principles for the regulation of multiple developmental pathways by a versatile transcriptional factor, BLIMP1. *Nucleic Acids Research*, *45*, 12152–12169.

Miyauchi, H., Ohta, H., Nagaoka, S., Nakaki, F., Sasaki, K., Hayashi, K., et al. (2017). Bone morphogenetic protein and retinoic acid synergistically specify female germ-cell fate in mice. *The EMBO Journal*, *36*, 3100–3119.

Miyauchi, H., Ohta, H., & Saitou, M. (2018). Induction of fetal primary oocytes and the meiotic prophase from mouse pluripotent stem cells. *Methods in Cell Biology*, *144*, 409–429.

Miyoshi, N., Stel, J. M., Shioda, K., Qu, N., Odahima, J., Mitsunaga, S., et al. (2016). Erasure of DNA methylation, genomic imprints, and epimutations in a primordial germ-cell model derived from mouse pluripotent stem cells. *Proceedings of the National Academy of Sciences of the United States of America*, *113*, 9545–9550.

Molaro, A., Falciatori, I., Hodges, E., Aravin, A. A., Marran, K., Rafii, S., et al. (2014). Two waves of de novo methylation during mouse germ cell development. *Genes & Development*, *28*, 1544–1549.

Morohaku, K., Tanimoto, R., Sasaki, K., Kawahara-Miki, R., Kono, T., Hayashi, K., et al. (2016). Complete in vitro generation of fertile oocytes from mouse primordial germ cells. *Proceedings of the National Academy of Sciences of the United States of America*, *113*, 9021–9026.

Murakami, K., Gunesdogan, U., Zylicz, J. J., Tang, W. W., Sengupta, R., Kobayashi, T., et al. (2016). NANOG alone induces germ cells in primed epiblast in vitro by activation of enhancers. *Nature*, *529*, 403–407.

Nady, N., Gupta, A., Ma, Z., Swigut, T., Koide, A., Koide, S., et al. (2015). ETO family protein Mtgr1 mediates Prdm14 functions in stem cell maintenance and primordial germ cell formation. *eLife*, *4*, e10150.

Nady, N., Lemak, A., Walker, J. R., Avvakumov, G. V., Kareta, M. S., Achour, M., et al. (2011). Recognition of multivalent histone states associated with heterochromatin by UHRF1 protein. *The Journal of Biological Chemistry*, *286*, 24300–24311.

Nakaki, F., Hayashi, K., Ohta, H., Kurimoto, K., Yabuta, Y., & Saitou, M. (2013). Induction of mouse germ-cell fate by transcription factors in vitro. *Nature*, *501*, 222–226.

Nakamura, T., Okamoto, I., Sasaki, K., Yabuta, Y., Iwatani, C., Tsuchiya, H., et al. (2016). A developmental coordinate of pluripotency among mice, monkeys and humans. *Nature*, *537*, 57–62.

Nakamura, T., Yabuta, Y., Okamoto, I., Aramaki, S., Yokobayashi, S., Kurimoto, K., et al. (2015). SC3-seq: A method for highly parallel and quantitative measurement of single-cell gene expression. *Nucleic Acids Research*, *43*, e60.

Nakamura, T., Yabuta, Y., Okamoto, I., Sasaki, K., Iwatani, C., Tsuchiya, H., et al. (2017). Single-cell transcriptome of early embryos and cultured embryonic stem cells of cynomolgus monkeys. *Science Data*, *4* 170067.

Nashun, B., Hill, P. W., Smallwood, S. A., Dharmalingam, G., Amouroux, R., Clark, S. J., et al. (2015). Continuous histone replacement by Hira is essential for normal transcriptional regulation and de novo DNA methylation during mouse oogenesis. *Molecular Cell*, *60*, 611–625.

Ng, J. H., Kumar, V., Muratani, M., Kraus, P., Yeo, J. C., Yaw, L. P., et al. (2013). In vivo epigenomic profiling of germ cells reveals germ cell molecular signatures. *Developmental Cell*, *24*, 324–333.

Ohinata, Y., Ohta, H., Shigeta, M., Yamanaka, K., Wakayama, T., & Saitou, M. (2009). A signaling principle for the specification of the germ cell lineage in mice. *Cell*, *137*, 571–584.

Ohinata, Y., Payer, B., O'Carroll, D., Ancelin, K., Ono, Y., Sano, M., et al. (2005). Blimp1 is a critical determinant of the germ cell lineage in mice. *Nature*, *436*, 207–213.

Ohta, H., Kurimoto, K., Okamoto, I., Nakamura, T., Yabuta, Y., Miyauchi, H., et al. (2017). In vitro expansion of mouse primordial germ cell-like cells recapitulates an epigenetic blank slate. *The EMBO Journal*, *36*, 1888–1907.

Popp, C., Dean, W., Feng, S., Cokus, S. J., Andrews, S., Pellegrini, M., et al. (2010). Genome-wide erasure of DNA methylation in mouse primordial germ cells is affected by AID deficiency. *Nature*, *463*, 1101–1105.

Rando, O. J. (2016). Intergenerational transfer of epigenetic information in sperm. *Cold Spring Harbor Perspectives in Medicine*, *6*, a022988.

Rottach, A., Frauer, C., Pichler, G., Bonapace, I. M., Spada, F., & Leonhardt, H. (2010). The multi-domain protein Np95 connects DNA methylation and histone modification. *Nucleic Acids Research*, *38*, 1796–1804.
Rulands, S., Lee, H. J., Clark, S. J., Angermueller, C., Smallwood, S. A., Krueger, F., et al. (2018). Genome-scale oscillations in DNA methylation during exit from pluripotency. *Cell Systems*, *7*, 63–76.
Sachs, M., Onodera, C., Blaschke, K., Ebata, K. T., Song, J. S., & Ramalho-Santos, M. (2013). Bivalent chromatin marks developmental regulatory genes in the mouse embryonic germline in vivo. *Cell Reports*, *3*, 1777–1784.
Saitou, M., Barton, S. C., & Surani, M. A. (2002). A molecular programme for the specification of germ cell fate in mice. *Nature*, *418*, 293–300.
Saitou, M., & Kurimoto, K. (2014). Paternal nucleosomes: Are they retained in developmental promoters or gene deserts? *Developmental Cell*, *30*, 6–8.
Samans, B., Yang, Y., Krebs, S., Sarode, G. V., Blum, H., Reichenbach, M., et al. (2014). Uniformity of nucleosome preservation pattern in mammalian sperm and its connection to repetitive DNA elements. *Developmental Cell*, *30*, 23–35.
SanMiguel, J. M., Abramowitz, L. K., & Bartolomei, M. S. (2018). Imprinted gene dysregulation in a *Tet1* null mouse model is stochastic and variable in the germline and offspring. *Development*, *145* dev160622.
Sasaki, K., Nakamura, T., Okamoto, I., Yabuta, Y., Iwatani, C., Tsuchiya, H., et al. (2016). The germ cell fate of cynomolgus monkeys is specified in the nascent amnion. *Developmental Cell*, *39*, 169–185.
Sasaki, K., Yokobayashi, S., Nakamura, T., Okamoto, I., Yabuta, Y., Kurimoto, K., et al. (2015). Robust in vitro induction of human germ cell fate from pluripotent stem cells. *Cell Stem Cell*, *17*, 178–194.
Seisenberger, S., Andrews, S., Krueger, F., Arand, J., Walter, J., Santos, F., et al. (2012). The dynamics of genome-wide DNA methylation reprogramming in mouse primordial germ cells. *Molecular Cell*, *48*, 849–862.
Seki, Y., Hayashi, K., Itoh, K., Mizugaki, M., Saitou, M., & Matsui, Y. (2005). Extensive and orderly reprogramming of genome-wide chromatin modifications associated with specification and early development of germ cells in mice. *Developmental Biology*, *278*, 440–458.
Seki, Y., Yamaji, M., Yabuta, Y., Sano, M., Shigeta, M., Matsui, Y., et al. (2007). Cellular dynamics associated with the genome-wide epigenetic reprogramming in migrating primordial germ cells in mice. *Development*, *134*, 2627–2638.
Sharif, J., Endo, T. A., Nakayama, M., Karimi, M. M., Shimada, M., Katsuyama, K., et al. (2016). Activation of endogenous retroviruses in Dnmt1(-/-) ESCs involves disruption of SETDB1-mediated repression by NP95 binding to hemimethylated DNA. *Cell Stem Cell*, *19*, 81–94.
Sharif, J., Muto, M., Takebayashi, S., Suetake, I., Iwamatsu, A., Endo, T. A., et al. (2007). The SRA protein Np95 mediates epigenetic inheritance by recruiting Dnmt1 to methylated DNA. *Nature*, *450*, 908–912.
Shirane, K., Kurimoto, K., Yabuta, Y., Yamaji, M., Satoh, J., Ito, S., et al. (2016). Global landscape and regulatory principles of DNA methylation reprogramming for germ cell specification by mouse pluripotent stem cells. *Developmental Cell*, *39*, 87–103.
Shirane, K., Toh, H., Kobayashi, H., Miura, F., Chiba, H., Ito, T., et al. (2013). Mouse oocyte methylomes at base resolution reveal genome-wide accumulation of non-CpG methylation and role of DNA methyltransferases. *PLoS Genetics*, *9* e1003439.
Stewart, K. R., Veselovska, L., & Kelsey, G. (2016). Establishment and functions of DNA methylation in the germline. *Epigenomics*, *8*, 1399–1413.

Stewart, K. R., Veselovska, L., Kim, J., Huang, J., Saadeh, H., Tomizawa, S., et al. (2015). Dynamic changes in histone modifications precede de novo DNA methylation in oocytes. *Genes & Development*, *29*, 2449–2462.

Sugawa, F., Arauzo-Bravo, M. J., Yoon, J., Kim, K. P., Aramaki, S., Wu, G., et al. (2015). Human primordial germ cell commitment in vitro associates with a unique PRDM14 expression profile. *The EMBO Journal*, *34*, 1009–1024.

Tachibana, M., Matsumura, Y., Fukuda, M., Kimura, H., & Shinkai, Y. (2008). G9a/GLP complexes independently mediate H3K9 and DNA methylation to silence transcription. *The EMBO Journal*, *27*, 2681–2690.

Tang, W. W., Dietmann, S., Irie, N., Leitch, H. G., Floros, V. I., Bradshaw, C. R., et al. (2015). A unique gene regulatory network resets the human germline epigenome for development. *Cell*, *161*, 1453–1467.

Tu, S., Narendra, V., Yamaji, M., Vidal, S. E., Rojas, L. A., Wang, X., et al. (2016). Co-repressor CBFA2T2 regulates pluripotency and germline development. *Nature*, *534*, 387–390.

Vertesy, A., Arindrarto, W., Roost, M. S., Reinius, B., Torrens-Juaneda, V., Bialecka, M., et al. (2018). Parental haplotype-specific single-cell transcriptomics reveal incomplete epigenetic reprogramming in human female germ cells. *Nature Communications*, *9*, 1873.

von Meyenn, F., Berrens, R. V., Andrews, S., Santos, F., Collier, A. J., Krueger, F., et al. (2016). Comparative principles of DNA methylation reprogramming during human and mouse in vitro primordial germ cell specification. *Developmental Cell*, *39*, 104–115.

von Meyenn, F., Iurlaro, M., Habibi, E., Liu, N. Q., Salehzadeh-Yazdi, A., Santos, F., et al. (2016). Impairment of DNA methylation maintenance is the main cause of global demethylation in naive embryonic stem cells. *Molecular Cell*, *62*, 848–861.

Weber, M., Hellmann, I., Stadler, M. B., Ramos, L., Paabo, S., Rebhan, M., et al. (2007). Distribution, silencing potential and evolutionary impact of promoter DNA methylation in the human genome. *Nature Genetics*, *39*, 457–466.

Yabuta, Y., Kurimoto, K., Ohinata, Y., Seki, Y., & Saitou, M. (2006). Gene expression dynamics during germline specification in mice identified by quantitative single-cell gene expression profiling. *Biology of Reproduction*, *75*, 705–716.

Yamaguchi, S., Hong, K., Liu, R., Inoue, A., Shen, L., Zhang, K., et al. (2013). Dynamics of 5-methylcytosine and 5-hydroxymethylcytosine during germ cell reprogramming. *Cell Research*, *23*, 329–339.

Yamaguchi, S., Shen, L., Liu, Y., Sendler, D., & Zhang, Y. (2013). Role of Tet1 in erasure of genomic imprinting. *Nature*, *504*, 460–464.

Yamaji, M., Seki, Y., Kurimoto, K., Yabuta, Y., Yuasa, M., Shigeta, M., et al. (2008). Critical function of Prdm14 for the establishment of the germ cell lineage in mice. *Nature Genetics*, *40*, 1016–1022.

Yamaji, M., Ueda, J., Hayashi, K., Ohta, H., Yabuta, Y., Kurimoto, K., et al. (2013). PRDM14 ensures naive pluripotency through dual regulation of signaling and epigenetic pathways in mouse embryonic stem cells. *Cell Stem Cell*, *12*, 1–15.

Yamashiro, C., Hirota, T., Kurimoto, K., Nakamura, T., Yabuta, Y., Nagaoka, S. I., et al. (2016). Persistent requirement and alteration of the key targets of PRDM1 during primordial germ cell development in mice. *Biology of Reproduction*, *94*, 7.

Yamashiro, C., Sasaki, K., Yabuta, Y., Kojima, Y., Nakamura, T., Okamoto, I., et al. (2018). Generation of human oogonia from induced pluripotent stem cells in vitro. *Science*, *362*, 356–360.

Ying, Q. L., Wray, J., Nichols, J., Batlle-Morera, L., Doble, B., Woodgett, J., et al. (2008). The ground state of embryonic stem cell self-renewal. *Nature*, *453*, 519–523.

Yokobayashi, S., Liang, C. Y., Kohler, H., Nestorov, P., Liu, Z., Vidal, M., et al. (2013). PRC1 coordinates timing of sexual differentiation of female primordial germ cells. *Nature*, *495*, 236–240.

Yokobayashi, S., Okita, K., Nakagawa, M., Nakamura, T., Yabuta, Y., Yamamoto, T., et al. (2017). Clonal variation of human induced pluripotent stem cells for induction into the germ cell fate. *Biology of Reproduction*, *96*, 1154–1166.

Zhang, M., Leitch, H. G., Tang, W. W. C., Festuccia, N., Hall-Ponsele, E., Nichols, J., et al. (2018). Esrrb complementation rescues development of nanog-null germ cells. *Cell Reports*, *22*, 332–339.

Zhang, J., Zhang, M., Acampora, D., Vojtek, M., Yuan, D., Simeone, A., et al. (2018). OTX2 restricts entry to the mouse germline. *Nature*, *562*, 595–599.

Zhang, B., Zheng, H., Huang, B., Li, W., Xiang, Y., Peng, X., et al. (2016). Allelic reprogramming of the histone modification H3K4me3 in early mammalian development. *Nature*, *537*, 553–557.

Zhou, Q., Wang, M., Yuan, Y., Wang, X., Fu, R., Wan, H., et al. (2016). Complete meiosis from embryonic stem cell-derived germ cells in vitro. *Cell Stem Cell*, *18*, 330–340.

CHAPTER FOUR

From worm to germ: Germ cell development and regeneration in planarians

Melanie Issigonis[a], Phillip A. Newmark[a,b,c,*]
[a]Regenerative Biology, Morgridge Institute for Research, Madison, WI, United States
[b]Howard Hughes Medical Institute, Chevy Chase, MD, United States
[c]Department of Integrative Biology, University of Wisconsin, Madison, WI, United States
*Corresponding author: e-mail address: pnewmark@morgridge.org

Contents

1. Introduction 128
2. Planarian biology: Some background about the "other" worm 129
3. Germ cell-intrinsic regulation 133
   3.1 Reproductive system regeneration 133
   3.2 Asexuals "do it" too 135
   3.3 An unbiased screen for germ cell regulators 136
4. Extrinsic regulation via local cues 138
   4.1 The niche is back: The somatic gonad and germ cell regeneration 138
5. Extrinsic regulation via systemic cues 139
   5.1 Sex on the brain: A role for neuropeptides 139
   5.2 A welcome reception: Roles for small-molecule receptors in the soma 141
6. A blurred line between neoblasts and germ cells 143
References 145

Abstract

The specification and proper differentiation of germ cells ensure the propagation of sexually reproducing species. Studies of a wide range of organisms have uncovered several important, conserved features of germ cell development, including the critical roles played by localized niches and somatically derived systemic cues. The planarian *Schmidtea mediterranea* is an excellent model to study fundamental aspects of germ cell development. Planarians are well known for their remarkable regenerative abilities and can regenerate whole animals from small tissue fragments. This amazing ability is bestowed by neoblasts, pluripotent somatic stem cells that are maintained throughout the planarian's lifetime. Advances in functional genomic methodologies have made planarians a powerful model to investigate the molecular mechanisms underlying germ cell development and reproductive maturation. Here we review recent studies that have led to the discovery of several germ cell-intrinsic factors and somatically derived extrinsic signals important for regulating various aspects of germ cell development.

Current Topics in Developmental Biology, Volume 135
ISSN 0070-2153
https://doi.org/10.1016/bs.ctdb.2019.04.001

In addition to revealing deep conservation of mechanisms that intrinsically regulate germ cells, these studies also uncover an important function for neuroendocrine control of planarian reproduction as well as novel roles for GPCR signaling in the somatic gonadal niche.

1. Introduction

All sexually reproducing organisms are made up of two main cell types: somatic cells that comprise the body and germ cells that produce gametes. In vivo, germ cells are unipotent and only give rise to differentiated oocytes and sperm. Yet, the union of these highly specialized cells creates a totipotent zygote, which produces *all* the cells of an organism. During development, a new germ cell lineage must be specified and germ cells must differentiate properly in order to ensure continuity of the species. How do germ cells maintain totipotency throughout development? In recent years, in vitro models have recapitulated the germ cell cycle, providing key insights into the origin and development of the germline. Germ cells can be derived from embryonic stem (ES) cells and pluripotent stem cells (Geijsen et al., 2004; Hayashi, Ohta, Kurimoto, Aramaki, & Saitou, 2011; Hikabe et al., 2016; Hübner et al., 2003; Ishikura et al., 2016; Qing et al., 2007; Toyooka, Tsunekawa, Akasu, & Noce, 2003; Yamashiro et al., 2018; Zhou et al., 2016). Furthermore, germ cells can be reprogrammed back to an embryonic-like pluripotent state (Guan et al., 2006; Kanatsu-Shinohara et al., 2004; Seandel et al., 2007). Therefore, it seems that the developmental potential of germ cells is unlimited, and that it becomes restricted by regulatory mechanisms in vivo. There is profound interest in defining the principles governing germ cell specification and development, and studies of diverse organisms—from invertebrates to mammals—have uncovered conserved features of germ cell development.

In sexually reproducing species, two apparently distinct mechanisms specify germ cell fate: determinate or inductive specification. In determinate specification, specialized maternal cytoplasm known as germ plasm is asymmetrically partitioned in the oocyte or zygote and specifies the new germ cell lineage. Determinate specification has been studied extensively in traditional model organisms such as *Drosophila*, *C. elegans*, zebrafish, and *Xenopus* (Extavour, 2007; Extavour & Akam, 2003; Seydoux & Braun, 2006; Strome & Updike, 2015). Inductive specification of germ cells occurs later in embryogenesis via extrinsic signals from surrounding tissues.

Examination of germ cell specification strategies throughout metazoans has revealed that inductive specification is the ancestral and most prevalent mode. Despite this, inductive specification has only been studied mechanistically in mice, and more recently, in axolotls and the cricket *Gryllus bimaculatus* (Ewen-Campen, Donoughe, Clarke, & Extavour, 2013; Extavour & Akam, 2003; Johnson et al., 2003; Seydoux & Braun, 2006; Strome & Updike, 2015). Interestingly, BMP signaling appears to be a phylogenetically conserved mechanism for inducing germ cells in all these species (Chang & Matzuk, 2001; Chu, Dunn, Anderson, Oxburgh, & Robertson, 2004; de Sousa Lopes et al., 2004; Donoughe et al., 2014; Hayashi et al., 2002; Johnson et al., 2003; Lawson et al., 1999; Tremblay, Dunn, & Robertson, 2001; Ying, Liu, Marble, Lawson, & Zhao, 2000). Current models for inductive germ cell specification represent only two of the major bilaterian superphyla: Deuterostomia (mice and axolotls) and Ecdysozoa (crickets), but not Lophotrochozoa (which includes planarians). Studies on animals from this underexplored clade will provide important insight into the evolution of mechanisms specifying germ cell fate and regulating germline development. Here, we introduce regenerative planarians as an excellent model in which to study germ cell development; knowledge gained from studying this versatile animal will help elucidate the nature of totipotency and further our understanding of stem cell and reproductive biology.

2. Planarian biology: Some background about the "other" worm

Planarians are free-living flatworms of the phylum Platyhelminthes (from the Greek "platy" meaning flat, and "helminth" meaning worm). They are among the simplest animals that are comprised of three germ layers (ectoderm, mesoderm, and endoderm), yet they possess a complex anatomy comprised of organ systems with striking molecular and physiological conservation to mammalian tissues (for a review, see Roberts-Galbraith & Newmark, 2015). The planarian central nervous system consists of a bilobed brain at the anterior end of the animal made up of dozens of distinct neural subtypes, interspersed by glia (Collins et al., 2010; Nishimura et al., 2007a, 2007b; Nishimura et al., 2008; Nishimura, Kitamura, Taniguchi, & Agata, 2010; Nishimura, Kitamura, Umesono, et al., 2008; Roberts-Galbraith, Brubacher, & Newmark, 2016; Wang, Lapan, Scimone, Clandinin, & Reddien, 2016). The planarian brain extends lateral projections, or brain branches, that connect to chemoreceptors along the margins of the planarian head, which relay sensory information back to the brain (MacRae, 1967). The brain also connects to two ventral nerve cords, which run along the entire length of the animal and connect to a peripheral nerve plexus that innervates the planarian's body (Agata et al., 2009; Baguñà & Ballester, 1978; Chien & Koopowitz, 1977; Koopowitz & Chien, 1975). The planarian epidermis contains a single layer of epidermal cells overlaying a complex body wall musculature composed of layers of circular, diagonal, and longitudinal muscle fibers (Cebrià et al., 1996; Hay & Coward, 1975; Kobayashi, Kobayashi, Orii, Watanabe, & Agata, 2009; MacRae, 1965; Morita, 1965). The planarian digestive system consists of a highly branched intestine comprised of absorptive phagocytes that take up food particles and secretory goblet cells that produce digestive enzymes (Bowen, Ryder, & Thompson, 1974; Corrales & Gamo, 1986; Garcia-Corrales & Gamo, 1988; Ishii, 1965). The intestine connects to the pharynx, a muscular feeding organ responsible for ingesting food and expelling waste (Ishii, 1965; MacRae, 1963). A network of broadly distributed tubules, the protonephridia, serves as the excretory system and regulates osmotic balance in the animal (Ishii, 1980a, 1980b; McKanna, 1968a, 1968b; Rink, Vu, & Alvarado, 2011; Scimone, Srivastava, Bell, & Reddien, 2011; Thi-Kim Vu et al., 2015). Surrounding the planarian's organs is the mesenchyme, called the parenchyma, which includes pluripotent stem cells called neoblasts.

Planarians reproduce by two distinct strategies: asexual and sexual reproduction (Hyman, 1951). Asexual planarians reproduce clonally by fissioning and subsequently regenerating two genetically identical animals. Sexual planarians reproduce as cross-fertilizing hermaphrodites that lay embryo-containing egg capsules. Some planarian species can alternate between asexual and sexual reproductive modes in response to various environmental and physiological stimuli such as seasonal changes or population density. Alternating between reproductive strategies allows some planarians to benefit from asexual reproduction when mating partners are scarce and from sexual reproduction, which generates genetically diverse offspring.

Planarians are well known for their extraordinary regenerative capabilities and can regenerate whole animals from small tissue fragments. This seemingly magical ability to regenerate has fascinated naturalists and scientists for centuries. Armed with nothing more than curiosity, scientists performed various amputations, injuries, and transplantation experiments—testing the virtually boundless limits of planarian regeneration in hopes of unraveling the secret to this creature's regenerative prowess. However, limited by research tools, studies on these animals lagged for the better part of a century and the mysteries of planarian regeneration remained unexplained.

With the advent of the modern era of molecular biology, planarians have experienced a renaissance as a laboratory model, with *Schmidtea mediterranea* becoming the species of choice (Newmark & Sánchez Alvarado, 2002). Two distinct strains of *S. mediterranea* exist in nature: an asexual strain and a sexual strain, which are genetically distinguishable by a chromosomal translocation present in the asexuals (Benazzi, Baguná, Ballester, Puccinelli, & Papa, 1975). The *S. mediterranea* genome has been sequenced and annotated (Grohme et al., 2018; Robb, Ross, & Sánchez Alvarado, 2008), and there is an arsenal of cellular, molecular, and transcriptomic tools available to study planarian biology. Neoblasts are abundant and easily accessible, comprising ~25% of all planarian cells (Baguñà & Romero, 1981). Dividing neoblasts can be specifically labeled with thymidine analogs (i.e., bromodeoxyuridine and fluoro-5-ethynyluridine) (Newmark & Sánchez Alvarado, 2000; Zhu, Hallows, Currie, Xu, & Pearson, 2015), which have proved useful in pulse-chase experiments and defining differentiation lineages in planarians. In situ hybridization and immunohistochemistry techniques allow for visualization of gene expression patterns with single-cell resolution and have already identified a myriad of markers for neoblasts and other cell types, including the reproductive system (Chong, Stary, Wang, & Newmark, 2011; Forsthoefel, Ross, Newmark, & Zayas, 2018; Forsthoefel, Waters, & Newmark, 2014; King & Newmark, 2018; Pearson et al., 2009). Gene function can be specifically inhibited by RNA interference (RNAi) by microinjecting animals or feeding them double stranded RNA (dsRNA) (Newmark, Reddien, Cebrià, & Sánchez Alvarado, 2003; Rouhana et al., 2013; Sánchez Alvarado & Newmark, 1999). Recent advances in single-cell sequencing technology have allowed for the classification and characterization of virtually all cell types in the adult planarian and have further refined neoblast subclasses and defined differentiation trajectories (Fincher, Wurtzel, de Hoog, Kravarik, & Reddien, 2018; Plass et al., 2018; Swapna, Molinaro, Lindsay-Mosher, Pearson, & Parkinson, 2018; Zeng et al., 2018). The availability of functional genomic tools greatly facilitates our ability to query the genetic programs that control neoblast behavior and drive fundamental developmental processes such as regeneration and reproduction.

Classic experiments demonstrated that neoblasts are the cells responsible for driving the remarkable regenerative feats seen in planarians. Since neoblasts are the only dividing cells in the planarian soma, they can be specifically ablated by irradiation; planarians devoid of neoblasts cannot sustain new tissue generation and ultimately die. However, inevitable lethality in irradiated hosts can be reversed simply by transplanting tissue grafts from unirradiated donors (Wolff, 1962). Transplantations of cell preparations enriched in neoblasts (but not differentiated cells) are also capable of rescuing lethality in irradiated planarians indicating that, as a population, neoblasts are pluripotent (Baguñà, Saló, & Auladell, 1989). Finally, groundbreaking work from the Reddien lab definitively confirmed the existence of a truly pluripotent stem cell; transplantation of single neoblasts, called clonogenic neoblasts, rescue stem cell-depleted planarians from certain death (Wagner, Wang, & Reddien, 2011).

Most studies examining the cellular and molecular mechanisms underlying regeneration and neoblast regulation have focused exclusively on the asexual strain of *S. mediterranea*. However, the sexual strain boasts equivalent regenerative capacity in response to tissue injury and similarly depends on neoblasts for homeostatic tissue maintenance, turnover, and regeneration. Interestingly, the planarian's regenerative capacity is not restricted to somatic tissues—they can also regenerate new germ cells.

Sexual *S. mediterranea* are simultaneous hermaphrodites harboring both male and female reproductive organs (Fig. 1). These animals contain dozens of dorsolaterally distributed testes, which are connected by sperm ducts to the seminal vesicles and copulatory apparatus. Individual testes contain several presumptive germline stem cells (GSCs) dispersed along the testis periphery, which give rise to an outer layer of spermatogonia that undergo three rounds of mitosis with incomplete cytokinesis to form 2-, 4-, and 8-cell spermatogonial cysts. These differentiate further to form primary spermatocytes, which undergo meiosis to generate 32 round spermatids. As spermatogenesis progresses, round spermatids give rise to elongated spermatids that undergo spermiogenesis and ultimately form sperm with elongated nuclei. Sexual planarians also contain a pair of ventrally located ovaries that are nestled at the base of the bilobed brain (Fig. 1). Ovaries produce oocytes, which are fertilized internally in a specialized compartment of the oviduct called the tuba. After oocyte fertilization, one or more zygotes are packaged together with yolk produced by vitelline glands as they are transported posteriorly along the ciliated oviducts and deposited into the genital atrium. Egg capsules are laid through the gonopore and hatchlings emerge from these egg capsules approximately 2 weeks later.

In this review, we will explore current advances in our understanding of planarian reproductive development by considering both intrinsic and

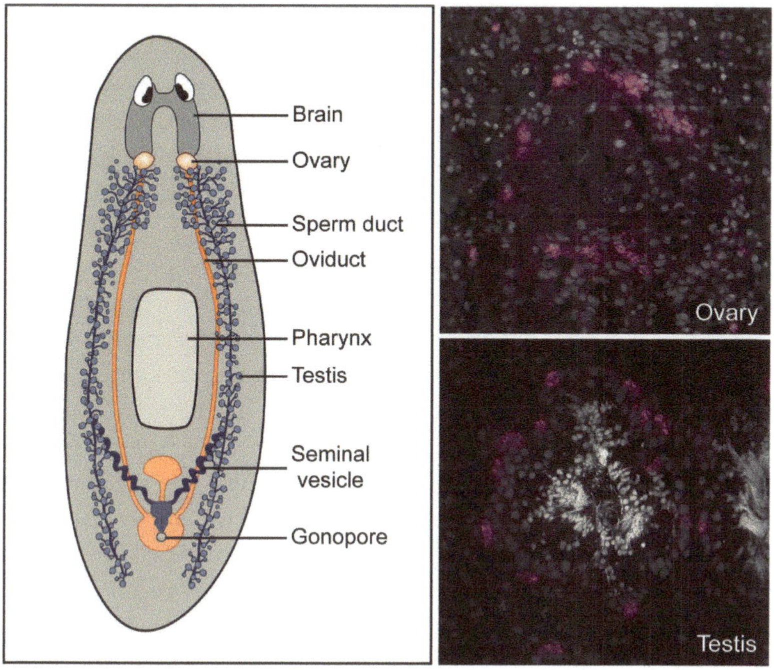

Fig. 1 Diagram depicting the brain, pharynx, and various reproductive organs in adult sexual *S. mediterranea* (left). Confocal image showing *nanos* (magenta) expression by fluorescent in situ hybridization in an ovary (top) and testis (bottom). All nuclei are counterstained with DAPI (gray). The most undifferentiated *nanos*$^{+}$ germ cells are located along the periphery of the ovary and testis. Oogenesis and spermatogenesis proceed from the periphery toward the center (lumen) of the gonads where large oocytes and elongated sperm are located (right).

extrinsic factors that regulate germ cell specification, maintenance, and differentiation. Extrinsic control of germ cells includes both local signaling from the somatic gonadal niche as well as long-range systemic cues. Finally, we will discuss the similarities between germ cells and pluripotent stem cells, and how the distinction between these two cell types may not be as rigid as it seems.

3. Germ cell-intrinsic regulation

3.1 Reproductive system regeneration

Planarians are an excellent model to understand the molecular mechanisms underlying inductive germ cell specification. Astonishingly, they can regenerate germ cells and reproductive tissues de novo! The earliest evidence that planarians can inductively specify new germ cells comes from an experiment performed over a century ago by T. H. Morgan wherein amputated head fragments lacking any reproductive tissues (i.e., composed entirely of somatic cells) ultimately regenerated a reproductive system (Morgan, 1902). New germ cells are most likely specified from pluripotent neoblasts, which give rise to all tissues in the planarian. Rather than relying on maternally supplied germ plasm to form germ cells, planarians specify germ cells late in embryogenesis and develop reproductive organs post-embryonically (Curtis, 1902; Davies et al., 2017).

To understand how pluripotent somatic cells acquire germ cell fate in *S. mediterranea*, a candidate-based approach was used to identify and functionally characterize the planarian homolog of *nanos* (Wang, Zayas, Guo, & Newmark, 2007). *nanos* encodes a deeply conserved RNA-binding protein, orthologs of which have been observed in germ cells in almost all taxa examined (Extavour & Akam, 2003). In *Drosophila*, *nanos* is required maternally for abdominal segmentation in embryos, and is essential for proper GSC maintenance and germline function in adults (Bhat, 1999; Deshpande, Calhoun, Yanowitz, & Schedl, 1999; Forbes & Lehmann, 1998; Hayashi, Hayashi, & Kobayashi, 2004; Kobayashi, Yamada, Asaoka, & Kitamura, 1996; Wang & Lehmann, 1991; Wang & Lin, 2004). *nanos* orthologs also play conserved roles in germ cell development in *C. elegans* (Subramaniam & Seydoux, 1999), zebrafish (Köprunner, Thisse, Thisse, & Raz, 2001), and mouse (Sada, Suzuki, Suzuki, & Saga, 2009; Suzuki, Tsuda, & Saga, 2007; Tsuda et al., 2003).

Newly hatched planarians express *nanos* dorsolaterally, in the testis primordia. *nanos* expression has not been reported in the ovary primordia in hatchlings, consistent with their protoandrous reproductive development. Irradiation of hatchlings eliminates the presence of these *nanos*$^+$ germ cells, suggesting that dividing neoblasts are the likely source of newly specified germ cells (Wang et al., 2007).

At later stages of sexual development, *nanos* expression is detected in testes and ovaries of juvenile and mature planarians (Fig. 1). Hatchling planarians fed dsRNA targeting *nanos* for 2–3 months (the time it takes planarians to reach sexual maturity) do not form ovaries or testes, indicating that *nanos* is required for postembryonic development of planarian gonads (Wang et al., 2007).

Armed with *nanos* as the earliest marker for planarian germ cells, classic amputation experiments were revisited. Transverse amputation *anterior* to the ovaries results in germline-devoid fragments that regenerate new *nanos*$^+$

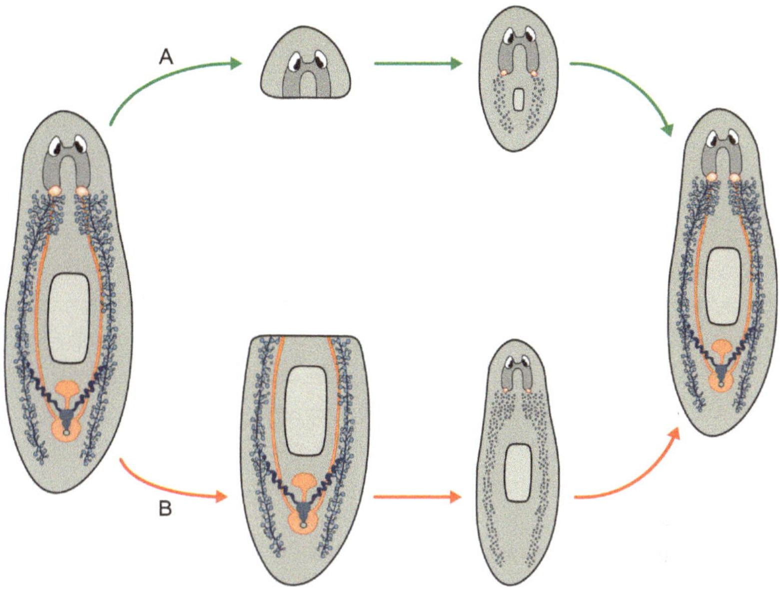

Fig. 2 Image depicting different amputation schemes in sexual planarians. (A) Amputation anterior to the ovaries produces a head fragment devoid of all reproductive tissues. This head fragment will regenerate a new tail and will specify new germ cells de novo. (B) Amputation posterior to the ovaries produces a tail fragment containing reproductive structures. The regenerating planarian will regress to a juvenile-like state (undifferentiated gonads, immature reproductive system) and will subsequently develop into a sexually mature adult.

cells within 2 weeks post-amputation (Fig. 2). Therefore, *nanos*$^+$ germ cells can be specified post-embryonically. Transection *posterior* to the ovaries leads to testis regression in the tail fragments; only small clusters of *nanos*$^+$ early germ cells remain (Fig. 2). Upon regeneration of the head and brain, the testes regenerate; *nanos*$^+$ germ cells resume differentiation and all stages of spermatogenesis reappear. *nanos* function is required for testis regeneration and development as *nanos* RNAi followed by amputation posterior to the ovaries results in animals that regenerate their bodies, but fail to form or maintain testes and accessory reproductive organs (Wang et al., 2007).

3.2 Asexuals "do it" too

Surprisingly, asexual planarians, which reproduce exclusively by fissioning, also express *nanos*. Small clusters of *nanos*$^+$ cells are distributed dorsolaterally in a pattern reminiscent of testis primordia in sexual planarians (Handberg-Thorsager & Saló, 2007; Sato et al., 2006; Wang et al., 2007). These *nanos*$^+$

presumptive germ cells do not differentiate but do require *nanos* for their maintenance and/or specification: *nanos* RNAi in both intact and regenerating asexual planarians results in germ cell loss (Wang et al., 2007).

3.3 An unbiased screen for germ cell regulators

How are *nanos*$^+$ germ cells maintained in planarians? What molecular program drives germ cell development and differentiation? To address these questions, an unbiased transcriptomic approach was used to identify germ cell-intrinsic genes. Transcriptional profiles were compared between animals with and without germ cells (control versus *nanos(RNAi)* asexuals and juvenile sexuals). Hundreds of genes were found to be down regulated following germ cell loss. These genes encode proteins involved in transcription, translation, RNA processing, and signal transduction. Due to the massive overrepresentation of testes compared to ovaries in planarians, the vast majority of down-regulated genes were expressed in testes; only five ovary-expressed genes were found using this whole-animal approach.

Several genes are expressed in specific testis germ cell compartments: for example, *glyceraldehyde-3-phosphate dehydrogenase, spermatogenic* and *rap55* (*RNA-associated protein 55*) are specifically expressed in spermatogonial cells, including *nanos*$^+$ germ cells; *heterogeneous nuclear ribonucleoprotein A2* (*hnRNPA2*) is enriched in spermatogonia and early spermatocytes; *poly(A)-binding protein, cytoplasmic* (*PABPC*) expression is detected in spermatocytes and spermatids; and *nuclear factor Y, subunit B* (*NF-YB*) is expressed in all testis germ cells, except mature sperm (Wang, Stary, Wilhelm, & Newmark, 2010).

RNAi targeting positive hits from the screen led to the discovery of 13 new genes required for distinct stages of germline development. For example, knockdown of several genes disrupted meiotic progression (*embryonic-lethal abnormal vision-1* and *-2, PABPC, secondary spermatocyte accumulation-1*) or spermatid elongation (*hnRNAP2, msy4, C3H-zinc finger-1, rap55, spermatid elongation-defective-1, Brahma-related gene 1*). RNAi knockdown of several genes also led to complete testis ablation (*Bicaudal-C* [*Bic-C*], *eukaryotic initiation factor 3c, NF-YB*); these genes are required for proper maintenance of *nanos*$^+$ male germ cells in planarians (Wang et al., 2010).

NF-YB knockdown animals displayed a particularly intriguing testis degeneration phenotype. *NF-YB* is not required for specifying *nanos*$^+$ germ cells (Iyer, Collins, & Newmark, 2016). Over time, *NF-YB(RNAi)* animals progressively lose their spermatogonia, including *nanos*$^{+}$ GSCs. This loss is due to a defect in spermatogonial proliferation and GSC self-renewal. Differentiation is unaffected; as GSCs are lost, they differentiate into spermatocytes that progress normally through meiosis and produce spermatids and sperm. Without GSCs to sustain spermatogenesis, however, the testes ultimately degenerate (Iyer, Collins, & Newmark, 2016; Wang et al., 2010). The gradual loss of spermatogonia and resulting infertility observed in *NF-YB* knockdown planarians is remarkably similar to the testis defects observed in *Plzf* (*promyelocytic leukemia zinc finger*) and *TAF4b* (*TATA-box binding protein associated factor 4b*) mutant mice; neonatal mice initially produce sperm but then progressively lose their germ cells over time due to defects in spermatogonial stem cell self-renewal (Buaas et al., 2004; Costoya et al., 2004; Falender et al., 2005).

Many of the genes from the screen that resulted in RNAi-mediated testis defects encode conserved post-transcriptional regulators with known germline functions. For example, the Y Box-containing translational regulator MSY-4, which is required for spermatid elongation in planarians, plays a role in mouse spermatid differentiation (Giorgini, Davies, & Braun, 2002). Importantly, these studies have the potential to inform our understanding of mammalian germ cell biology. *Bic-C* and *rap55* (*trailer hitch*) are both essential for egg patterning during *Drosophila* oogenesis (Mahone, Saffman, & Lasko, 1995; Wilhelm, Buszczak, & Sayles, 2005). *Bic-C* is a regulator of poly(A) tail length (Chicoine et al., 2007) and RAP55 is a component of P bodies, sites of mRNA degradation and translational repression (Tanaka et al., 2006; Yang, Yu, Gulick, Bloch, & Bloch, 2006). In planarians, *Bic-C* is required for maintenance of *nanos*$^{+}$ male germ cells and *rap55* plays a role in spermatid elongation (Wang et al., 2010). Intriguingly, mouse *Bic-C* is highly expressed in spermatogonia. Furthermore, mouse RAP55 protein is expressed in spermatocytes and spermatids, and the human *rap55* ortholog, *LSM14*, is significantly upregulated during spermatogenesis (Chalmel et al., 2007; Pepling, Wilhelm, O'Hara, Gephardt, & Spradling, 2007). Even though genome-wide transcriptomic analyses are straightforward in the mouse model, gene expression mapping and functional characterization of differentially expressed genes in a high-throughput manner is not feasible. Given the deep conservation of many germ cell regulators, characterizing the factors involved in planarian germ cell development could help guide future experiments on mammalian systems.

4. Extrinsic regulation via local cues

4.1 The niche is back: The somatic gonad and germ cell regeneration

One conserved feature of germ cell development is the regulatory role played by germ cell-extrinsic factors from somatic tissues (Clarke, 2018; Greenspan, de Cuevas, & Matunis, 2015; Kimble & Seidel, 2013; Svingen & Koopman, 2013). Somatic cells of the gonad act as specialized microenvironments, or niches, that provide important germ cell regulatory signals. The highly conserved DM domain-containing family of transcription factors has been implicated in sex determination and gonadal differentiation in a wide range of species—from cnidarians to mammals (Kopp, 2012; Matson & Zarkower, 2012; Picard et al., 2015). Despite their importance in regulating sexual development across metazoans, mechanistic studies on DM domain genes have focused heavily on traditional models—*Drosophila* (*doublesex, dsx*), *C. elegans* (*male abnormal-3, mab-3*), and mouse (*Dsx- and mab-3-related transcription factor 1, Dmrt1*)—and our understanding of them outside these few species is lacking.

Planarians offer an excellent system to understand the ancient and conserved role played by DM domain genes. The *S. mediterranea* genome encodes four genes with conserved DM domains, one of which, *dmd-1*, is highly expressed in the planarian reproductive system. *dmd-1* transcripts are detected in male, but not female, reproductive structures. Specifically, *dmd-1* is expressed in somatic gonadal cells of the testes and in male accessory reproductive organs (i.e., the seminal vesicles, sperm ducts, and penis papilla). *dmd-1* is also expressed in sparse cells throughout the dorsal mesenchyme; these cells co-express the stem cell marker *piwi-1* and represent differentiating neoblast progeny that are committed to become testis somatic gonadal cells. A small number of neurons in the planarian brain also express *dmd-1*, raising the intriguing possibility that sexually dimorphic neurons may exist in this simultaneous hermaphrodite. *dmd-1* expression is detected in newly hatched sexual planarians, and *nanos*$^+$ primordial germ cells are always found in close proximity to these *dmd-1*-expressing cells. Interestingly, asexual planarians express *dmd-1* as well, in somatic cells that are intimately associated with *nanos*$^+$ germ cells in the testis primordia. Do *dmd-1*$^+$ somatic gonadal cells behave as a "niche" for *nanos*$^+$ germ cells?

RNAi-mediated knockdown of *dmd-1* in regenerating sexual planarians as well as in intact animals results in the loss of *nanos*$^+$ male germ cells; testes are ablated while ovaries remain unaffected, consistent with a male-specific role for *dmd-1*. *dmd-1* is also required for the initial post-embryonic development of planarian germ cells. Animals fed *dmd-1* dsRNA from hatching until adulthood do not form testes or any male accessory reproductive organs. Notably, ovaries fail to develop in these planarians, indicating that *dmd-1* is required for the development of both male *and* female reproductive tissues. This suggests that the presence of male reproductive structures in planarians may be a prerequisite for the proper development of female ones (Chong, Collins, Brubacher, Zarkower, & Newmark, 2013).

As described above, planarian germ cells are inductively specified from somatic neoblasts not only during development, but also during regeneration. Amputated planarian fragments comprised solely of somatic cells regenerate all missing tissues including new *nanos*$^+$ germ cells (Morgan, 1902; Wang et al., 2007). However, when combined with *dmd-1* RNAi, these tissue fragments fail to specify new germ cells. Therefore, *dmd-1* is required cell non-autonomously for the specification of germ cells (Chong et al., 2013). Questions for future work include: How do *dmd-1*$^+$ "niche" cells direct the specification and maintenance of male germ cells in planarians? What are the transcriptional targets of Dmd-1 and how do Dmd-1-dependent signals control development of male germ cells and reproductive organs?

5. Extrinsic regulation via systemic cues

5.1 Sex on the brain: A role for neuropeptides

It has long been appreciated in vertebrates that a neuroendocrine axis between the brain and gonads exists; this axis involves neuropeptide-mediated signaling from the brain to the gonads (via the pituitary) to control reproduction in both males and females. Recently, insulin-like peptides have been shown to regulate GSC proliferation in nematodes and flies (LaFever & Drummond-Barbosa, 2005; Michaelson, Korta, Capua, & Hubbard, 2010; Ueishi, Shimizu, & H Inoue, 2009). Do invertebrate peptide hormones regulate other aspects of reproductive system development and maturation? In other words, does a more expansive, "vertebrate-like" role for neurally derived peptide hormones exist in these "lower" organisms?

The first clue that brain-derived systemic signals influence planarian sexual development comes from experimental observations made decades ago. Transverse amputations in which the head and brain are completely removed result in testis regression. Gonadal regeneration and germ cell differentiation occur only once the brain has completely regenerated (Fedecka-Bruner, 1967). A role for neuropeptide signaling in planarian reproduction was recently defined by investigating the function of planarian *prohormone convertase 2* (*pc2*), the gene responsible for proteolytically cleaving prohormone precursors to mature neuropeptides. In sexual planarians, *pc2* is expressed in the central nervous system, mesenchyme, pharynx, and in the testes and copulatory apparatus. RNAi knockdown of *pc2* in sexually mature planarians and in juvenile animals reared to adulthood results in regressed testes composed of small undifferentiated clusters of *nanos*$^+$/*germinal histone 4*$^+$ spermatogonia (Collins et al., 2010).

Which *pc2*-dependent neuropeptide(s) mediate sexual development in planarians? The *S. mediterranea* genome encodes 51 prohormone genes, which are predicted to be processed into several distinct peptides. The existence of at least 142 distinct hormone peptides has been biochemically confirmed by mass spectrometry. Capitalizing on the existence of the distinct sexual and asexual strains of *S. mediterranea*, a comparative analysis of prohormone gene expression uncovered a single neuropeptide Y family member, *npy-8*, uniquely expressed in sexuals (Collins et al., 2010).

npy-8 prohormone gene expression is undetectable in asexual animals, but is expressed at high levels in a subset of cells in the sexual planarian CNS, ventral nerve cords, nerve plexus, pharyngeal neurons, and neurons innervating the testes (Collins et al., 2010; Saberi, 2016). *npy-8* knockdown animals fail to produce sperm and lack copulatory organs. Moreover, *npy-8* RNAi animals partially phenocopy the *pc2* RNAi phenotype and are characterized by regressed testes comprised of undifferentiated germ cells. Since the *npy-8* RNAi phenotype is less severe than that observed in *pc2* RNAi animals, it is possible that *npy-8* is not the only neuropeptide involved in planarian sexual development. This is supported by the fact that several prohormones are expressed in the planarian reproductive system, including the testes, oviducts, copulatory apparatus, and gland cells surrounding the copulatory apparatus (Collins et al., 2010). Future experiments will uncover the full neuropeptide complement involved in different aspects of planarian reproduction (e.g., ovarian development, egg-laying, and behaviors such as copulation).

A study examining the function of Cytoplasmic Polyadenylation Element Binding (CPEB) proteins in planarian sexual development found that *CPEB2* is expressed in the planarian nervous system and testes, and similar to *npy-8*, is required for proper spermatogenesis and development of copulatory organs. Biochemical analyses revealed that *CPEB2*(*RNAi*) leads to a reduction in *npy-8* transcript levels and NPY-8 peptide abundance. Therefore, *CPEB2* is essential for proper reproductive development in planarians and acts, at least in part, by promoting translation of NPY-8 (Rouhana, Tasaki, Saberi, & Newmark, 2017).

To elucidate the mechanism underlying NPY-8-mediated peptide hormone control of planarian reproduction and to identify its cognate receptor, a genome-wide analysis of planarian G-protein coupled receptors (GPCRs) was performed. Transcriptomic data and sensitive pattern discovery algorithms were used to define a complete *S. mediterranea* "GPCRome" comprising 566 predicted GPCRs. Sixteen planarian NPY receptors (NPYRs) were identified based on their homology to human NPY receptors; most of these *npyr* genes are expressed in the planarian CNS and testes. A single NPYR, encoded by *npyr-1*, is activated in response to NPY-8 peptide in an in vitro cell-based receptor-activation assay. Furthermore, knockdown of *npyr*-1, but not other *npyr* genes, phenocopies *npy-8* loss of function; *npyr-1(RNAi)* animals exhibit defects in spermatogonial and oogonial differentiation and a loss of accessory reproductive organs. Therefore, NPYR-1 is the bona fide receptor for NPY-8 (Saberi, Jamal, Beets, Schoofs, & Newmark, 2016).

How does NPY-8/NPYR-1 signaling regulate germ cell differentiation and accessory reproductive organ development in planarians? A clue lies in the expression pattern of *npyr-1*. *npyr-1* is not expressed in gonads or accessory reproductive tissues, as its phenotype might suggest. Instead *npyr-1* transcripts are detected in a subset of $pc2^+$ peptidergic neurons in the CNS and ventral nerve cords. *npyr-1* and *npy-8* are not co-expressed in the same $pc2^+$ neuroendocrine cells, invoking a paracrine NPY-8/NPYR-1 neuronal circuit which systemically influences reproductive development via unknown signals (Saberi et al., 2016). Future experiments will uncover the downstream factors that mediate NPY-8/NPYR-1-mediated neuroendocrine control on reproductive function.

5.2 A welcome reception: Roles for small-molecule receptors in the soma

What other long-range endocrine signals besides peptide hormones influence planarian reproductive development? Lipophilic hormones and their receptors—ligand-binding transcription factors called nuclear hormone receptors (NHRs)—have been shown to play important roles in reproduction from invertebrates (Allen & Spradling, 2008; Asahina et al., 2000; Buszczak et al., 1999; Gissendanner et al., 2008) to mammals

(Chappell, Lydon, Conneely, O'Malley, & Levine, 1997; Walker & Korach, 2004; Wang, Yeh, Tzeng, & Chang, 2009). The *S. mediterranea* genome encodes 23 NHRs, one of which, *nhr-1*, is dramatically upregulated in sexual planarians compared to the asexual strain. In adult sexual planarians, *nhr-1* is expressed in both male (sperm ducts, seminal vesicles, and penis papilla) and female (tuba, oviduct, and copulatory bursa) accessory reproductive organs.

RNAi-mediated knockdown of *nhr-1* in sexual planarians results in the loss of all *nhr-1*-expressing accessory reproductive organs in addition to sex organs lacking detectable levels of *nhr-1* (cement glands and gonopore). These results indicate a previously underappreciated level of coordination in the development of different components that make up the accessory reproductive system in planarians. Intriguingly, loss of *nhr-1* also leads to a failure in germ cell differentiation in both ovaries and testes; only the most undifferentiated *nanos*$^{+}$/*gh4*$^{+}$ germ cells remain. Taken together, these results suggest that during planarian sexual maturation, somatically expressed *nhr-1* is required cell autonomously for proper development of male and female sex organs and that long-range hormonal soma-germline signaling is critical for coordinated differentiation of the germline (Tharp, Collins, & Newmark, 2014). NHRs can bind to a diverse range of ligands including steroid hormones, retinoic acid, fatty acids, and phospholipids—the identity and the source of the *nhr-1* ligand remain to be determined.

How do planarian gonads integrate/respond to these long-range signals? Germ cells either respond directly to systemic cues or indirectly via their somatic niche. GPCRs represent the largest class of cell-surface receptors and, therefore, are a likely candidate for receiving external stimuli and mediating downstream signaling events required for germ cell development. A comparative analysis of GPCR expression was performed between sexual and asexual *S. mediterranea* (Saberi et al., 2016), since genes enriched in the reproductive system account for the majority of transcriptional differences between the two planarian strains (Chong et al., 2011). Of the full GPCR complement, 8% (46/566) are more than four-fold upregulated in sexual planarians. A targeted functional screen of these sexually upregulated GPCRs revealed that one gene, *ophis*, is required for germ cell differentiation and sexual maturation; after knockdown of *ophis*, ovaries and testes are developmentally stunted and contain only undifferentiated *nanos*$^{+}$ germ cells, and accessory reproductive organs fail to develop (Saberi et al., 2016).

ophis is expressed in a variety of reproductive tissues; specifically, in somatic gonadal cells in both testes and ovaries, as well as in accessory reproductive organs (tuba, oviducts, yolk-producing vitelline glands, and the copulatory apparatus). *ophis* is co-expressed in *dmd-1*$^{+}$ progenitor cells in the dorsal mesenchyme as well as in *dmd-1*$^{+}$ testis somatic gonadal cells juxtaposed to *nanos*$^{+}$ GSCs. In fact, in sexual planarians, *nanos*$^{+}$ germ cells are invariably found in close proximity to *dmd-1*$^{+}$/*ophis*$^{+}$ somatic gonadal cells (Saberi et al., 2016). These data suggest that germ cells are only specified once the appropriate microenvironment to sustain them has been established. Alternatively, since *nanos* is the earliest (and only) known marker for planarian GSCs, it is possible that the earliest timepoint for GSC specification is not being detected until, under the influence of somatically derived signals, *nanos* is expressed. This scenario is reminiscent of mouse germ cell devlopment, in which *Nanos2* is expressed in primordial germ cells only after they have migrated to and colonized the genital ridge (Tsuda et al., 2003).

Unlike *dmd-1,* which is essential for specifying new germ cells and maintaining GSCs in the testes, *ophis* is required for promoting germ cell differentiation. The dichotomous role of *dmd-1*$^{+}$/*ophis*$^{+}$ somatic gonadal niche cells in the planarian testes is similar to that of mammalian Sertoli cells. Specialized domains within Sertoli cells support the maintenance of spermatogonial stem cells and facilitate their proper differentiation throughout spermatogenesis (Chen & Liu, 2015). *ophis*$^{+}$ somatic gonadal cells in the ovary are similarly required for female germ cell differentiation (Saberi et al., 2016). What are the niche signals that promote germ cell maintenance in the ovaries? What is the identity and the cellular source of the *ophis* ligand? What downstream signaling events are triggered by *ophis* receptor activation? Answers to these questions will require a greater understanding of the male and female somatic gonadal niche cells in planarians.

6. A blurred line between neoblasts and germ cells

During development, a new germ cell lineage must be established and differentiate properly to ensure propagation of the species. Despite their crucial role, much remains to be learned about how germ cells are specified and how they escape the mortal fate of the soma. Most animals accomplish these feats by segregating the germline away from the soma during development, regardless of whether they specify their germ cells with germ plasm (e.g., *Drosophila, C. elegans*, zebrafish, *Xenopus*) or via inductive signals (e.g., mice) (Extavour, 2007; Extavour & Akam, 2003; Seydoux & Braun, 2006; Strome & Updike, 2015). In all these animals, primordial germ cells are established once during embryogenesis. Embryonic germline segregation from the soma is a widely employed strategy, but not a universal one (Extavour, 2007; Juliano, Swartz, & Wessel, 2010).

In sexual planarians, germ cells are specified late in embryogenesis, but then continue to be generated throughout its seemingly limitless lifetime. This is possible due to the presence of neoblasts that persist into adulthood, and have the capacity to give rise to both somatic cells and new germ cells. Consequently, neoblasts are perpetually poised to decide between somatic versus germ cell fates. What molecular mechanisms underlie this choice? In the quest to uncover unique germ cell programs, planarian biologists have surprisingly uncovered more similarities than differences.

In transmission electron microscope images, neoblasts are indistinguishable from germ cells in asexual planarians of the species *Dugesia japonica* (Sato et al., 2006). In fact, neoblasts and germ cells share several important features. They are both small cells characterized by a large nucleus with little heterochromatin, surrounded by scant cytoplasm. Neoblasts and germ cells are also the only planarian cells that display distinctive chromatoid bodies: large, electron-dense, ribonucleoprotein complexes found in the cytoplasm by the nuclear membrane. Chromatoid bodies are highly reminiscent of germ granules found in germ cells of other species (Auladell, Valero, & Baguñà, 1993; Coward, 1974; Hay & Coward, 1975; Le Moigne, 1967; Morita, Best, & Noel, 1969; Pedersen, 1959; Sato et al., 2006; Sauzin, 1968; Yoshida-Kashikawa, Shibata, Takechi, & Agata, 2007).

Somatic neoblasts and germ cells even share a strikingly similar molecular signature. Neoblasts express several genes with conserved roles in germ cell development in other organisms, including homologs of *piwi*, *vasa*, *pumilio*, and *bruno*. These "germ cell genes" that are typically associated with germ granules and are essential for proper germ cell function play important roles in planarian neoblasts: RNAi knockdown of planarian *piwis* (Nakagawa, Ishizu, Hasegawa, Kobayashi, & Matsumoto, 2012; Palakodeti, Smielewska, Lu, Yeo, & Graveley, 2008; Reddien, Oviedo, Jennings, Jenkin, & Sánchez Alvarado, 2005; Shibata et al., 2016), *bruno-like* (Guo, Peters, & Newmark, 2006), *pumilio* (Iyer, Issigonis, Sharma, Extavour, & Newmark, 2016; Salvetti et al., 2005), *tudor* (Solana, Lasko, & Romero, 2009), and *vasa* (Rouhana, Shibata, Nishimura, & Agata, 2010; Wagner, Ho, & Reddien, 2012) all result in neoblast maintenance or differentiation defects.

The resemblances between neoblasts and germ cells beg the question—are these separate cell states? Classic experiments have indicated that the soma-germline boundary is far more blurred (and germ cells far less committed) than originally thought. A mixoploid planarian strain (*Schmidtea polychroa*) which possesses triploid somatic cells, hexaploid oogonia, and diploid spermatogonia, allowed Gremigni and colleagues to distinguish between these cell types based on DNA content and to perform what is essentially an "old school" lineage tracing experiment. Capitalizing on this species' unique mosaicism, the fates of somatic versus germ cells could be tracked. Following amputation, hexaploid and/or diploid cells (originating from germ cells) were observed among the triploid somatic cells of the regeneration blastema and could generate differentiated somatic tissues (e.g., pharyngeal cells) (Gremigni, Miceli, & Picano, 1980; Gremigni, Miceli, & Puccinelli, 1980). This result suggests that germ cells in planarians are not only unipotent (generating gametes), but retain pluripotency and can give rise to somatic lineages in vivo.

Planarian germ cells are morphologically and molecularly similar to neoblasts, and can even function as pluripotent somatic stem cells. This is reminiscent of the developmental plasticity observed in mammalian in vitro studies: pluripotent ES cells can produce germ cells, and germ cells can be reprogrammed back to a pluripotent state. In planarians, this germ cell cycle occurs in vivo *without* genetic manipulation. Planarian germ cells retain the ability to revert to pluripotency while concomitantly allowing for germ cell commitment to make gametes (a necessity for species continuity). What mechanisms govern this remarkable inherent plasticity? Planarians present a unique opportunity to investigate this important question.

References

Agata, K., Soejima, Y., Kato, K., Kobayashi, C., Umesono, Y., & Watanabe, K. (2009). Structure of the planarian central nervous system (CNS) revealed by neuronal cell markers. *Zoological Science*, *15*(3), 433–440.

Allen, A. K., & Spradling, A. C. (2008). The *Sf1*-related nuclear hormone receptor *Hr39* regulates *Drosophila* female reproductive tract development and function. *Development*, *135*(2), 311–321.

Asahina, M., Ishihara, T., Jindra, M., Kohara, Y., Katsura, I., & Hirose, S. (2000). The conserved nuclear receptor Ftz-F1 is required for embryogenesis, moulting and reproduction in *Caenorhabditis elegans*. *Genes to Cells: Devoted to Molecular & Cellular Mechanisms*, *5*(9), 711–723.

Auladell, C., Valero, J. G., & Baguñà, J. (1993). Ultrastructural localization of RNA in the chromatoid bodies of undifferentiated cells (neoblasts) in planarians by the RNase–gold complex technique. *Journal of Morphology*, *216*(3), 319–326.

Baguñà, J., & Ballester, R. (1978). The nervous system in planarians: Peripheral and gastrodermal plexuses, pharynx innervation, and the relationship between central nervous system structure and the acoelomate organization. *Journal of Morphology*, *155*(2), 237–252.

Baguñà, J., & Romero, R. (1981). Quantitative analysis of cell types during growth, degrowth and regeneration in the planarians *Dugesia mediterranea* and *Dugesia tigrina*. *Hydrobiologia*, *84*(1), 181–194.

Baguñà, J., Saló, E., & Auladell, C. (1989). Regeneration and pattern formation in planarians. III. Evidence that neoblasts are totipotent stem cells and the source of blastema cells. *Development*, *107*(1), 77–86.

Benazzi, M., Baguná, J., Ballester, R., Puccinelli, I., & Papa, R. D. (1975). Further contribution to the taxonomy of the "*Dugesia lugubris-Polychroa* group" with description of *Dugesia mediterranea* n.sp. (Tricladida, Paludicola). *Bolletino Di Zoologia*, *42*(1), 81–89.

Bhat, K. M. (1999). The posterior determinant gene *nanos* is required for the maintenance of the adult germline stem cells during Drosophila oogenesis. *Genetics*, *151*(4), 1479–1492.

Bowen, I. D., Ryder, T. A., & Thompson, J. A. (1974). The fine structure of the planarian *Polycelis tenuis* Iijima. *Protoplasma*, *79*(1–2), 1–17.

Buaas, F. W., Kirsh, A. L., Sharma, M., McLean, D. J., Morris, J. L., Griswold, M. D., et al. (2004). Plzf is required in adult male germ cells for stem cell self-renewal. *Nature Genetics*, *36*(6), 647–652.

Buszczak, M., Freeman, M. R., Carlson, J. R., Bender, M., Cooley, L., & Segraves, W. A. (1999). Ecdysone response genes govern egg chamber development during mid-oogenesis in *Drosophila*. *Development*, *126*(20), 4581–4589.

Cebrià, F., Vispo, M., Bueno, D., Carranza, S., Newmark, P., & Romero, R. (1996). Myosin heavy chain gene in *Dugesia* (*G.*) *tigrina*: A tool for studying muscle regeneration in planarians. *The International Journal of Developmental Biology*, (Suppl. 1), 177S–178S.

Chalmel, F., Rolland, A. D., Niederhauser-Wiederkehr, C., Chung, S. S. W., Demougin, P., Gattiker, A., et al. (2007). The conserved transcriptome in human and rodent male gametogenesis. *Proceedings of the National Academy of Sciences of the United States of America*, *104*(20), 8346–8351.

Chang, H., & Matzuk, M. M. (2001). *Smad5* is required for mouse primordial germ cell development. *Mechanisms of Development*, *104*(1–2), 61–67.

Chappell, P. E., Lydon, J. P., Conneely, O. M., O'Malley, B. W., & Levine, J. E. (1997). Endocrine defects in mice carrying a null mutation for the progesterone receptor gene. *Endocrinology*, *138*(10), 4147–4152.

Chen, S.-R., & Liu, Y.-X. (2015). Regulation of spermatogonial stem cell self-renewal and spermatocyte meiosis by sertoli cell signaling. *Reproduction*, *149*(4), R159–R167.

Chicoine, J., Benoit, P., Gamberi, C., Paliouras, M., Simonelig, M., & Lasko, P. (2007). Bicaudal-C recruits CCR4-NOT deadenylase to target mRNAs and regulates oogenesis, cytoskeletal organization, and its own expression. *Developmental Cell*, *13*(5), 691–704.

Chien, P. K., & Koopowitz, H. (1977). Ultrastructure of nerve plexus in flatworms. III. The infra-epithelial nervous system. *Cell and Tissue Research*, *176*(3), 335–347.

Chong, T., Collins, J. J., Brubacher, J. L., Zarkower, D., & Newmark, P. A. (2013). A sex-specific transcription factor controls male identity in a simultaneous hermaphrodite. *Nature Communications*, *4*, 1814.

Chong, T., Stary, J. M., Wang, Y., & Newmark, P. A. (2011). Molecular markers to characterize the hermaphroditic reproductive system of the planarian *Schmidtea mediterranea*. *BMC Developmental Biology*, *11*, 69.

Chu, G. C., Dunn, N. R., Anderson, D. C., Oxburgh, L., & Robertson, E. J. (2004). Differential requirements for Smad4 in TGFβ-dependent patterning of the early mouse embryo. *Development*, *131*(15), 3501–3512.

Clarke, H. J. (2018). Regulation of germ cell development by intercellular signaling in the mammalian ovarian follicle. *Wiley Interdisciplinary Reviews: Developmental Biology*, 7(1), e294.

Collins, J. J., III, Hou, X., Romanova, E. V., Lambrus, B. G., Miller, C. M., Saberi, A., et al. (2010). Genome-wide analyses reveal a role for peptide hormones in planarian germline development. *PLoS Biology*, *8*(10), e1000509.
Corrales, P. G., & Gamo, J. (1986). The ultrastructure of the gastrodermal gland cells in the freshwater planarian *Dugesia gonocephala* s.l. *Acta Zoologica*, *67*(1), 43–51.
Costoya, J. A., Hobbs, R. M., Barna, M., Cattoretti, G., Manova, K., Sukhwani, M., et al. (2004). Essential role of Plzf in maintenance of spermatogonial stem cells. *Nature Genetics*, *36*(6), 653–659.
Coward, S. J. (1974). Chromatoid bodies in somatic cells of the planarian: Observations on their behavior during mitosis. *The Anatomical Record*, *180*(3), 533–545.
Curtis, W. C. (1902). The life history, the normal fission and the reproductive organs of *Planaria maculata*. *Proceedings of the Boston Society of Natural History*, *30*(7), 515–559.
Davies, E. L., Lei, K., Seidel, C. W., Kroesen, A. E., McKinney, S. A., Guo, L., et al. (2017). Embryonic origin of adult stem cells required for tissue homeostasis and regeneration. *eLife*, *6*, e21052.
de Sousa Lopes, S. M. C., Roelen, B. A. J., Monteiro, R. M., Emmens, R., Lin, H. Y., Li, E., et al. (2004). BMP signaling mediated by ALK2 in the visceral endoderm is necessary for the generation of primordial germ cells in the mouse embryo. *Genes & Development*, *18*(15), 1838–1849.
Deshpande, G., Calhoun, G., Yanowitz, J. L., & Schedl, P. D. (1999). Novel functions of *nanos* in downregulating mitosis and transcription during the development of the *Drosophila* germline. *Cell*, *99*(3), 271–281.
Donoughe, S., Nakamura, T., Ewen-Campen, B., Green, D. A., Henderson, L., & Extavour, C. G. (2014). BMP signaling is required for the generation of primordial germ cells in an insect. *Proceedings of the National Academy of Sciences of the United States of America*, *111*(11), 4133–4138.
Ewen-Campen, B., Donoughe, S., Clarke, D. N., & Extavour, C. G. (2013). Germ cell specification requires zygotic mechanisms rather than germ plasm in a basally branching insect. *Current Biology*, *23*(10), 835–842.
Extavour, C. G. M. (2007). Evolution of the bilaterian germ line: Lineage origin and modulation of specification mechanisms. *Integrative and Comparative Biology*, *47*(5), 770–785.
Extavour, C. G., & Akam, M. (2003). Mechanisms of germ cell specification across the metazoans: Epigenesis and preformation. *Development*, *130*(24), 5869–5884.
Falender, A. E., Freiman, R. N., Geles, K. G., Lo, K. C., Hwang, K., Lamb, D. J., et al. (2005). Maintenance of spermatogenesis requires TAF4b, a gonad-specific subunit of TFIID. *Genes & Development*, *19*(7), 794–803.
Fedecka-Bruner, B. (1967). Études sur la régénération des organes genitaux chez la planaire *Dugesia lugubris*. I. Régénération des testicules après destruction. *Bulletin Biologique de la France et de la Belgique*, *101*(4), 255–319.
Fincher, C. T., Wurtzel, O., de Hoog, T., Kravarik, K. M., & Reddien, P. W. (2018). Cell type transcriptome atlas for the planarian *Schmidtea mediterranea*. *Science*, *360*(6391), eaaq1736.
Forbes, A., & Lehmann, R. (1998). Nanos and pumilio have critical roles in the development and function of *Drosophila* germline stem cells. *Development*, *125*(4), 679–690.
Forsthoefel, D. J., Ross, K. G., Newmark, P. A., & Zayas, R. M. (2018). Fixation, processing, and immunofluorescent labeling of whole mount planarians. *Methods in Molecular Biology*, *1774*(2), 353–366.
Forsthoefel, D. J., Waters, F. A., & Newmark, P. A. (2014). Generation of cell type-specific monoclonal antibodies for the planarian and optimization of sample processing for immunolabeling. *BMC Developmental Biology*, *14*(1), 45.

Garcia-Corrales, P., & Gamo, J. (1988). Ultrastructural changes in the gastrodermal phagocytic cells of the planarian *Dugesia gonocephala* s.l. during food digestion (Plathelminthes). *Zoomorphology, 108*(2), 109–117.
Geijsen, N., Horoschak, M., Kim, K., Gribnau, J., Eggan, K., & Daley, G. Q. (2004). Derivation of embryonic germ cells and male gametes from embryonic stem cells. *Nature, 427*(6970), 148–154.
Giorgini, F., Davies, H. G., & Braun, R. E. (2002). Translational repression by MSY4 inhibits spermatid differentiation in mice. *Development, 129*(15), 3669–3679.
Gissendanner, C. R., Kelley, K., Nguyen, T. Q., Hoener, M. C., Sluder, A. E., & Maina, C. V. (2008). The *Caenorhabditis elegans* NR4A nuclear receptor is required for spermatheca morphogenesis. *Developmental Biology, 313*(2), 767–786.
Greenspan, L. J., de Cuevas, M., & Matunis, E. (2015). Genetics of gonadal stem cell renewal. *Annual Review of Cell and Developmental Biology, 31*, 291–315.
Gremigni, V., Miceli, C., & Picano, E. (1980). On the role of germ cells in planarian regeneration. II. Cytophotometric analysis of the nuclear Feulgen-DNA content in cells of regenerated somatic tissues. *Journal of Embryology and Experimental Morphology, 55*, 65–76.
Gremigni, V., Miceli, C., & Puccinelli, I. (1980). On the role of germ cells in planarian regeneration. I. A karyological investigation. *Journal of Embryology and Experimental Morphology, 55*, 53–63.
Grohme, M. A., Schloissnig, S., Rozanski, A., Pippel, M., Young, G. R., Winkler, S., et al. (2018). The genome of *Schmidtea mediterranea* and the evolution of core cellular mechanisms. *Nature, 554*(7690), 56–61.
Guan, K., Nayernia, K., Maier, L. S., Wagner, S., Dressel, R., Lee, J. H., et al. (2006). Pluripotency of spermatogonial stem cells from adult mouse testis. *Nature, 440*(7088), 1199–1203.
Guo, T., Peters, A. H. F. M., & Newmark, P. A. (2006). A *bruno-like* gene is required for stem cell maintenance in planarians. *Developmental Cell, 11*(2), 159–169.
Handberg-Thorsager, M., & Saló, E. (2007). The planarian *nanos*-like gene *Smednos* is expressed in germline and eye precursor cells during development and regeneration. *Development Genes and Evolution, 217*(5), 403–411.
Hay, E. D., & Coward, S. J. (1975). Fine structure studies on the planarian, *Dugesia*. I. Nature of the "neoblast" and other cell types in noninjured worms. *Journal of Ultrastructure Research, 50*(1), 1–21.
Hayashi, Y., Hayashi, M., & Kobayashi, S. (2004). Nanos suppresses somatic cell fate in *Drosophila* germ line. *Proceedings of the National Academy of Sciences of the United States of America, 101*(28), 10338–10342.
Hayashi, K., Kobayashi, T., Umino, T., Goitsuka, R., Matsui, Y., & Kitamura, D. (2002). SMAD1 signaling is critical for initial commitment of germ cell lineage from mouse epiblast. *Mechanisms of Development, 118*(1–2), 99–109.
Hayashi, K., Ohta, H., Kurimoto, K., Aramaki, S., & Saitou, M. (2011). Reconstitution of the mouse germ cell specification pathway in culture by pluripotent stem cells. *Cell, 146*(4), 519–532.
Hikabe, O., Hamazaki, N., Nagamatsu, G., Obata, Y., Hirao, Y., Hamada, N., et al. (2016). Reconstitution *in vitro* of the entire cycle of the mouse female germ line. *Nature, 539*(7628), 299–303.
Hübner, K., Fuhrmann, G., Christenson, L. K., Kehler, J., Reinbold, R., La, F., et al. (2003). Derivation of oocytes from mouse embryonic stem cells. *Science, 300*(5623), 1251–1256.
Hyman, L. H. (1951). The invertebrates. *Platyhelminthes and Rhynchocoela, the acoelomate Bilateria: Vol 2* New York: McGraw-Hill
Ishii, S. (1965). Electron microscopic observations on the planarian tissues. II. The intestine. *Fukushima Journal of Medical Science, 12*(1), 67–87.

Ishii, S. (1980a). The ultrastructure of the protonephridial flame cell of the freshwater planarian *Bdellocephala brunnea*. *Cell and Tissue Research*, *206*(3), 441–449.
Ishii, S. (1980b). The ultrastructure of the protonephridial tubules of the freshwater planarian *Bdellocephala brunnea*. *Cell and Tissue Research*, *206*(3), 451–458.
Ishikura, Y., Yabuta, Y., Ohta, H., Hayashi, K., Nakamura, T., Okamoto, I., et al. (2016). In vitro derivation and propagation of spermatogonial stem cell activity from mouse pluripotent stem cells. *Cell Reports*, *17*(10), 2789–2804.
Iyer, H., Collins, J. J., & Newmark, P. A. (2016). NF-YB regulates spermatogonial stem cell self-renewal and proliferation in the planarian *Schmidtea mediterranea*. *PLoS Genetics*, *12*(6), e1006109.
Iyer, H., Issigonis, M., Sharma, P. P., Extavour, C. G., & Newmark, P. A. (2016). A premeiotic function for *boule* in the planarian *Schmidtea mediterranea*. *Proceedings of the National Academy of Sciences of the United States of America*, *113*(25), E3509–E3518.
Johnson, A. D., Crother, B., White, M. E., Patient, R., Bachvarova, R. F., Drum, M., et al. (2003). Regulative germ cell specification in axolotl embryos: A primitive trait conserved in the mammalian lineage. *Philosophical Transactions of the Royal Society of London. Series B, Biological Sciences*, *358*(1436), 1371–1379.
Juliano, C. E., Swartz, S. Z., & Wessel, G. M. (2010). A conserved germline multipotency program. *Development*, *137*(24), 4113–4126.
Kanatsu-Shinohara, M., Inoue, K., Lee, J., Yoshimoto, M., Ogonuki, N., Miki, H., et al. (2004). Generation of pluripotent stem cells from neonatal mouse testis. *Cell*, *119*(7), 1001–1012.
Kimble, J., & Seidel, H. (2013). *C. elegans* germline stem cells and their niche. StemBook, ed. In *The Stem Cell Research Community, StemBook*. https://doi.org/10.3824/stembook.1.95.1.
King, R. S., & Newmark, P. A. (2018). Whole-mount in situ hybridization of planarians. *Methods in Molecular Biology*, *1774*(8), 379–392.
Kobayashi, C., Kobayashi, S., Orii, H., Watanabe, K., & Agata, K. (2009). Identification of two distinct muscles in the planarian *Dugesia japonica* by their expression of myosin heavy chain genes. *Zoological Science*, *15*(6), 861–869.
Kobayashi, S., Yamada, M., Asaoka, M., & Kitamura, T. (1996). Essential role of the posterior morphogen *nanos* for germline development in *Drosophila*. *Nature*, *380*(6576), 708–711.
Koopowitz, H., & Chien, P. (1975). Ultrastructure of nerve plexus in flatworms. II. Sites of synaptic interactions. *Cell and Tissue Research*, *157*(2).
Kopp, A. (2012). *Dmrt* genes in the development and evolution of sexual dimorphism. *Trends in Genetics*, *28*(4), 175–184.
Köprunner, M., Thisse, C., Thisse, B., & Raz, E. (2001). A zebrafish *nanos*-related gene is essential for the development of primordial germ cells. *Genes & Development*, *15*(21), 2877–2885.
LaFever, L., & Drummond-Barbosa, D. (2005). Direct control of germline stem cell division and cyst growth by neural insulin in *Drosophila*. *Science*, *309*(5737), 1071–1073.
Lawson, K. A., Dunn, N. R., Roelen, B. A., Zeinstra, L. M., Davis, A. M., Wright, C. V., et al. (1999). *Bmp4* is required for the generation of primordial germ cells in the mouse embryo. *Genes & Development*, *13*(4), 424–436.
Le Moigne, A. (1967). Présence d'émissions nucléaires fréquemment associées à des mitochondries dans les cellules embryonnaires des planaires. *Comptes Rendus des Séances de la Société de Biologie et de ses Filiales*, *161*, 508–511.
MacRae, E. K. (1963). Observations on the fine structure of pharyngeal muscle in the planarian *Dugesia tigrina*. *The Journal of Cell Biology*, *18*(3), 651–662.
MacRae, E. K. (1965). The fine structure of muscle in a marine turbellarian. *Zeitschrift für Zellforschung und Mikroskopische Anatomie*, *68*(3), 348–362.

MacRae, E. K. (1967). The fine structure of sensory receptor processes in the auricular epithelium of the planarian, *Dugesia tigrina*. *Zeitschrift für Zellforschung und Mikroskopische Anatomie*, *82*(4), 479–494.

Mahone, M., Saffman, E. E., & Lasko, P. F. (1995). Localized *Bicaudal-C* RNA encodes a protein containing a KH domain, the RNA binding motif of FMR1. *The EMBO Journal*, *14*(9), 2043–2055.

Matson, C. K., & Zarkower, D. (2012). Sex and the singular DM domain: Insights into sexual regulation, evolution and plasticity. *Nature Reviews. Genetics*, *13*(3), 163–174.

McKanna, J. A. (1968a). Fine structure of the protonephridial system in planaria. I. Flame cells. *Zeitschrift Für Zellforschung Und Mikroskopische Anatomie*, *92*(4), 509–523.

McKanna, J. A. (1968b). Fine structure of the protonephridial system in planaria. II. Ductules, collecting ducts, and osmoregulatory cells. *Zeitschrift für Zellforschung und Mikroskopische Anatomie*, *92*(4), 524–535.

Michaelson, D., Korta, D. Z., Capua, Y., & Hubbard, E. J. A. (2010). Insulin signaling promotes germline proliferation in *C. elegans*. *Development*, *137*(4), 671–680.

Morgan, T. H. (1902). Growth and regeneration in *Planaria lugubris*. *Archiv Für Entwicklungsmechanik Der Organismen*, *13*(1–2), 179–212.

Morita, M. (1965). Electron microscopic studies on planaria. I. Fine structure of muscle fiber in the head of the planarian *Dugesia dorotocephala*. *Journal of Ultrastructure Research*, *13*(5–6), 383–395.

Morita, M., Best, J. B., & Noel, J. (1969). Electron microscopic studies of planarian regeneration. I. Fine structure of neoblasts in *Dugesia dorotocephala*. *Journal of Ultrastructure Research*, *27*(1), 7–23.

Nakagawa, H., Ishizu, H., Hasegawa, R., Kobayashi, K., & Matsumoto, M. (2012). *Drpiwi-1* is essential for germline cell formation during sexualization of the planarian *Dugesia ryukyuensis*. *Developmental Biology*, *361*(1), 167–176.

Newmark, P. A., Reddien, P. W., Cebrià, F., & Sánchez Alvarado, A. (2003). Ingestion of bacterially expressed double-stranded RNA inhibits gene expression in planarians. *Proceedings of the National Academy of Sciences of the United States of America*, *100*(Suppl. 1), 11861–11865.

Newmark, P. A., & Sánchez Alvarado, A. (2000). Bromodeoxyuridine specifically labels the regenerative stem cells of planarians. *Developmental Biology*, *220*(2), 142–153.

Newmark, P. A., & Sánchez Alvarado, A. (2002). Not your father's planarian: A classic model enters the era of functional genomics. *Nature Reviews. Genetics*, *3*(3), 210–219.

Nishimura, K., Kitamura, Y., Inoue, T., Umesono, Y., Sano, S., Yoshimoto, K., et al. (2007a). Reconstruction of dopaminergic neural network and locomotion function in planarian regenerates. *Developmental Neurobiology*, *67*(8), 1059–1078.

Nishimura, K., Kitamura, Y., Inoue, T., Umesono, Y., Yoshimoto, K., Takeuchi, K., et al. (2007b). Identification and distribution of tryptophan hydroxylase (TPH)-positive neurons in the planarian *Dugesia japonica*. *Neuroscience Research*, *59*(1), 101–106.

Nishimura, K., Kitamura, Y., Inoue, T., Umesono, Y., Yoshimoto, K., Taniguchi, T., et al. (2008). Characterization of tyramine β-hydroxylase in planarian *Dugesia japonica*: Cloning and expression. *Neurochemistry International*, *53*(6–8), 184–192.

Nishimura, K., Kitamura, Y., Taniguchi, T., & Agata, K. (2010). Analysis of motor function modulated by cholinergic neurons in planarian *Dugesia japonica*. *Neuroscience*, *168*(1), 18–30.

Nishimura, K., Kitamura, Y., Umesono, Y., Takeuchi, K., Takata, K., Taniguchi, T., et al. (2008). Identification of glutamic acid decarboxylase gene and distribution of GABAergic nervous system in the planarian *Dugesia japonica*. *Neuroscience*, *153*(4), 1103–1114.

Palakodeti, D., Smielewska, M., Lu, Y.-C., Yeo, G. W., & Graveley, B. R. (2008). The PIWI proteins SMEDWI-2 and SMEDWI-3 are required for stem cell function and piRNA expression in planarians. *RNA*, *14*(6), 1174–1186.

Pearson, B. J., Eisenhoffer, G. T., Gurley, K. A., Rink, J. C., Miller, D. E., & Alvarado, A. S. (2009). Formaldehyde-based whole-mount in situ hybridization method for planarians. *Developmental Dynamics*, *238*(2), 443–450.
Pedersen, K. J. (1959). Cytological studies on the planarian neoblast. *Zeitschrift Fur Zellforschung Und Mikroskopische Anatomie (Vienna, Austria: 1948)*, *50*, 799–817.
Pepling, M. E., Wilhelm, J. E., O'Hara, A. L., Gephardt, G. W., & Spradling, A. C. (2007). Mouse oocytes within germ cell cysts and primordial follicles contain a Balbiani body. *Proceedings of the National Academy of Sciences of the United States of America*, *104*(1), 187–192.
Picard, M. A.-L., Cosseau, C., Mouahid, G., Duval, D., Grunau, C., Toulza, È., et al. (2015). The roles of *Dmrt* (Double sex/Male-abnormal-3 Related Transcription factor) genes in sex determination and differentiation mechanisms: Ubiquity and diversity across the animal kingdom. *Comptes Rendus Biologies*, *338*(7), 451–462.
Plass, M., Solana, J., Wolf, F. A., Ayoub, S., Misios, A., Glažar, P., et al. (2018). Cell type atlas and lineage tree of a whole complex animal by single-cell transcriptomics. *Science*, *360*(6391), eaaq1723.
Qing, T., Shi, Y., Qin, H., Ye, X., Wei, W., Liu, H., et al. (2007). Induction of oocyte-like cells from mouse embryonic stem cells by co-culture with ovarian granulosa cells. *Differentiation*, *75*(10), 902–911.
Reddien, P. W., Oviedo, N. J., Jennings, J. R., Jenkin, J. C., & Sánchez Alvarado, A. (2005). SMEDWI-2 is a PIWI-like protein that regulates planarian stem cells. *Science*, *310*(5752), 1327–1330.
Rink, J. C., Vu, H. T.-K., & Alvarado, A. S. (2011). The maintenance and regeneration of the planarian excretory system are regulated by EGFR signaling. *Development*, *138*(17), 3769–3780.
Robb, S. M. C., Ross, E., & Sánchez Alvarado, A. (2008). SmedGD: The *Schmidtea mediterranea* genome database. *Nucleic Acids Research*, *36*, D599–D606.
Roberts-Galbraith, R. H., Brubacher, J. L., & Newmark, P. A. (2016). A functional genomics screen in planarians reveals regulators of whole-brain regeneration. *eLife*, *5*, e17002.
Roberts-Galbraith, R. H., & Newmark, P. A. (2015). On the organ trail: Insights into organ regeneration in the planarian. *Current Opinion in Genetics & Development*, *32*, 37–46.
Rouhana, L., Shibata, N., Nishimura, O., & Agata, K. (2010). Different requirements for conserved post-transcriptional regulators in planarian regeneration and stem cell maintenance. *Developmental Biology*, *341*(2), 429–443.
Rouhana, L., Tasaki, J., Saberi, A., & Newmark, P. A. (2017). Genetic dissection of the planarian reproductive system through characterization of *Schmidtea mediterranea* CPEB homologs. *Developmental Biology*, *426*(1), 43–55.
Rouhana, L., Weiss, J. A., Forsthoefel, D. J., Lee, H., King, R. S., Inoue, T., et al. (2013). RNA interference by feeding in vitro-synthesized double-stranded RNA to planarians: Methodology and dynamics. *Developmental Dynamics*, *242*(6), 718–730.
Saberi, A. (2016). *Somatic regulation of germ cell development in the planarian Schmidtea mediterranea: The role of systemic and local factors*. Doctoral thesis. Urbana, IL: University of Illinois at Urbana-Champaign.
Saberi, A., Jamal, A., Beets, I., Schoofs, L., & Newmark, P. A. (2016). GPCRs direct germline development and somatic gonad function in planarians. *PLoS Biology*, *14*(5), e1002457.
Sada, A., Suzuki, A., Suzuki, H., & Saga, Y. (2009). The RNA-binding protein NANOS2 is required to maintain murine spermatogonial stem cells. *Science*, *325*(5946), 1394–1398.
Salvetti, A., Rossi, L., Lena, A., Batistoni, R., Deri, P., Rainaldi, G., et al. (2005). *DjPum*, a homologue of *Drosophila pumilio*, is essential to planarian stem cell maintenance. *Development*, *132*(8), 1863–1874.

Sánchez Alvarado, A., & Newmark, P. A. (1999). Double-stranded RNA specifically disrupts gene expression during planarian regeneration. *Proceedings of the National Academy of Sciences of the United States of America*, *96*(9), 5049–5054.
Sato, K., Shibata, N., Orii, H., Amikura, R., Sakurai, T., Agata, K., et al. (2006). Identification and origin of the germline stem cells as revealed by the expression of *nanos*-related gene in planarians. *Development, Growth & Differentiation*, *48*(9), 615–628.
Sauzin, M. J. (1968). Présence d'émissions nucléaires dans les cellules différenciées et en différenciation de la planaire adulte *Dugesia gonocephala*. *Comptes Rendus de l'Académie des Sciences*, *267*, 1146–1148.
Scimone, M. L., Srivastava, M., Bell, G. W., & Reddien, P. W. (2011). A regulatory program for excretory system regeneration in planarians. *Development*, *138*(20), 4387–4398.
Seandel, M., James, D., Shmelkov, S. V., Falciatori, I., Kim, J., Chavala, S., et al. (2007). Generation of functional multipotent adult stem cells from GPR125^{+} germline progenitors. *Nature*, *449*(7160), 346–350.
Seydoux, G., & Braun, R. E. (2006). Pathway to totipotency: Lessons from germ cells. *Cell*, *127*(5), 891–904.
Shibata, N., Kashima, M., Ishiko, T., Nishimura, O., Rouhana, L., Misaki, K., et al. (2016). Inheritance of a nuclear PIWI from pluripotent stem cells by somatic descendants ensures differentiation by silencing transposons in planarian. *Developmental Cell*, *37*(3), 226–237.
Solana, J., Lasko, P., & Romero, R. (2009). *Spoltud-1* is a chromatoid body component required for planarian long-term stem cell self-renewal. *Developmental Biology*, *328*(2), 410–421.
Strome, S., & Updike, D. (2015). Specifying and protecting germ cell fate. *Nature Reviews Molecular Cell Biology*, *16*(7), 406–416.
Subramaniam, K., & Seydoux, G. (1999). *nos-1* and *nos-2*, two genes related to *Drosophila nanos*, regulate primordial germ cell development and survival in *Caenorhabditis elegans*. *Development*, *126*(21), 4861–4871.
Suzuki, A., Tsuda, M., & Saga, Y. (2007). Functional redundancy among Nanos proteins and a distinct role of Nanos2 during male germ cell development. *Development*, *134*(1), 77–83.
Svingen, T., & Koopman, P. (2013). Building the mammalian testis: Origins, differentiation, and assembly of the component cell populations. *Genes & Development*, *27*(22), 2409–2426.
Swapna, L. S., Molinaro, A. M., Lindsay-Mosher, N., Pearson, B. J., & Parkinson, J. (2018). Comparative transcriptomic analyses and single-cell RNA sequencing of the freshwater planarian *Schmidtea mediterranea* identify major cell types and pathway conservation. *Genome Biology*, *19*(1), 124.
Tanaka, K. J., Ogawa, K., Takagi, M., Imamoto, N., Matsumoto, K., & Tsujimoto, M. (2006). RAP55, a cytoplasmic mRNP component, represses translation in *Xenopus* oocytes. *The Journal of Biological Chemistry*, *281*(52), 40096–40106.
Tharp, M. E., Collins, J. J., & Newmark, P. A. (2014). A lophotrochozoan-specific nuclear hormone receptor is required for reproductive system development in the planarian. *Developmental Biology*, *396*(1), 150–157.
Thi-Kim Vu, H., Rink, J. C., McKinney, S. A., McClain, M., Lakshmanaperumal, N., Alexander, R., et al. (2015). Stem cells and fluid flow drive cyst formation in an invertebrate excretory organ. *eLife*, *4*, e10501.
Toyooka, Y., Tsunekawa, N., Akasu, R., & Noce, T. (2003). Embryonic stem cells can form germ cells *in vitro*. *Proceedings of the National Academy of Sciences of the United States of America*, *100*(20), 11457–11462.
Tremblay, K. D., Dunn, N. R., & Robertson, E. J. (2001). Mouse embryos lacking Smad1 signals display defects in extra-embryonic tissues and germ cell formation. *Development*, *128*(18), 3609–3621.
Tsuda, M., Sasaoka, Y., Kiso, M., Abe, K., Haraguchi, S., Kobayashi, S., et al. (2003). Conserved role of nanos proteins in germ cell development. *Science*, *301*(5637), 1239–1241.

Ueishi, S., Shimizu, H., & H Inoue, Y. (2009). Male germline stem cell division and spermatocyte growth require insulin signaling in *Drosophila*. *Cell Structure and Function*, *34*(1), 61–69.

Wagner, D. E., Ho, J. J., & Reddien, P. W. (2012). Genetic regulators of a pluripotent adult stem cell system in planarians identified by RNAi and clonal analysis. *Cell Stem Cell*, *10*(3), 299–311.

Wagner, D. E., Wang, I. E., & Reddien, P. W. (2011). Clonogenic neoblasts are pluripotent adult stem cells that underlie planarian regeneration. *Science*, *332*(6031), 811–816.

Walker, V. R., & Korach, K. S. (2004). Estrogen receptor knockout mice as a model for endocrine research. *Institute for Laboratory Animal Research Journal*, *45*(4), 455–461.

Wang, I. E., Lapan, S. W., Scimone, M. L., Clandinin, T. R., & Reddien, P. W. (2016). Hedgehog signaling regulates gene expression in planarian glia. *eLife*, *5*, e16996.

Wang, C., & Lehmann, R. (1991). Nanos is the localized posterior determinant in *Drosophila*. *Cell*, *66*(4), 637–647.

Wang, Z., & Lin, H. (2004). *Nanos* maintains germline stem cell self-renewal by preventing differentiation. *Science*, *303*(5666), 2016–2019.

Wang, Y., Stary, J. M., Wilhelm, J. E., & Newmark, P. A. (2010). A functional genomic screen in planarians identifies novel regulators of germ cell development. *Genes & Development*, *24*(18), 2081–2092.

Wang, R.-S., Yeh, S., Tzeng, C.-R., & Chang, C. (2009). Androgen receptor roles in spermatogenesis and fertility: Lessons from testicular cell-specific androgen receptor knockout mice. *Endocrine Reviews*, *30*(2), 119–132.

Wang, Y., Zayas, R. M., Guo, T., & Newmark, P. A. (2007). *nanos* function is essential for development and regeneration of planarian germ cells. *Proceedings of the National Academy of Sciences of the United States of America*, *104*(14), 5901–5906.

Wilhelm, J. E., Buszczak, M., & Sayles, S. (2005). Efficient protein trafficking requires Trailer Hitch, a component of a ribonucleoprotein complex localized to the ER in *Drosophila*. *Developmental Cell*, *9*(5), 675–685.

Wolff, E. (1962). Recent researches on the regeneration of planaria. In D. Rudnick (Ed.), *Vol. 20*. Regeneration: 20th Growth Symposium (pp. 53–84). New York: The Roland Press.

Yamashiro, C., Sasaki, K., Yabuta, Y., Kojima, Y., Nakamura, T., Okamoto, I., et al. (2018). Generation of human oogonia from induced pluripotent stem cells in vitro. *Science*, *362*(6412), 356–360.

Yang, W.-H., Yu, J. H., Gulick, T., Bloch, K. D., & Bloch, D. B. (2006). RNA-associated protein 55 (RAP55) localizes to mRNA processing bodies and stress granules. *RNA*, *12*(4), 547–554.

Ying, Y., Liu, X. M., Marble, A., Lawson, K. A., & Zhao, G. Q. (2000). Requirement of *Bmp8b* for the generation of primordial germ cells in the mouse. *Molecular Endocrinology*, *14*(7), 1053–1063.

Yoshida-Kashikawa, M., Shibata, N., Takechi, K., & Agata, K. (2007). DjCBC-1, a conserved DEAD box RNA helicase of the RCK/p54/Me31B family, is a component of RNA-protein complexes in planarian stem cells and neurons. *Developmental Dynamics*, *236*(12), 3436–3450.

Zeng, A., Li, H., Guo, L., Gao, X., McKinney, S., Wang, Y., et al. (2018). Prospectively isolated Tetraspanin^{+} neoblasts are adult pluripotent stem cells underlying planaria regeneration. *Cell*, *173*(7), 1593–1608.e20.

Zhou, Q., Wang, M., Yuan, Y., Wang, X., Fu, R., Wan, H., et al. (2016). Complete meiosis from embryonic stem cell-derived germ cells in vitro. *Cell Stem Cell*, *18*(3), 330–340.

Zhu, S. J., Hallows, S. E., Currie, K. W., Xu, C., & Pearson, B. J. (2015). A *mex3* homolog is required for differentiation during planarian stem cell lineage development. *eLife*, *4*, e07025.

CHAPTER FIVE

Heterogeneity of primordial germ cells

Daniel H. Nguyen†, Rebecca G. Jaszczak†, Diana J. Laird*

Department of Obstetrics, Gynecology and Reproductive Science, Center for Reproductive Sciences, Eli and Edythe Broad Center for Regeneration Medicine and Stem Cell Research, University of California, San Francisco, CA, United States

*Corresponding author: e-mail address: diana.laird@ucsf.edu

Contents

1. Introduction 156
2. Heterogeneous phenotypes of primordial germ cells 157
   2.1 Primordial germ cell specification and heterogeneity 159
   2.2 Primordial germ cell migration and heterogeneity 164
   2.3 Proliferation of primordial germ cells and heterogeneity 166
   2.4 Primordial germ cell survival and heterogeneity 168
   2.5 Differentiation of PGCs and heterogeneity 171
3. Determinants of heterogeneity 173
   3.1 Genetic determinants 174
   3.2 Extrinsic sources of heterogeneity: Environment, niche, and timing 182
   3.3 Buffers of heterogeneity 183
4. Selection and heterogeneity 184
   4.1 Defining selection in the developing germline 185
   4.2 How does PGC fitness relate to fitness of gametes or individuals? 186
5. Conclusion 188
References 189

Abstract

Primordial germ cells (PGCs) must complete a complex and dynamic developmental program during embryogenesis to establish the germline. This process is highly conserved and involves a diverse array of tasks required of PGCs, including migration, survival, sex differentiation, and extensive epigenetic reprogramming. A common theme across many organisms is that PGC success is heterogeneous: only a portion of all PGCs complete all these steps while many other PGCs are eliminated from further germline contribution. The differences that distinguish successful PGCs as a population are not well understood. Here, we examine variation that exists in PGCs as they navigate the

† Equal contribution.

Current Topics in Developmental Biology, Volume 135
ISSN 0070-2153
https://doi.org/10.1016/bs.ctdb.2019.04.009

many stages of this developmental journey. We explore potential sources of PGC heterogeneity and their potential implications in affecting germ cell behaviors. Lastly, we discuss the potential for PGC development to function as a multistage selection process that assesses heterogeneity in PGCs to refine germline quality.

1. Introduction

Sexual reproduction culminates with the fusion of two gametes to generate a zygote, briefly returning life to a fleeting, single-celled state. This simple cellular arithmetic, however, belies the momentous journeys undertaken by each gametic partner to enable this union. As we consider the epic reproductive development that precedes the production of this unicellular zygote, it is most appropriate that our focus begin, humbly, at the level of the single cell. Though the rise of metazoans introduced the dazzling complexity of diversified cell types and tissues, all multicellular systems are still fundamentally reducible to single cells—the smallest unit of biological structure.

While recent advances in single-cell technologies promise to reveal powerful insights about the identities of individual cells that compose a lineage or tissue, much of biology has begun by asking questions of single cells. From Anton van Leeuwenhoek at his microscope to Paul Ehrlich and his histological dyes, we have long analyzed differences that exist from cell to individual cell. The advantage of whole-mount or section immunofluorescence imaging lies in the ability to resolve multiple cells in a tissue while preserving individual differences for measurement (Levsky & Singer, 2003). Such powerful resolution permits a more comprehensive understanding of cellular interactions and a sensitivity to cell-to-cell variation. In contrast, analytical methods like Western blots, PCR, or bulk sequencing provide only population-level insights under the assumption that behavior is uniform. Even within single lineages, it is increasingly understood that significant variation can exist among seemingly identical cells (Altschuler & Wu, 2010; Raj & van Oudenaarden, 2008) and that this heterogeneity can have far-ranging functional consequences.

With this principle in mind, we can begin tracing the steps that generate gametes—themselves single-celled carriers of the genome. Each germ cell must endure a lengthy and complex development that begins shortly after fertilization when the germline is set aside from the somatic lineages with the formation of primordial germ cells (PGCs). The path from PGC to gamete is highly conserved, demonstrating the evolutionary importance of reproductive development (Nieuwkoop & Sutasurya, 1979, 1981). Early PGC development is especially dynamic and requires transition through differentiated states, interaction with diverse cellular environments, and processing a multitude of signals. This complexity can amplify differences among individual cells to yield varied cell fates. For germ cells that are tasked with creating gametes, those that survive this developmental crucible secure the ultimate biological prize: propagation of their genetic identity.

The concept of selective events acting on variation within a population has long been appreciated in biology, albeit on the level of organisms in Darwinian natural selection. These same principles are also applicable at the cellular level as well, and we argue that they can govern the fate of PGC heterogeneity during development (Buss, 1988; Laird, Chang, Weissman, & Lauzon, 2005; Weissman, 2015). The diversity of challenges posed to PGCs in this period may function as developmental selection that acts upon cellular variation in the germline in a manner analogous to natural selection with organisms. Here, we consider how heterogeneity in PGCs provides source material for selection by PGC development. This review presents a comprehensive perspective on early events in germ cell development with a particular emphasis on how heterogeneity manifests in germ cells, how it arises, and its impact on the germline and reproductive fitness.

2. Heterogeneous phenotypes of primordial germ cells

Heterogeneity is a fundamental property of biological systems consisting of multiple units such as cells. Even though cells may share a common lineage, it is increasingly recognized that substantial variation exists even within cell types. As our ability to measure cellular phenotypes and behaviors improves, we can distinguish rare cell subpopulations in tissues such the intestinal crypt (Grün et al., 2015) as well as canvass the broad diversity that exists in chaotic intratumoral settings (McGranahan & Swanton, 2017). In hematopoiesis, these subpopulations can represent increasingly smaller organizational units that possess distinctive behaviors and fates (Schroeder, 2010).

While heterogeneity has been studied in many homeostatic contexts such as the adult stem cells of the intestine or blood, its role in germline development is a unique and interesting consideration. The establishment of PGCs at the outset of embryogenesis, as the first lineage in many different organisms, mandates a lengthy process of development through which a subset of germ cells become gametes. The dynamic and multifaceted challenges encountered by PGCs during fetal development present numerous opportunities to assess heterogeneity that may exist in nascent PGCs. As PGCs go on to establish the germline and generate gametes, how heterogeneity is resolved by developmental selection can prove especially influential for transgenerational inheritance.

This section aims to establish a conceptual framework to define and identify heterogeneity as it pertains to the germline and reproductive fitness (see Box 1). We will examine evidence for heterogeneity in specific aspects

BOX 1 Clarification of terminology

Cell type: Although cell type has been defined by many criteria such as morphology, location, lineage history, and distinct transcriptional profile (Altschuler & Wu, 2010; Muller-Sieburg, Sieburg, Bernitz, & Cattarossi, 2012; Seoane, 2017), we treat the germ cell as a functional entity theoretically capable of giving rise to gametes of either sex (Fig. 1). This definition encompasses changes in morphology, location, and gene expression involved in PGC development, as well as the different modes of specification and routes of PGC migration between organisms. We define PGCs as the committed but yet sexually undifferentiated embryonic precursor of the gamete. Although we acknowledge that PGC embryonic origins may have evolved multiple times, we draw comparisons between PGCs in different organisms based on their operational similarities and parallel constraints.

Cell state: A similarly nebulous term, we consider cell state to be a stable but transient set of activities, a potency, a distinct transcriptional profile, and/or a set of epigenetic marks that defines a subpopulation of cells within a particular cell type. Similar to the poised, active, and silenced states of embryonic stem cells associated with distinct sets of epigenetic marks and gene expression (Young, 2011), we consider that PGCs can occupy different states as well as transition back and forth between them. Cell state and cell type are interdependent; PGCs are a cell type, but PGCs in the embryo are a collection of cells in multiple states. PGC differentiation to male gonocyte or oocyte involves a change of cell type that may occur through a series of cell state transitions.

Heterogeneity (of cells): Synonymous with cell individuality and cell diversity, cell heterogeneity is treated here as a biological difference between cells of the same type (as opposed to noise that arises from experimental or technical factors). Heterogeneity can be detected as distinct cellular behaviors, including morphology or function (such as proliferation, capacity for differentiation, or migration); these behaviors may be reflected in transcriptional state or epigenetic profile. Determinants or causes of cellular heterogeneity may include genetic changes, epigenetic changes (which is relevant for PGCs more than other cells), non-genetic changes that may or may not be stochastic, or extrinsically induced changes.

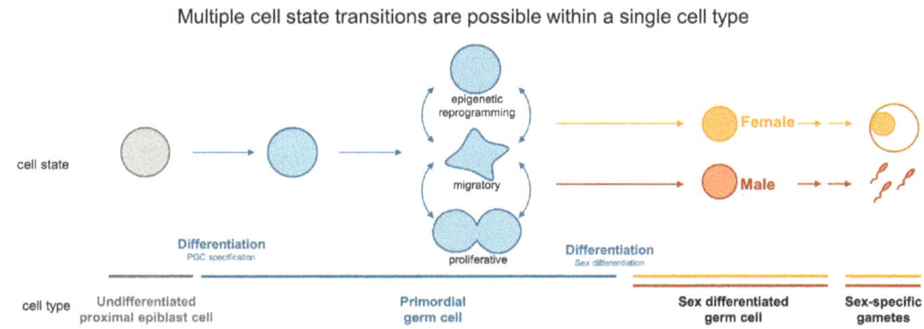

Fig. 1 PGCs transit through multiple cell states during development but remain a single cell type defined by an uncommitted sexual identity. Cell states are not necessarily discrete nor mutually exclusive. We consider germ cells to encompass all cell types with gametogenic potential, including PGCs and sex-differentiated oogonia and prospermatogonia.

of PGC development in several model organisms to uncover general principles that underlie the unique challenges and adaptations utilized by germ cells to address variation (schematized in Fig. 2).

2.1 Primordial germ cell specification and heterogeneity

This section considers the potential for heterogeneity to arise between PGCs during their specification, as well as specific examples. Among metazoans, there are two modes by which PGCs are specified: preformation and induction. Also termed epistasis, preformation confers a germline fate by cytoplasmic inheritance of maternally derived organelles composed of proteins and RNAs termed germ plasm, germ granules, polar or P granules; mechanisms of germline preformation have been reviewed extensively by others (Extavour, 2003; Saitou, Payer, O'Carroll, Ohinata, & Surani, 2005; Solter, 2016). Of the model organisms commonly studied, *Caenorhabditis elegans*, *Drosophila melanogaster*, *Xenopus laevis*, and *Danio rerio* all generate PGCs through early preformation. By contrast, mammals, urodele amphibians and crickets specify germline by induction from somatic cells, which involves signaling from the surrounding niche that leads to a binary fate decision (Ewen-Campen, Donoughe, Clarke, & Extavour, 2013; Lawson et al., 1999; Nieuwkoop, 1947; Tam & Zhou, 1996); underlying molecular mechanisms are well reviewed elsewhere (Solter, 2016).

Does the mode of specification limit the extent of heterogeneity that is possible between PGCs? It is not unreasonable to assume that the germ plasm promotes *homogeneity* from the inception of the PGC lineage by preformation. The inclusion of proteins and transcripts in the germ plasm

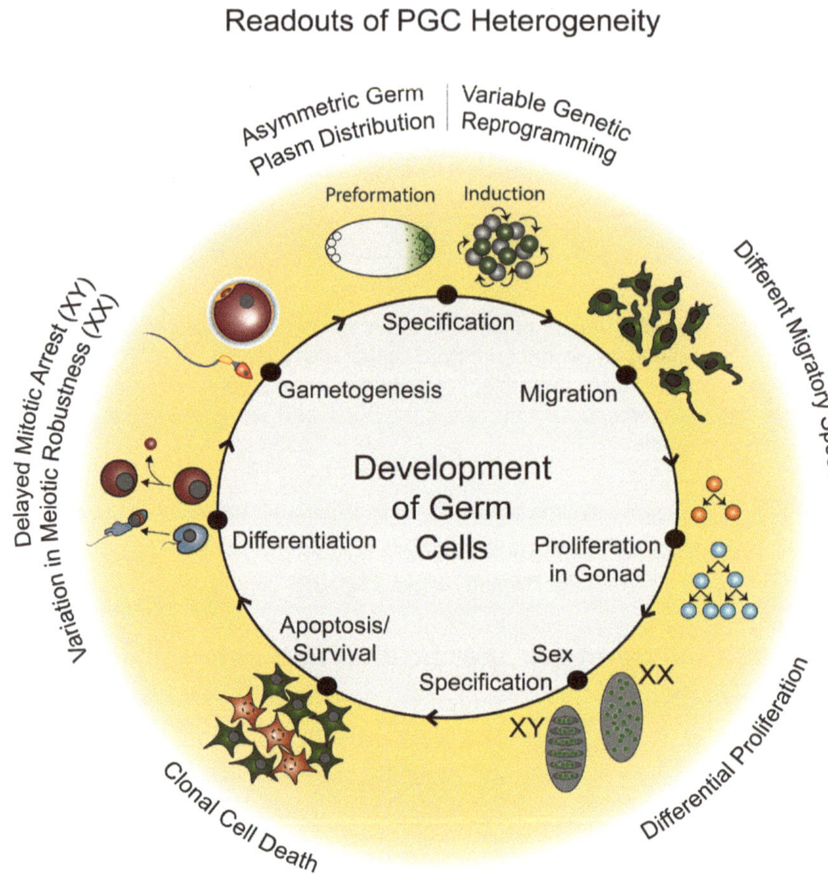

Fig. 2 Readouts of PGC Heterogeneity. Snapshots of the reproductive life cycle evidence potential readouts of heterogeneity. These readouts can be broadly surveyed across many model organisms. Key phases of potential heterogeneity occur during specification, migration, proliferation, sex determination, survival, differentiation, and gametogenesis. Specific readouts may be applicable several cell states.

regulates the levels of both in the cytoplasm, particularly since PGC specification precedes zygotic transcription in model organisms with preformation, and transcription remains suppressed in nascent PGCs even after beginning in the soma (Strome & Lehmann, 2007). It might be expected that germ plasm promotes PGC homogeneity on a continual basis given the capacity for movement through specialized cytoplasmic bridges between germ cells (Extavour, 2003). During PGC specification, it has been proposed that niche functions to shield PGCs from somatic fate signals; this idea is reinforced by the observation that PGCs arise via preformation or induction paradigms in sequestered locations, such as the posterior pole in fly embryos (Eddy, 1975), near the extraembryonic tissues in mammals and birds (Ginsburg & Eyal-Giladi, 1986, 1987; Ginsburg, Snow, & McLaren, 1990), or on the periphery of zebrafish embryos (Raz, 2002). However, in the context of this discussion, we might speculate that the niche for specification instead protects PGCs from signals that would promote heterogeneity. On the other hand, the inductive mode is arguably more conducive to producing heterogeneous PGCs since the process of fate conferral by signaling can occur over a more protracted period of time in principle. The comparative effects of these two mechanisms of PGC fate determination may have relevance in an evolutionary context. Examining patterns of PGC specification across many species precipitated the notion that preformation is derived whereas induction is ancestral (Extavour, 2003). We might ask, then—if germ plasm indeed promotes uniformity, whereas induction promotes greater diversity of PGCs—what benefit does reduced heterogeneity confer upon those species that acquired preformation?

2.1.1 PGC specification by preformation

We discuss several examples of heterogeneous behaviors of PGCs associated with specification by preformation. In this mode of development, germline establishment requires asymmetrical distribution of germ plasm in the oocyte and zygote (Hird, Paulsen, & Strome, 1996), followed by asymmetric inheritance during early divisions into the germline precursors and PGCs (Eddy, 1975; Mahowald, 1962, 2001). However, subsequent development of PGCs involves a switch from asymmetric to symmetric inheritance of germ plasm to daughter cells. A key question is whether minor asymmetries in either the allocation of germ plasm between founder PGCs or the subsequent allocation to PGC daughter cells have phenotypic consequences. A recent study in Drosophila suggests that indeed some founder PGCs are short-changed in their inheritance. Quantifiable variation was observed in the levels of maternal mRNAs *Nanos* and *Wunen* associated with the germ plasm between individual PGCs. Importantly, the levels of both were not arbitrary, but highly correlated with the probability of subsequent germ cell death, suggesting that Wunen is an indicator of germ plasm quality (Slaidina & Lehmann, 2017). Similarly, in *Caenorhabditis elegans*, the asymmetric segregation of cytoplasmic determinants known as P granules occurs in the P4 ancestor cell of the two PGCs known as Z2 and Z3 (Sulston, Schierenberg, White, & Thomson, 1983). Surprisingly, PGCs were specified and fertility was maintained in a mutant with symmetric distribution of

P granules, suggesting that the P granules themselves are not sufficient for the worm germline (Gallo, Wang, Motegi, & Seydoux, 2010).

Further insight into the critical components of germ plasm for conferring a germ cell fate and potentially generating diversity of PGCs comes from recent work in Drosophila. The germ plasm component *oskar* is both necessary and sufficient for specifying PGCs in flies, and plays a deterministic role in the number of precursor pole cells (Ephrussi & Lehmann, 1992; Ewen-Campen, Srouji, Schwager, & Extavour, 2012; Smith, Wilson, & Macdonald, 1992). A surprising function of one isoform of *oskar* is to recruit mitochondria to nascent PGCs via the actin cytoskeleton; a reduction in the number of PGCs and diminished downstream oogenesis in mutants lacking this *oskar* long isoform implicates mitochondrial inheritance and number in the competence and normal development of PGCs (Hurd et al., 2016). In zebrafish, the maternal gene *bucky ball* functions in a parallel manner to *oskar* via interactions with microtubule networks to assemble the germ plasm, which involves aggregating mitochondria into a structure known as the Balbiani body. Ectopic PGCs are produced by ectopic expression of *bucky ball*, but how they may differ from properly localized germ cells is unclear (Bontems et al., 2009; Marlow, 2015). Similarly, it is not known how mislocalized PGCs that result from ectopic expression of *oskar* in flies might be different than those specified according to established patterning, but an avenue of inquiry will be interrogate potential differences in mitochondria between individual PGCs as related to subsequent fate and function.

2.1.2 PGC specification by induction

Of the organisms that specify their germline through induction and signaling from somatic cells, *Mus musculus* is one of the most studied. Despite this fact, much remains unknown about how heterogeneity may impact murine reproductive potential. Early fate mapping studies of single cells from E4.5 the mouse embryo produced an important negative result: no single cell of the inner cell mass exclusively gave rise to PGCs (Gardner & Rossant, 1979). Mouse PGCs are induced in the epiblast by BMPs from the surrounding extraembryonic tissues (Fujiwara, Dunn, & Hogan, 2001; Lawson et al., 1999; Ying, Liu, Marble, Lawson, & Zhao, 2000). The earliest known marker of the PGC fate is *Blimp1* (also known as *Prdm1*), which first appears in the epiblast at E6.25. Although other genes of PGC commitment do not come on until ~E7, lineage tracing with Blimp1-Cre suggests that all cells in the region of the allantois that have expressed *Blimp1* become PGCs (Ohinata et al., 2005). This approach does not distinguish between cells that descend from the first to express *Blimp1* and those derived from later onset of *Blimp1* in subsequent recruits from the epiblast, hence the number of PGCs specified directly from epiblast in mice remains elusive. Several groups have attempted to identify the number of founder PGCs which actually contribute to germ cell lineages. Examination of the earliest *Blimp1* transcript found four to eight Blimp1-positive cells in most proximal epiblast which grow into a tight cluster of approximately 20 cells at the mid-streak (MS) stage (Saitou et al., 2005). Approaches using retroviruses or generating chimeras with different-colored embryonic stem cells delivered to the blastocyst have estimated that all PGCs derive from a founder population of three or four cells (Soriano & Jaenisch, 1986; Ueno, Turnbull, & Weissman, 2009). It is unclear whether the growth of cell number from 4 to 5 to the observed BLIMP1+ cells in the epiblast to the ~40 PGCs observed at e7 (Ginsburg et al., 1990) occurs through induction of new cells into the germline from the epiblast, or proliferation of the previously induced 4–5 PGCs. The mouse germline undoubtedly derives from a narrow set of ancestral founders, and the yet unclear size of this founder population has potential implications for heterogeneity of PGCs (Zheng, Luebeck, Byers, & Moolgavkar, 2005).

Further insight into the question of whether all PGC founders are equivalent will emerge from more sophisticated lineage tracing of the germline, such as genetic barcoding. A glimpse into potential heterogeneity comes from knockout studies. Genetic ablation of *Blimp1* leads to the specification of PGCs with incomplete repression of somatic genes and activation of pluripotency genes, but this phenotype varies in severity depending on the mouse genetic background (Avilion et al., 2003; Yabuta, Kurimoto, Ohinata, Seki, & Saitou, 2006). These results raise the possibility that variable levels of *Blimp1* between early mouse PGCs could lead to differences in their capacity to fully realize their fate, which could potentially impact their behavior and function. More recently, the homeobox transcription factor Otx2 was identified as an earlier repressor of the PGC fate that turns off 12–24 h before the appearance of *Blimp1*. Whereas overexpression of Otx2 decreases the number of PGCs, the absence of Otx2 augments the number of PGCs, indicating a negative correlation between PGCs and Otx2 expression (Zhang et al., 2018). Although these experiments lacked the sensitivity to detect low levels of Otx2, the question remains whether PGC precursors that incompletely extinguish Otx2 become PGCs with low levels of somatic gene expression that parlay into measurable differences in behavior (Laird, 2018).

2.2 Primordial germ cell migration and heterogeneity

In many organisms, PGCs are specified far from their permanent tissue niche, and must walk a long distance through growing and developing somatic compartments to find the gonad primordia, which will furnish a niche that promotes their survival and differentiation. To accomplish this feat, PGCs must be responsive to external migration cues as well as able to navigate their way through temporary migratory niches, recognize the gonad niche, and stop migrating when they arrive. Navigation of these challenges by each individual germ cell may amplify initially small differences, such that the migratory process reveals heterogeneity.

From the single germline progenitor P4 in *C. elegans* emerges the two PGCs: Z2 and Z3 (Sulston et al., 1983). While Z2 and Z3 themselves do not migrate per se, their development includes two relevant processes. First, cells of the somatic niche derived from Z1 and Z4 (called the distal tip cells) undertake a migration to reach the PGCs. Distal tip cells signal through Notch pathway to control proliferation of the germline (Rohrschneider & Nance, 2013), although how migration shapes their signaling abilities is unknown. Second, development of the surrounding embryo pulls the PGCs from the surface to the center of the embryo in order to reach their destination (Rohrschneider & Nance, 2013). Analogous to PGCs in other species, it remains unknown which mechanisms signal cessation to the moving cells (in this case the somatic gonad progenitors Z1 and Z4), although the presence of PGCs is required for the final positioning of Z1 and Z4 (Rohrschneider & Nance, 2013).

D. melanogaster PGCs, or pole cells, are specified in the posterior extreme of the embryo. Their migration begins by passive conveyance to the interior of the embryo during gastrulation. Upon reaching the gut epithelium, the PGCs exhibit amoeboid migratory movements triggered by gut tissue (Grubb, 2006); it is not known whether these gut-derived signals may vary depending on location or intrinsic signaling capacities of PGCs. However, functional studies of individual PGCs and their ability to respond to cues or enact an amoeboid migratory phenotype could begin to address this question. Active migration through the mesoderm niche takes PGCs to the forming gonads (Santos & Lehmann, 2004). Bifurcation from the midline toward the opposing gonads involves repulsive as well as attractive cues and takes individual PGCs on varying path lengths at varying speeds, but the extent or significance of this potential heterogeneity is not known.

The study of Zebrafish PGC migration reveals several interesting windows where heterogeneity may play a role in the development of the germline. After being specified in four random locations along the developmental axis, Zebrafish PGC morphology is smooth and round (Richardson & Lehmann, 2010). PGCs in Zebrafish must undergo a round of mitosis and a round of polarization before active migration initiates (Blaser et al., 2005), at which time they begin a directed, active movement under the guidance of chemokines (Raz & Reichman-Fried, 2006). This awakening occurs on an individual cell basis, emphasizing the potential for heterogeneity among the initial cell divisions and polarization processes (Raz & Reichman-Fried, 2006). Migratory initiation depends on the capability to mount an entirely new transcriptional program (Richardson & Lehmann, 2010), such that variations in transcription may correlate with variations in migration. Zebrafish PGCs exhibit a curious cycling of active "running" migration when they are polarized, and a "tumbling" pausing phase where they lose cell polarity and are able to make course corrections (Reichman-Fried, Minina, & Raz, 2004). The ability of an individual PGC to cycle properly between these stages is a necessity for survival, as those failing to complete these exercises become ectopic and undergo apoptosis outside of the range of survival signals. Interestingly, Zebrafish PGCs switch between individualized migration in initial phases to movement toward the gonads in clusters following aggregation at concentrated locations of the chemoattractant Sdf-1. This later stage of group migration may serve to mitigate differences in chemokine responsiveness or motility that arise earlier in Zebrafish PGC development.

Mouse PGCs begin their migration from their site of specification on the posterior side of the epiblast near the base of the extraembryonic structure known as the allantois (Ginsburg et al., 1990). This 3–4 day journey takes PGCs from the allantois, through the primitive gut and surrounding mesentery, and into the gonadal ridge. Migration is not synchronous, but spread out in space and time; some PGCs "lead" migration while others "lag" behind. Time-lapse imaging captures the individualistic behavior of PGCs as well as the fraction that migrate improperly, failing to reach the next milepost or moving into ectopic locations (McDole et al., 2018; Molyneaux, Stallock, Schaible, & Wylie, 2001). Also evidenced in the live movies are differing levels of activity from E9.0 to E9.5, when germ cells are active and motile, but not yet exhibiting the directed migration which causes them to exit the gut epithelium. What distinguishes the earliest "pioneer" germ cells first exiting the gut from other germ cells that follow them is a heterogeneity profile as of yet unknown (Cantú, Altshuler-Keylin, & Laird, 2016; Gomperts, Garcia-Castro, Wylie, & Heasman, 1994). Another point at which heterogeneous PGC behaviors emerge is during the last stages of migration at E10.5, when approximately 10% of PGCs remain outside of the gonads, with little hope of completing their migration as determined by their persistence at E13.5 (Cantú et al., 2016; D.H. Nguyen & D.J. Laird, unpublished data). Motility of mouse PGCs is regulated by multiple signaling mechanisms, including the non-canonical Wnt signaling pathway via Wnt5a (Laird, Altshuler-Keylin, Kissner, Zhou, & Anderson, 2011), which attenuates the mitogenic beta-catenin dependent arm of the pathway (Cantú et al., 2016). Although migratory mouse PGC behavior is influenced by signals from different niches, which certainly promotes heterogeneous migratory outcomes, there also appears to be intrinsic heterogeneity of PGCs.

Cultured ex vivo in defined conditions, mouse PGCs mounted variable signaling responses to cues such as Wnt5a (Cantú et al., 2016). PGCs from embryos have also been observed to be heterogeneous in expression of important signaling receptors such as cKit, which together with Kit ligand regulate migration and survival (Morita-Fujimura, Tokitake, & Matsui, 2009). Finally, it is also possible that as the mesentery niche changes and develops around them, the PGCs in the lagging group lose contact with the proper migratory signals which would enable them to move forward. Even more intriguing is the idea that migration acts as a selection metric, to prevent germ cells with deleterious or detrimental heterogeneities from reaching the gonad and colonizing. This will be discussed more in the final section of this chapter.

2.3 Proliferation of primordial germ cells and heterogeneity

PGC development involves the expansion of a small handful of founder cells to a cohort of sex-undifferentiated germ cells in the gonads, which range in number from tens in flies and worms to thousands in mice. The proliferative histories of individual PGCs have not been examined. However, discrepancies in the number of self-renewing divisions could potentially amplify any inherent heterogeneity between PGCs by over-representing or under-representing particular founders. Alternatively, unequal divisions could disproportionately increase the opportunity for replication-associated damage in subsets of germ cells, thereby introducing new heterogeneity.

We present examples in several organisms of heterogeneous behavior of PGCs during proliferation. The two *C. elegans* founder PGCs Z2 and Z3 defer proliferation until larval stages, and rely upon regulation by somatic gonad precursors (Kimble & White, 1981). Germline development is acutely responsive to nutritional cues in worms (Hubbard, Korta, & Dalfo, 2013). Upon hatching, unless larvae eat, they enter a state of L1 developmental arrest called diapause, accompanied by an arrest in the proliferation of PGCs and their somatic support cells (Fukuyama, Rougvie, & Rothman, 2006). During L1 diapause, PGCs halt their cell cycle in G2 in a precisely controlled process under the auspices of Pten and the insulin/IGF-like signaling pathway. Release from L1 diapause and G2 arrest occurs upon feeding and involves cell cycle re-entry of only one of the two PGCs, which is believed to reflect communication between the PGCs via yet unknown mechanisms (Fukuyama et al., 2006). Although it is further unknown which of the two PGCs proliferates and the basis for this decision, this proliferative asymmetry in the *C. elegans* germline could amplify any heterogeneity that exists between Z2 and Z3.

Similarly, the regulation of Drosophila PGCs during development involves differential proliferation that could introduce or augment preexisting heterogeneity. The 12 PGCs that arrive at each of the gonads double their number every 24 h during the first and second instar larval stages before curbing their expansion. Mutants possessing severely reduced numbers of embryonic PGCs—such as *oskar* and *germcell-less* which average at 2—reach the middle of 3rd larval instar with the correct number of PGCs. These mutants revealed a cross-talk between PGCs and somatic cells of the ovary via EGF signaling that senses and adjusts the number of PGCs (Gilboa & Lehmann, 2006). As in other organs, the size of the PGC pool in the larval fly ovary relies upon Hippo signaling between PGCs and somatic cells that form their niche (Sarikaya & Extavour, 2015). These compensatory mechanisms for PGC proliferation indicate that the Drosophila germline can be built from different numbers of founder PGCs and through differing numbers of divisions.

Such mechanisms for detecting and adjusting cell numbers similarly govern the proliferation of PGCs in chicken, where compensatory expansion was observed after treatment with the anti-mitogen busulfan (Lee et al., 2013). In mice, PGCs were deemed to undergo approximately eight divisions based on a population doubling time of 16 h, with compensation for early losses in the number of PGCs occurring during the period of expansion in the fetal gonads (Tam & Snow, 1981). During their migration, the proliferation of PGCs in the mouse embryo increases with each successive niche through which they transit (Cantú et al., 2016). PGC mitosis is regulated by beta-catenin-dependent Wnt signaling pathway but instructed by the niche.

This extrinsic and niche-specific control of PGC proliferation, coupled with the asynchrony in PGC migration, raises the possibility that those migrating first or fastest would undergo more cycles of expansion, leading to a clonal dominance over those migrating last or slowest (Cantú et al., 2016). Although the number of cell divisions has been proposed to be deterministic in a number of processes—including the decision to enter meiosis (McLaren, 2003; McLaren & Southee, 1997; Ohkubo, Shirayoshi, & Nakatsuji, 1996), the probability of acquiring mutations (Ohno, 2018), and the quality of the oocyte (Henderson & Edwards, 1968)—the functional impact of proliferative history on the germline and gamete remains to be determined. The advent of molecular "flight recorders" that track cell divisions in vivo hold promise for revealing the extent and the consequences of heterogeneous proliferation in PGCs.

2.4 Primordial germ cell survival and heterogeneity

As PGCs progress through fetal development, many fail to complete the entire journey and are eliminated from contributing to the future germline. This elimination occurs mainly via developmentally programmed apoptosis at several stages in the fetal period. Importantly, this apoptosis is separate from PGC death in response to cytotoxic insults, such as irradiation or chemical exposure. Consistent PGC apoptosis during development suggests that apoptosis may act as a selective event. The PGCs that do survive fetal development represent the founding population that establishes the adult germline. Differences in PGC survival during this fetal period are therefore critical to determining the composition of gamete-producing cells in adulthood. Given that only a subset of the initial PGC population completes fetal development, survival represents a heterogeneous outcome that may reflect important differences between survivors and eliminated PGCs that are relevant to germ cell function.

Apoptosis is an invariant feature of *C. elegans* embryo development, with 131 out of 1090 somatic cells eliminated in a lineage-specific manner (Gartner, Boag, & Blackwell, 2008). In the germline, apoptosis occurs exclusively in the adult during oogenesis and eliminates a variable number of potential oocytes that does not exceed 50% (Gumienny, Lambie, Hartwieg, Horvitz, & Hengartner, 1999). Importantly, germline apoptosis differs from somatic apoptosis by virtue of the single oocyte lineage containing two variably sized but distinctly fated populations: surviving versus eliminated cells. In contrast, somatic apoptosis reproducibly eliminates precisely the same 131 cells.

Oocyte apoptosis can be classified into two categories: stress-induced and physiological. Stress-induced apoptosis occurs in response to insults such as environmental stress, DNA damage, or bacterial infection. In contrast, physiological apoptosis affects a fraction of cells during oogenesis in the absence of any stress and differs from stress-induced apoptosis by not relying on pro-apoptotic BH3 proteins. Physiological apoptosis is restricted to oocytes in late pachytene that occupy a particular region of the gonad. Inhibition of germline apoptosis results in poorer quality oocytes as characterized by decreased egg viability and difficulty producing full-sized oocytes (Andux & Ellis, 2008). These reproductive defects may be due to intrinsic faults in the normally eliminated population that are not present in the oocytes that survive physiological apoptosis. Alternatively, apoptosis may not reflect intrinsic cellular differences between oocytes in the survivor versus dying populations but instead functions as a resource-management strategy. *C. elegans* germ cells are linked as a syncytia, allowing apoptotic cells to act as nurse cells and direct their cytoplasmic contents toward surviving cells (Wolke, Jezuit, & Priess, 2007). These two models are non-exclusive; cellular differences stemming from oocyte heterogeneity can serve as the basis for determining whether an oocyte acquires a nurse cell or survivor fate.

Drosophila germ cell development shares many conserved elements with higher-order organisms such as migration. Drosophila PGCs, also termed pole cells, that fail to migrate accurately undergo programmed cell death in the early embryo (Coffman et al., 2002). As many as half of the initial cohort of PGCs mismigrate and are ultimately eliminated (Underwood, Caulton, Allis, & Mahowald, 1980). Ectopic PGC removal does not require apoptosis genes *grim*, *reaper*, or *hid* and is caspase-independent (Sano, Renault, & Lehmann, 2005). Several studies have demonstrated that elimination of ectopic PGCs instead requires a pathway involving both *p53* and *out* (Coffman et al., 2002; Yamada, Davis, & Coffman, 2008). Overexpression of *p53* induces programmed cell death only in ectopic PGCs, suggesting that the threshold for elimination varies across subpopulations of PGCs that differ by migratory success. Differentiation is another criterion that can distinguish surviving PGCs from eliminated ones. In studying Nanos (*nos*), an evolutionarily conserved protein critical for many aspects of germline development, Hayashi et al. discovered that *nos*-mutant PGCs inappropriately acquire somatic identities and are subsequently removed through apoptosis (Hayashi, Hayashi, & Kobayashi, 2004). Interestingly, despite the central role of *nanos* in securing germline identity, *nos*-mutant PGCs only display a partially penetrant phenotype of somatic expression (Schaner, Deshpande,

Schedl, & Kelly, 2003), suggesting that heterogeneity exists among PGCs in regard to retaining somatic competence. Given the role of *nanos* in regulating histone modifications, heterogeneous chromatin states among individual cells may explain this variation in differentiation propensity. To summarize, programmed cell death in Drosophila operates through both an apoptosis-independent and dependent mechanism, with apoptosis being associated with PGCs that inappropriately differentiate to the somatic lineage.

In Zebrafish, PGC apoptosis also is linked to migration, with disrupted migration leading to apoptosis. Weidinger et al. injected morpholinos targeting *dead end* mRNA, a component of germ plasm in zebrafish, and observed an increase in ectopic PGCs that would ultimately be eliminated by apoptosis (Weidinger et al., 2003). Apoptosis was preceded by downregulation of key germ cell maintenance genes such as *nos-1*, indicating that mismigration disrupts germ cell identity in ectopic PGCs, leading to downstream apoptosis. Apoptosis is the eventual fate of ectopic PGCs in other migratory mutants (Köprunner, Thisse, Thisse, & Raz, 2001; Ramasamy, Wang, Quach, & Sampath, 2006) and may also be due to loss of *nos-1* in ectopic PGCs. Importantly, in these migratory mutants some PGCs do correctly localize to the gonads, maintain *nos-1* and its protein product Nos1, and survive, suggesting that correct migration supports the maintenance of germ cell identity and provides a survival advantage.

PGC survival in mice follows a similar paradigm to that in Drosophila and Zebrafish. Correct PGC migration is tightly linked to PGC survival. This coupling is primarily accomplished through genes that have dual roles in chemotaxis as well as survival. In mice, mutants for Dead-end (Dnd) and KitL (Steel) result in increased numbers of ectopic PGCs, which subsequently undergo apoptosis (Runyan et al., 2006; Youngren et al., 2005). KitL is downregulated in the midline after PGCs migrate through the tissue by E10.5, enforcing a narrow pro-survival niche (Runyan et al., 2006). The stringent connection between survival and migratory signals resolves migratory heterogeneity in favor of PGCs that can accurately navigate to the gonad. Because ectopic PGCs can give rise to germ cell tumors, this linking of migratory success to survival can prevent deleterious differentiation from mismigrated PGCs (Runyan, Gu, Shoemaker, Looijenga, & Wylie, 2008).

After reaching the gonads, mouse PGC apoptosis occurs primarily through a *Bax*-mediated mechanism peaking by E13.5 in both males and females (Coucouvanis, Sherwood, Carswell-Crumpton, Spack, & Jones, 1993; Wang, Nakane, & Koji, 1998). PGCs begin to upregulate transcripts of *Bax*, increasing their apoptotic potential between E10.5 through E13.5 (Stallock, Molyneaux, Schaible, Knudson, & Wylie, 2003). The Bcl-2 family members such as Bcl-x (Rucker et al., 2000) further regulate PGC survival. Variation in the expression and balancing of these apoptotic and survival pathways can modulate cell death propensity and could give rise to the apoptotic fraction observed even in wild-type backgrounds. Interestingly, successful differentiation is associated with increased survival. Knockout of *Nanos3*, a germline transcription factor important for germline differentiation, results in PGC depletion by *Bax*-dependent and -independent mechanisms (Suzuki & Saga, 2008). However, a subset of *Nanos3* null PGCs are capable of surviving and differentiating in male and female adult gonads, suggesting that a subset of PGCs is more robust and capable of survival despite the loss of pro-differentiation components.

While studies of mutant PGCs have illuminated much of the differentiation and survival networks operating in the fetal period, the fraction of wild-type PGCs that undergo apoptosis indicate that not all PGCs are equivalent in their likelihood to contribute to gametogenesis. The identities of apoptotic germ cells on a wild-type background are beginning to be understood through techniques such as single-cell RNA sequencing and clonal labeling, both of which can elucidate subpopulation-level differences among germ cells that bulk analyses cannot. In particular, clonal labeling allows for direct comparisons among potentially heterogeneous clonal subpopulations. Multicolor clonal labeling revealed that apoptosis in the fetal male germline is clonal, suggesting that mitotically heritable differences among subpopulations—particularly in the expression of epigenetically regulated genes that control sex differentiation—can determine whether germ cells survive or are eliminated (Nguyen & Laird, 2019).

2.5 Differentiation of PGCs and heterogeneity

After specification, the next major differentiation event in PGC development is sex differentiation. Of the common model organisms, *C. elegans* is unique in featuring hermaphroditic gametogenesis, although genera-principles regarding somatic regulation of germline differentiation can apply. For example, *C. elegans* somatic cells at the distal tips of the gonads maintain mitotic potential, and meiosis only occurs as germ cells move away (Ellis, 2008). In *Drosophila* and mice, the somatic gonadal sexual identity significantly influences the sex differentiation of resident germ cells

(Casper & Van Doren, 2006; Ewen & Koopman, 2010). This section will primarily discuss differentiation of male and female germ cells in mice, with a focus on meiotic initiation and mitotic arrest.

Upon arriving in the gonad, PGCs undergo sexually dimorphic differentiation into the highly specialized male and female lineages. The signal for meiotic initiation in mice is retinoic acid (RA), which is secreted by somatic cells in the mesonephros (Bowles et al., 2006; Koubova et al., 2006). As RA must diffuse to germ cells in the ovary, the diffusion gradient could produce variation in both strength and timing of RA signaling. Meiosis in the mouse ovary is observed to occur in an anterior to posterior wave that may reflect asynchronous initiation (Menke, Koubova, & Page, 2003). Interestingly, this anteroposterior wave does not align with the axis of diffusion between the mesonephros and the ovary, suggesting that other signals may modulate RA signaling within the ovary. RA has also been discovered to be produced by ovarian cells (Bowles et al., 2016; Childs, Cowan, Kinnell, Anderson, & Saunders, 2011) and the Aldh1a1 enzyme. Meiotic progression involves Stra8, which is activated by Dmrt binding to the Stra8 promoter (Matson et al., 2010). *Stra8* mutants display greatly reduced follicle formation, but a small population of *Dmrt1*-null oocytes can continue to form functional follicles and express the appropriate markers of later meiosis, including Stra8, Sycp3, and Msy2. This meiotically competent subpopulation represents more robust oocytes that do not require Dmrt1 to complete meiosis, possibly due to a compensatory meiotic program that is only activated in a subset of oocytes. Double knockout mice that lack both *Stra8* and *Msx1/2*, another activator of *Stra8*, also contain a subpopulation of oocytes that can still complete meiosis (Le Bouffant et al., 2011). The variation in meiotic robustness suggests that heterogeneity exists during female sex differentiation to endow certain oocytes with greater meiotic potential.

Male sex differentiation in mice is initiated at E12 by supportive somatic cells in the nascent testis called Sertoli cells (Ohta et al., 2012). Sertoli cell primarily regulates production of female-inhibitory and male-inducing factors such as Fgf9 and Cyp26b1 to suppress the female meiotic fate (Bowles et al., 2010, 2006). Male germ cells continue to proliferate in the testis but undergo mitotic arrest as early as E12.5. This timing is asynchronous, as some germ cells are still proliferating by E14.5 (Western, Miles, van den Bergen, Burton, & Sinclair, 2008). Mitotic arrest is accompanied by expression of *Nanos2*, a key male sex differentiation marker (Suzuki & Saga, 2008). Delayed mitotic arrest is associated with aberrant maintenance of pluripotency, decreased *Nanos2*, and an increased susceptibility for teratoma formation (Dawson et al., 2018; Heaney et al., 2012). Heterogeneity in the timing of mitotic arrest can expose later-arresting male germ cells to these deleterious fates. Mitotic arrest is part of many processes regulated by complex networks of secreted factors such as Tgfβ and Nodal (Spiller, Burnet, & Bowles, 2017; Spiller, Koopman, & Bowles, 2017) active in germ cells during this period. Heterogeneous expression of these factors as well as of their receptors on germ cells can significantly modulate the differentiation states of male germ cells, leading to vastly asynchronous or divergent cell fates.

An overarching theme during sex differentiation is the variation in both position and timing of these events. Many secreted factors regulate this process in males and females, and the precise signaling environments remain to be defined. Heterogeneous progression through sex differentiation can also be due to PGCs themselves differing in their capacity to respond to differentiation cues. Sex differentiation has been shown to be facilitated by epigenetically regulated genes that are activated by demethylation (Hill et al., 2018). Individual variation in the extent of demethylation particularly at these loci could lead to highly variable differentiation potentials. Thus, initial cellular differences in epigenetic state could be amplified as epigenetically heterogeneous subpopulations of PGCs proceed through sex differentiation asynchronously, or fail to do so at all. The later impact of differentiation heterogeneity can be significant, affecting reproductive and tumorigenic potential (Hunt & Hassold, 2008).

3. Determinants of heterogeneity

In this section, we consider the potential causes for diverse behaviors among PGCs described above. We separate these causes into those which are genetically encoded, those which arise from traditional epigenetic marks such as DNA methylation, chromatin modifications, or small RNAs, and a third and separate category of non-genetic determinants (summarized in Fig. 3). Among these sources, we distinguish between those which are permanent (and therefore heritable to all cellular progeny), those which are transient and do not cause similar behaviors in cellular progeny, and those determinants which are semi-permanent, or lasting in daughter cells but not for infinite cellular generations. Finally, we discuss mechanisms that mitigate variation between germ cells, or buffers of heterogeneity. For an outstanding taxonomy of the sources of heterogeneity, refer to Huang (2009).

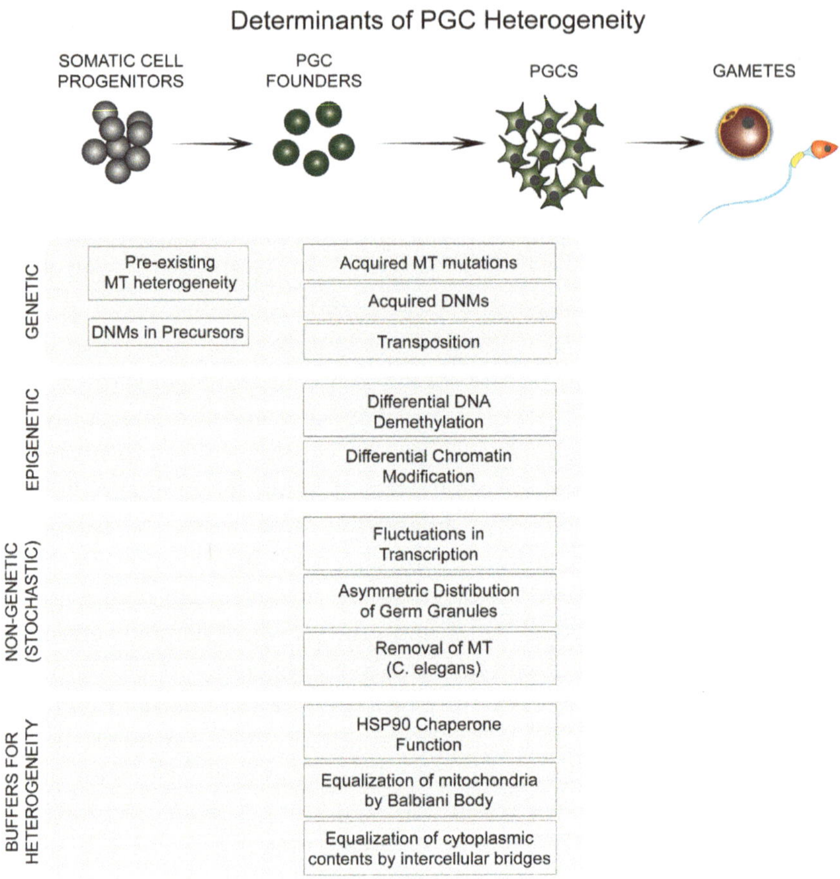

Fig. 3 Determinants of PGC Heterogeneity. For the various cell types, different underlying cell states can be causes for heterogeneity, impacting cell type transition as well as impacting the fitness of progeny.

3.1 Genetic determinants

3.1.1 Acquired genetic changes

De novo mutations (DNMs) in the DNA sequence typically originate during cell division. Although DNMs occur in all cells, only those in the germline are inherited to offspring. The Darwinian model of evolution by random mutation, causing a phenotypic change which undergoes natural selection in fact originates in germ cells, or potentially in the early embryo of animals such as mice which induce PGCs from somatic cells. Although the diversity of multicellular organisms is living testament to DNMs in the germline, the rate of mutation is low: approximately 10^{-9} to 10^{-10}

mutations per nucleotide pair per cell division in eukaryotic cells (Arana & Kunkel, 2010; Drake, Charlesworth, Charlesworth, & Crow, 1998). Studies across mice and humans suggest that DNMs arise in the germline at an even lower frequency than in somatic cells (Kohler et al., 1991; Milholland et al., 2017; Walter, Intano, McCarrey, McMahan, & Walter, 1998).

The observed rate of mutation is the output of an equation that includes incurred errors, DNA repair, and elimination of cells. For studies that infer germline mutations by examining progeny, gamete fitness is also included in the rate calculation. The recent generation of exome and whole genome data in human families has provided precise measurements of germline DNMs, with single-nucleotide variations (SNVs) arising at 1.2×10^{-8} per nucleotide per generation (Besenbacher et al., 2015; Campbell et al., 2012; Conrad et al., 2011; Jónsson et al., 2017; Kong et al., 2012; Lynch, 2010; Michaelson et al., 2012; Rahbari et al., 2016; Roach et al., 2010; Ségurel, Wyman, & Przeworski, 2014; Wong et al., 2016). The rate of germline mutation in humans as well as non-human primates (Pfeifer, 2017) is nearly an order of magnitude greater than that measured in mice (5.4×10^{-9} per nt/generation; Uchimura et al., 2015), birds (4.6×10^{-9}/nt/generation; Smeds, Qvarnstrom, & Ellegren, 2016), or Drosophila (2.8×10^{-9}/nt/generation; Keightley, Ness, Halligan, & Haddrill, 2014). This difference most likely stems from the greater number of cell divisions that occur in the human germline and specifically in the spermatogonial stem cells, which divide throughout life, in contrast to oocytes which cease replication in the fetus (Crow, 2000). Indeed, sequencing of parents and offspring revealed that the majority of human DNMs arise in the paternal germline and increase with paternal age by 1–2 per year (Campbell et al., 2012; Jónsson et al., 2017; Kong et al., 2012; Michaelson et al., 2012; Rahbari et al., 2016). However, with an estimated 10 cellular divisions from the zygote to PGC in mammals (Rahbari et al., 2016), the theoretical number of incurred mutations across the genome of each PGC ranges from ~2 to 15 for mice and ~4 to 38 for humans; hence DNMs do not account for the extent of heterogeneity in cellular behavior that is observed.

A second source of acquired mutations in PGCs is the mitochondrial genome. Separate from the nuclear genome, the circular mitochondrial DNA molecule produces 13 mRNAs, which all encode oxidative phosphorylation components, as well as 2 rRNAs and 22 tRNAs needed for mitochondrial translation of these mRNAs (Falkenberg, Larsson, & Gustafsson, 2007). Critically, the rate of mutation of the mitochondrial genome is 2- to 20-fold greater than the nuclear genome, due to replication errors (Allio, Donega, Galtier, & Nabholz, 2017). The presence of more than one mitochondrial haplotype within the same cell, or heteroplasmy, arises in cells because fidelity of the mitochondrial polymerase is poor and varies depending on biological condition; for example, the presence of reactive oxygen species (ROS) decreases the fidelity of the mitochondrial polymerase, while the presence of a proofreading subunit increases fidelity (Longley, Nguyen, Kunkel, & Copeland, 2001; Pinz, Shibutani, & Bogenhagen, 1995). Although mitochondrial replication is paused in the early mouse embryo until the time of gastrulation (Cao et al., 2007; Wai, Teoli, & Shoubridge, 2008), in the course of PGC development, estimates suggest the number of mitochondria per cell increases from 100 at E7.5 to over 6000 at E14.5 (Wai et al., 2008). Given the size of the mitochondrial genome, mutation rate, and rate of mitochondrial expansion, approximately 35 mitochondrial DNMs would accumulate by E14.5 in mice, although this does not account for unequal replication. However, as with nuclear mutations, this rate of acquired mutation in the mitochondrial genome is likely a very small contribution, at best, to observed heterogeneity of PGCs.

A third potential source of acquired mutation in PGCs is through retrotransposition. Transposable elements comprise a significant fraction of the genome and many of the evolutionarily young Alu and autonomous L1 (or Long Interspersed, LINE-1) elements remain active and capable of re-inserting into new sites in the genome (Kazazian & Moran, 2017). During development, the period of epigenetic reprogramming in PGCs de-silences transposable elements, leading to their expression and potential for transposition in the window of development before transposons are silenced by piRNAs (Brennecke et al., 2007; Teixeira et al., 2017). In humans, new L1 and Alu integrations were detected in ~1/100 and 1/20 births, respectively (Hancks & Kazazian, 2012). This rate is even higher in mice, where new L1 insertions occur in 1/8 births (Richardson et al., 2017). However, on a cellular level, given its rate of occurrence, transposition again makes an exceedingly minor contribution to PGC heterogeneity.

3.1.2 Pre-existing genetic changes

A special case for animals that specify their germline by induction is the potential for prior genetic mutations as a source for heterogeneity. In mice, DNMs arising in the epiblast would lead to mosaicism among the founder PGCs as well as in somatic lineages. Cells in the mouse epiblast undergo rapid divisions—measuring as little as 3h (Snow, 1977)—hence providing opportunity for errors in replication. Furnishing evidence of the elevated potential for damage and repair during this period of rapid division in the early mouse embryo, PGCs and cells of the epiblast have been shown to be hypersensitive to genotoxic stress as measured by DNA breaks, Atm and p53-dependent apoptosis (Heyer, MacAuley, Behrendtsen, & Werb, 2000). From genome sequencing of human families, it has been imputed that approximately 10 divisions occur between the zygote and the specification of PGCs (Ohno, 2018; Rahbari et al., 2016); and this number is probably similar in mice. Detection of DNMs in offspring that co-exist at low levels in the blood of one parent provides evidence that replication errors in cells of the early embryo before PGC specification are present in a subset of somatic and germline derivatives. Data in humans estimate that 0.2–0.6 DNMs accumulate per cell division in the pre-PGC embryo (Rahbari et al., 2016), which would suggest a theoretical maximum of six pre-existing mutations among PGC founders could contribute to heterogeneity.

A second type of pre-existing genetic mutation that could contribute to heterogeneity among PGCs is mitochondrial heteroplasmy (Mishra & Chan, 2014). Mitochondrial heteroplasmy and mutations in mitochondrial (mt) DNA arise in oocyte maturation due to the inherently higher error rate in mitochondrial polymerase (Allio et al., 2017; Khrapko et al., 1997; Madsen, Ghivizzani, & Hauswirth, 1993). During early embryogenesis, the replication of mtDNA and fission of mitochondria ceases (~E1.0–E3.0 in mice) (Cao et al., 2007; Wai et al., 2008), so that each cell division segregates existing mitochondria and produces an increasingly purer pool of mtDNA. Known as a mitochondrial bottleneck, this segregation limits the diversity of mitochondrial genomes inherited from the oocyte (Mishra & Chan, 2014; Stewart & Chinnery, 2015). As PGC specification in mice occurs during this bottleneck, the founders inherit different mtDNA cohorts. Replication of mitochondria resumes after PGC specification, amplifying the heterogeneity between cells that was created during the bottleneck. Together these phenomena may reveal mtDNA mutations by altered PGC phenotypes.

The potential for revealing a mutation in a mitochondrial genome increases as heteroplasmy from the oocyte trends to homoplasmy (similarity in mtDNA) in PGCs. Preexisting or de novo mtDNA mutations would be inherited in a minority of PGCs during specification or could arise during PGC development; either way, those cells with a high concentration of mitochondria bearing deleterious mutations would exhibit altered phenotypes. Concomitant with the onset of PGC migration, metabolism within the embryo shifts from glycolysis to oxidative phosphorylation, which provides a potential source of selective pressure for cells unable to make the switch (Folmes & Terzic, 2014). Deleterious mitochondrial DNA mutations within PGCs may be revealed as compromised ability to initiate oxidative phosphorylation or the bioenergetic response required for migration. Little is known about the cellular or molecular mechanisms for mitochondrial organelle quality control and selection, but the migratory time-point may be particularly important for early embryonic metabolism. A new area of inquiry concerns the selective destruction of mitochondria in PGCs by mechanisms such as mitophagy. Recent work describes an unprecedented means for removing mitochondria from PGCs in *C. elegans*. This unique cellular surgical procedure is carried out by somatic endoderm cells via extrusion of cytoplasmic lobes from the PGCs (Abdu, Maniscalco, Heddleston, Chew, & Nance, 2016). Although speculative, the effect could serve to concentrate other contents such as germ granules or to remove faulty mitochondria in order to increase cellular fitness, reduce oxidative damage, or alter metabolic capacity of PGCs.

3.1.3 Epigenetic determinants

Having focused on the genetic determinants of heterogeneity, we turn our attention to the determinants not directly linked to the genome itself (Brock, Chang, & Huang, 2009). Our interest in the epigenetic determinants of heterogeneity encompasses both transcription regulation through modification of histone and chromatin structure, as well as the molecular states defined by covalent modifications of DNA. We note that epigenetic determinants have the capacity for altering PGC states semi-permanently, as many of the mechanisms below are preserved through cell divisions.

Reprogramming of the epigenome occurs in two main stages in the fetal germline: first, during PGC migration during which methylation is maintained at specific regions; and second, after gonadal colonization when epigenetic marks are erased to create a hypomethylated epigenetic ground state (Hackett & Surani, 2012; Seisenberger et al., 2012). At day 8 of mouse development, PGCs begin a significant loss of genome-wide DNA methylation that is considered the first of three waves of demethylation before E13 (Bohacek & Mansuy, 2015; Hackett & Surani, 2012; Hajkova et al., 2002; Seisenberger et al., 2012). Changes in histone modifications of PGCs occur starting at E9.0 include loss of histone H3-lysine (K) 9 dimethylation and increase in H3K27 trimethylation (Seki et al., 2005, 2007). These epigenetic changes occur during migration (Nikolic, Volarevic, Armstrong, Lako, & Stojkovic, 2016) and before sex differentiation, and are thought to carry out the following: ensure that the germline genome is ready to receive parent-of-origin specific epigenetic marks; erase potential epimutations incurred in early development; and confer a state of pluripotency for the embryo (Hajkova et al., 2002; Seki et al., 2007).

As epigenomic studies to date have analyzed PGCs in bulk, the variation in epigenetic states of single cells is not known. In principle, differences in the timing of epigenetic reprogramming or failure to lose cytosine methylation or histone marks—epimutations—could lead to PGC heterogeneities. Such differences could be manifested as altered expression of genes critical to differentiation. In support of this idea, PGC maturation genes such as *Mvh* (Vasa) and the meiotic component *Sycp3* are regulated by methylation at their promoters (Maatouk et al., 2006) and would be sensitive to epimutations. Recent studies have shown that efficient sex differentiation is dependent upon demethylation at the loci of 45 germline reprogramming-responsive genes (GRRs; Hill et al., 2018). Any individual cell differences in the extent of demethylation at these GRRs could stratify PGCs by differentiation capacity and produce greater population variation as individual PGCs heterogeneously differentiate.

Demethylation during epigenetic reprogramming can also relieve inhibition of transposon sequences that were previously repressed by methylation. Retrotransposition activity from these unsuppressed elements can lead to genomic disruption and inappropriate gene expression, further exacerbating initial variation among germ cell subpopulations. Accordingly, germ cells that fail to repress transposons (Kuramochi-Miyagawa et al., 2004; Malki, van der Heijden, O'Donnell, Martin, & Bortvin, 2014; Soper et al., 2008) arrest prior to gametogenesis and are eliminated, indicating that dysregulated transposons are stringently selected against. Single-cell assessments of transposon expression have identified LINE-1 elements as differentially expressed within both mouse and human PGCs at individual timepoints, which could reflect cell-to-cell variability in repressive methylation (Nguyen & Laird, 2019; Reznik et al., 2019).

A second mechanism of non-genomic transcriptional regulation to be addressed as a potential source of PGC heterogeneity is small non-coding RNAs, including micro RNAs (miRNAs), endogenous small interfering RNAs (*endo*-siRNAs), and piwi-interacting RNAs (piRNAs). In *D. melanogaster* and *D. rerio*, miRNAs function to silence germline enriched mRNAs in cells which are destined to be a part of a somatic fate (Banisch, Goudarzi, & Raz, 2012). In the mouse, the suppression of the miRNA *let-7* by *lin-28* is necessary for expression of critical PGC specification genes like *Blimp1/Prdm1*, and without *lin-28* the germline fails to form (West et al., 2009); this kind of binary signal requirement during specification does not invite much heterogeneity. However, it is not known whether incremental differences of *line-28* can lead to phenotypic consequences. Recent work has begun to characterize miRNA expression in the mouse at later developmental times such as sex differentiation, finding different sexually and developmentally regulated populations of miRNAs (Fernández-Pérez, Brieño-Enríquez, Isoler-Alcaraz, Larriba, & Del Mazo, 2018). Examining miRNA heterogeneity at a single cell level is becoming possible with new technology (Hagemann-Jensen, Abdullayev, Sandberg, & Faridani, 2018), and variability in the expression of miRNAs has recently been speculated to lead to non-genetic cell-to-cell heterogeneity outside of the germline (Wang et al., 2019). Since dysregulation of miRNA expression has been linked to testicular cancer (McIver, Roman, Nixon, & McLaughlin, 2012), the possibility of heterogeneous expression of miRNAs in the early germ cell populations merits exploration.

piRNAs are small RNAs that are nearly exclusive to the germ line which silence transposable elements in the genome (Banisch et al., 2012). Although transposition events are rare and therefore not a significant source of germ cell heterogeneity, heterogeneous expression of transposable elements such as intracisternal-A particle (IAP) family of endogenous retroviruses could lead to phenotypic readouts. Recently identified additional functions for IAPs in regulating the transcription of neighboring genes (Vasiliauskaitė et al., 2018) raise the possibility that the de-repression of IAPs could lead to inappropriate gene expression in PGCs as well as their derivative gametes.

The wave of demethylation which germ cells undergo during their development potentially enables transposable elements relocation to deleterious effect. During this period, *endo*-siRNAs are considered a secondary defense to silence transposable elements by degrading transcripts. Endo-siRNAs have been studied in *C. elegans*, *D. melanogaster*, and mouse, and also function to prevent viral infection (Banisch et al., 2012). By nature, *endo*-siRNAs are heterogeneous in their response to an infection or transposition event, as these occur on a cell-to-cell basis. However, the ability of an individual germ cell to mount an *endo*-siRNAs has not yet been investigated, as the contribution of endo-siRNA to the suppression of transposons is just beginning to be appreciated (Berrens et al., 2017).

3.1.4 Non-genetic determinants

Distinct from alterations to the DNA or epigenetic mechanisms discussed above, non-genetic determinants can lead to measurable differences between cells of a single type. Bacteriologists studying genetically identical, clonal populations cultured in uniform conditions first described "non-genetic cell individuality" based on cell-to-cell variations in viral yield (known as burst size; Delbruck, 1945) and chemotactic behavior (Spudich & Koshland, 1976). These heterogeneities were ascribed to the interaction of stochastic fluctuations in molecules within the cell (Spudich & Koshland, 1976). In the field of cancer biology, non-genetic causes of cellular heterogeneity have been parsed into "population noise" and "temporal noise." Temporal noise derives from the fluctuations of protein levels within a cell over short durations. Population noise, on the other hand, reflects different states of cells at a given moment. The broad assumption is that for genetically identical cells, population noise merely reflects temporal noise, and that all cells experience a similar profile of fluctuations over time (Brock et al., 2009).

One such generator of fluctuation may be the very nature of transcription, which was revealed to be pulsatile, occurring in bursts that emerge at locus-specific rates (Chubb, Trcek, Shenoy, & Singer, 2006; Raj, Peskin, Tranchina, Vargas, & Tyagi, 2006; Suter et al., 2011). In germ cells, stochastic variation due to transcription may be particularly relevant, given that an elevated rate of global transcription was recently reported in mouse fetal germ cells (Percharde, Wong, & Ramalho-Santos, 2017). This result raises the possibility that non-deterministic effects due to timing and transcription may be magnified in the developing germline, although there are currently no studies that have rigorously compared the absolute levels of transcripts in germ cells across time to capture such cycles. It is not yet clear how variation in levels and rate of transcription relates to the level of proteins, although imminent developments in technologies for single cell proteomics will be revealing.

A key assumption of the stochastic processes that generate fluctuations in molecules is that they are not permanent; although a cell may remain in a particular state, its daughter cells will not remember that state. Remarkably, there are documented instances of cells in culture maintaining levels of particular proteins through mitosis. In one example, variations in the levels of apoptotic regulators were maintained through 1–2 cellular generations of human mammary epithelial cells, so that the response to extrinsic death signals was similar between mother and daughter cells (Spencer, Gaudet, Albeck, Burke, & Sorger, 2009). In the early mouse embryo, the levels of cMYC are variable between cells, and inherited through divisions. As cMyc levels determine the capacity of pluripotent cells in the epiblast to outcompete their neighbors by inducing programmed cell death, this observation indicates that cell fitness states are remembered transiently through mechanisms that remain to be elucidated (Diaz-Diaz et al., 2017). We can conceive of scenarios by which non-epigenetic, stochastic heterogeneity could lead to permanent genetic changes; for example, variability in the levels of DNA repair machinery could render some cells more susceptible or resistant to acquired mutations.

A final non-genetic source of heterogeneity we consider is the germ granule. A highly conserved feature of germ cells, these cytoplasmic granules consist of aggregates of mRNA, small RNA, and proteins (Seydoux & Braun, 2006). In *C. elegans*, *Drosophila*, and *Xenopus*, these granules (known as polar granules, P granules or germ plasm) segregate to PGCs during specification and are required for specification and function of germ cells (Gallo et al., 2010; Hay, Ackerman, Barbel, Jan, & Jan, 1988; Strome & Wood, 1983). In mammals, PGCs synthesize components of germ granules rather than inherit them prior to specification, although these germ granules are then passed on with each cell division. Mammalian germ granules resemble processing bodies in somatic cells, which contain mRNA decay machinery. By delivering granule-associated RNAs that can regulate gene expression, germ granules can produce heritable changes in recipient cells independent of changes to DNA sequence, DNA methylation, or chromatin state. Heterogeneity could be imparted by the composition of the germ granules that are transmitted. The contents of germ granules can be modulated according to different environmental conditions as a stress response mechanism (Snee & Macdonald, 2004, 2009). Alternatively, the number of germ granules can be asymmetrically divided between two daughter cells. In *Drosophila*, the amount of germ plasm as measured by *wun2* mRNA levels can vary within a field of cells, with central cells receiving more germ plasm than peripheral cells (Slaidina & Lehmann, 2017). This asymmetry provides central cells with relatively higher amounts of maternal mRNA and is demonstrated to provide a survival advantage over peripheral cells.

3.2 Extrinsic sources of heterogeneity: Environment, niche, and timing

The necessity of the tissue niche for proper development of cells has been widely recognized across biology, and the germline niche is no exception. Unlike the well-characterized cooperative signaling and maintenance of cell fate of the adult *Drosophila* stem cell niche, the notion that PGCs receive instruction from their niche during specification, migration, and colonization is less established. Across many organisms, the developing embryo changes drastically from the time of PGC specification to the time that functional gametes are produced. The idea that the landscape with which the developing germ cells must interact changes constantly is not new. It is known that fly PGCs receive different levels of signaling from different niches (Stepanik et al., 2016). How the changing niche may contribute to heterogeneity is only just beginning to be appreciated.

PGC migration in mice exhibits spread over space and time. Egress of PGCs from the hindgut into the surrounding mesoderm occurs asynchronously, with "pioneers" emerging first; chemotaxis toward the gonads occurs at varying speeds; and colonization of the gonads takes place over >24h, with late arrivals remaining in the mesentery (Molyneaux et al., 2001). A recent study found that the rate of proliferation of PGCs increases during migration and is regulated by the successive niches through which they transit (Cantú et al., 2016). The progressive increase of mitotic signals along the migratory route implies that those PGCs that move through the series of niches first would gain a proliferative advantage. Upon reaching the gonads, the earlier or faster PGCs would have generated a larger number of clonal progeny as compared to the later or slower migrants. It is here that the heterogeneity of PGCs may be influenced by the environment, as cells which are capable of efficient migration are rewarded by enhanced clonal expansion.

3.3 Buffers of heterogeneity

In contrast to the determinants discussed above, which potentially generate and reinforce differences between individual cells, we now consider the opposite: those mechanisms that could mitigate differences between PGCs. Although we propose that a certain level of PGC heterogeneity could increase the overall fitness of the germline through selection (as discussed below), there are several ways that germ cell diversity could be unfavorable. First, genetic mutations or epi-mutations in PGCs that potentially promote the initiation, growth, or metastasis of tumors by rendering PGCs more resistant to apoptosis, more proliferative, or able to migrate would be dangerous in gametes. Such PGC variants could out-compete other PGCs and even homogenize the gamete pool. Second, the phenotypes that increase the fitness of a PGC are not necessarily the same phenotypes that benefit a gamete or embryo. Finally, too much tolerance for heterogeneity may have consequences for the overall fertility of the organism over a lifetime, such as early loss of fertility or lack of consistency of gametes. In this light, we point to evidence for mechanisms that buffer heterogeneity in PGCs.

Hsp90 is a highly conserved chaperone that assists with the folding of proteins. It has been dubbed an "evolutionary capacitor" because its absence in a number of organisms leads to enhanced phenotypic variability that is believed to arise from cryptic protein variants that are normally degraded or chaperoned by Hsp90 (Rutherford & Lindquist, 1998; Specchia et al., 2010). Most cell types express Hsp90b and elevate expression in stressful conditions, but an A isoform is exclusive to germ cells (Nishikawa & Kinjo, 2018). Hsp90a was further linked to germ cell development through its interaction with the germ plasm component and piRNA biogenesis factor Piwi (Gangaraju et al., 2010). We posit that Hsp90a is a likely candidate to buffer heterogeneity in PGCs during their development.

Intercellular bridges are a deeply conserved cellular structure found in germ cells at various points in their development. Formed by incomplete cytokinesis, the bridges stably connect sister germ cells into "cysts" and permit the passage of cytoplasm and organelles such as mitochondria (Greenbaum, Iwamori, Buchold, & Matzuk, 2011). Intercellular bridges coordinate differentiation and allow sharing of resources between cells, particularly through self-sacrifice, when one member of a cyst commits apoptosis (Lei & Spradling, 2016). Although the formation of intercellular bridges between sister cells limits the potential difference between cells of a cyst, we note their potential to buffer phenotypic consequences of heterogeneity that arises in PGCs such as acquired DNMs or epi-mutations. The increased geographic spread of cysts compared to single PGCs also raises the possibility that they buffer differences introduced by the niche.

4. Selection and heterogeneity

Natural selection is an elegant force of evolution, working slowly through time to shape and change organisms to better suit their needs in changing environments and pressures. This concept has implications for species that reproduce and share heritable characteristics with their offspring. As organisms reproduce, their offspring inherit different traits with stochastic variation; if a random trait is beneficial to the offspring, it will have a competitive advantage, and potentially pass this trait along through its germline. Through many of the causes of heterogeneity that we have already discussed in this review, changes can be wrought within an organism. Changes within somatic cells, while impactful for the organism itself, pale in comparison to the multigenerational effect that heterogeneous changes can have if the germline is altered.

Whether through a genetic or non-genetic transmission, discrete traits are brought to prominence over time if placed under enough positive selective pressure. Natural selection within a population results from interaction between organisms and their environment, as the environment creates selection pressure among heterogeneous individuals, resulting in increased fitness of the population. Less fit animals are phased out of prominence as natural selection leads to increase in the frequency of certain alleles over time.

We propose that like natural selection acts at the organismal and population level, there is a similar kind of selection operating at the cellular level (Buss, 1988; Laird et al., 2005). At both the population and cellular level, alleles subjected to positive selection increase in frequency. Over time, variants are selected due to a competitive advantage they convey to potentially produce an organism or cell with greater fitness. The diversity of PGC development is a special, comprehensive series of selection events during which multiple types of fitnesses can be examined. We will consider how the totality of these developmental events act selectively on PGCs and discuss their implications on reproductive fitness.

4.1 Defining selection in the developing germline

In developmental selection, each stage of germ cell development provides an opportunity for the emergence of underlying heterogeneity to produce different outcomes for subpopulations of germ cells. As previously discussed, both non-genetic and genetic variation among individual cells can present distinct cell identities or states for selection to act upon. We consider a selective process to be one which alters the proportions of PGC subpopulations during development—by mechanisms such as apoptosis, proliferation, migration, or differentiation. Here, we will briefly explicate what we consider a "selected cell" and the processes by which this selection occurs.

Selection can be either negative or positive with respect to gametogenic potential. In negative selection, a germ cell subpopulation is functionally eliminated from further contributing to gametogenesis. In positive selection, a germ cell subpopulation is increased relative to other subpopulations. For the developing germline, we will consider how key events during the fetal period can potentially select for certain subpopulations and discuss the impact on the germline. For example, migration is one of the first challenges a PGC faces after specification at E7.0. During migration, PGCs must respond to precise chemotactic cues to remain on course as they navigate to the gonads. Owing to the distinctive signaling environments presented by each niche along the migratory route, the precise timing of migration may also represent another form of selection. Damaged or unfit PGCs might migrate inefficiently, resulting in reduced colonization of the gonad by their cellular progeny, which corresponds to negative selection. If PGCs that arrive first in the gonad receive different signals than later arriving PGCs, this could alter differentiation on the basis of migratory speed. Additionally, the duration of these differentiation steps could vary as early migrants would be exposed to signals for a longer period of time compared to late arrivals.

In contrast to spatially based selection during PGC migration, selection can also occur during differentiation. After reaching the gonad, PGCs continue maturation by epigenetic reprogramming as well as sex differentiation. The concomitant increase in the expression of apoptotic genes such as *Bax* sensitizes PGCs toward elimination. Differentiation downregulates these genes and protects successfully differentiated PGCs from apoptosis. PGCs are additionally eliminated based on clonal lineage, implying that apoptotic selection can work on heritable qualities such as epigenetic permissiveness for differentiation (Nguyen & Laird, 2019). The balancing of these two programs selects for differentiated PGCs and selects against PGCs that are incapable or inefficient in differentiating.

Selection can also occur through proliferation, rather than elimination. Germ cell subpopulations that are more proliferative will increase their proportional representation within the germline with each successive division. This may also induce competition for niches or access to signaling factors that continue the tightly regulated differentiation program essential for further germ cell development into optimal gamete-producing cells. Earlier studies utilizing clonal labeling have noted that a small subset of clonal populations in the fetal period is responsible for producing the small number of large monoclonal areas in testes that actually contribute to spermatogenesis (Ueno et al., 2009). Later spermatogenesis in the adult has also been observed to originate from a limited number of spermatogonial stem cell clones (Kanatsu-Shinohara, Naoki, & Shinohara, 2016). In examining fetal populations of germ cells, Lei and Spradling found clone sizes to be variable (Lei & Spradling, 2013), although a longitudinal study would clarify if larger clones resulted from or continued to enjoy a proliferative advantage relative to other germ cell clones.

4.2 How does PGC fitness relate to fitness of gametes or individuals?

While heterogeneities and selective pressure are seen during all phases of germline development, we have thus limited our discussion to competition among fellow cells of the same type. Considering how each competitive cell contributes to the gametes as a whole, providing variation and heterogeneity within a single organism, leads us to speculate on how PGC selection may have an impact on the rate of evolution. Heterogeneity within the germline may exert effects not only within one or two generations, but could also shape the population of a species on a longer evolutionary scale. The fine processes of generating and selecting upon PGC heterogeneity in the embryo may influence the processes of natural selection or genetic drift.

The relationship between processes of germline development and processes of evolution has been considered by others with PGC specification. Broad surveys of modes of germline formation across phylogeny have driven the conclusion that PGC preformation is not ancestral but rather has evolved convergently in many species (reviewed in Extavour, 2003; Johnson, Richardson, Bachvarova, & Crother, 2011). While one study found that species that form their germline through preformation enjoy a faster rate of sequence evolution (Evans, Wade, Chapman, Johnson, & Loose, 2014), follow-up work found instead that species with PGC specification by preformation and PGC induction were equally represented among species with rapidly evolving genomes (Whittle & Extavour, 2016). Since the rate of evolution may be influenced by the heterogeneity of the germline, perhaps organisms with less constraint on their PGCs and a more stringent selection would promote variation within the germline and faster rate of evolution through generations.

If selection during the fetal period can have multi-generational effect, then the fitness of cells which will contribute to these population effects must be multifaceted in their ability to respond to changing environments. Tests of germ cell selection may occur at many developmental timepoints to ensure greatest cell competency; germ cells that excel in one characteristic might be less fit for others. Take, for example, cells which excel dramatically at migration. How might supermigrators such as these compete in later developmental tests? Previous work has shown that as oocytes mature, their order of entry into meiosis corresponds to the order of follicle maturation (Hirshfield, 1992; Polani & Crolla, 1991), and later differentiation follows a "first in, first out" order of follicle maturation and ovulation (Reizel et al., 2012). While supermigrators have not been directly linked to order of entry into meiosis, one might speculate that the earliest migrants will also be the first to enter meiosis, and thus the first to initiate folliculogenesis (Arora et al., 2016). However, many of the earliest follicles release before puberty occurs, causing them to die due to a lack of supportive hormones environment (McGee, 2000). In this instance, a cell overly specialized for migration may fail later developmental checkpoints. The shifting selective pressures of germ cell development ensure that the fittest cells are the germ cells competent at facing many developmental challenges.

These selective processes that comprise germ cell development provide an opportunity for selfish cells to take advantage. Selection favoring a subpopulation of germ cells can eliminate aberrant cells to improve PGC fitness, but it also is vulnerable to cheating. For example, mutations in the FGF receptors can increase FGF ligand binding affinity to confer a proliferative advantage in spermatogenesis. This would lead to clonal expansion of FGFR mutants during development and increased representation in the gamete pool (Goriely, McVean, Röjmyr, Ingemarsson, & Wilkie, 2003). While advantageous for germ cell success, these mutations are detrimental to the organism and can lead to disorders in somatic tissues such as in Apert syndrome (Glaser et al., 2003). The potential for germ cell cheating perhaps illustrates the utility of a multipartite developmental examination of various germ cell behaviors and phenotypes. A single mutation that provides an advantage in one stage may be selected against in another. This cumulative developmental selection would secure diversity in the gamete pool, which is essential for variation at the organismal level to respond to environmental and evolutionary challenges.

5. Conclusion

The arduous journey to establish the germline is defined by variation—both in the challenges faced by PGCs as well as the heterogeneity of the PGCs themselves. The diversity of events during PGC development creates a gauntlet that is ideally positioned to assess heterogeneity in PGCs as they begin to expand and differentiate to establish the eventual germline. Applying the principles of natural selection to the level of germ cells, we see that multiple selective opportunities exist to alter the PGC population that continues onto gametogenesis. The ultimate effect that PGC selection can have on reproductive fitness and even the fitness of organisms will be a fascinating area of study.

The advent of many single-cell analytical techniques now enables us to more comprehensively describe the variation within PGCs. Early attempts at morphological phenotyping are bolstered by single-cell transcriptomics, epigenomics, and single-molecule imaging. In addition to evaluating differences among PGCs at each stage of development, complex barcoding can also extend our understanding how these differences change over developmental time. This is especially relevant to the germline given the dynamic challenges faced by PGCs and the varied capabilities assessed throughout their entire journey. Combining these technologies with a longitudinal perspective will clarify how each event during PGC development contributes to improving gamete quality.

Heterogeneity is a ubiquitous property of biology, but this is especially amplified when a population is faced with making a highly regulated and complex transition such as PGCs in the early embryo. Recent advances in reproductive technology have enabled the generation of mouse germ cells in vitro that produce gametes capable of fertilization at exceedingly low efficiency (Hikabe et al., 2016; Ishikura et al., 2016). These germ cells do not undergo the same complex in vivo development and it remains to be seen how bypassing this lengthy process affects the fidelity of their development. Understanding the events and criteria of selection during early embryonic PGC development can improve how we evaluate in vitro-derived gametes to ensure the highest quality ones will be chosen.

References

Abdu, Y., Maniscalco, C., Heddleston, J. M., Chew, T.-L., & Nance, J. (2016). Developmentally programmed germ cell remodelling by endodermal cell cannibalism. *Nature Cell Biology*, *18*, 1302. EP.

Allio, R., Donega, S., Galtier, N., & Nabholz, B. (2017). Large variation in the ratio of mitochondrial to nuclear mutation rate across animals: Implications for genetic diversity and the use of mitochondrial DNA as a molecular marker. *Molecular Biology and Evolution*, *34*, 2762–2772. https://doi.org/10.1093/molbev/msx197.

Altschuler, S. J., & Wu, L. F. (2010). Cellular heterogeneity: Do differences make a difference? *Cell*, *141*, 559–563. https://doi.org/10.1016/j.cell.2010.04.033.

Andux, S., & Ellis, R. E. (2008). Apoptosis maintains oocyte quality in aging *Caenorhabditis elegans* females. *PLoS Genetics*, *4*, e1000295. https://doi.org/10.1371/journal.pgen.1000295.

Arana, M. E., & Kunkel, T. A. (2010). Mutator phenotypes due to DNA replication in delity. *Seminars in Cancer Biology*, *20*, 304–311.

Arora, R., Abby, E., Ross, A. D. J., Cantú, A. V., Kissner, M. D., Castro, V., et al. (2016). Meiotic onset is reliant on spatial distribution but independent of germ cell number in the mouse ovary. *Journal of Cell Science*, *129*, 2493–2499. https://doi.org/10.1242/jcs.189910.

Avilion, A. A., Nicolis, S. K., Pevny, L. H., Perez, L., Vivian, N., & Lovell-Badge, R. (2003). Multipotent cell lineages in early mouse development depend on SOX2 function. *Genes & Development*, *17*, 126–140.

Banisch, T. U., Goudarzi, M., & Raz, E. (2012). Small RNAs in germ cell development. *Current Topics in Developmental Biology*, *99*, 79–113. https://doi.org/10.1016/B978-0-12-387038-4.00004-5.

Berrens, R. V., Andrews, S., Spensberger, D., Santos, F., Dean, W., Gould, P., et al. (2017). An endosiRNA-based repression mechanism counteracts transposon activation during global DNA demethylation in embryonic stem cells. *Cell Stem Cell*, *21*, 694–703.e7. https://doi.org/10.1016/j.stem.2017.10.004.

Besenbacher, S., Liu, S., Izarzugaza, J. M. G., Grove, J., Belling, K., Bork-Jensen, J., et al. (2015). Novel variation and *de novo* mutation rates in population-wide *de novo* assembled Danish trios. *Nature Communications*, *6*, 5969.

Blaser, H., Eisenbeiss, S., Neumann, M., Reichman-Fried, M., Thisse, B., Thisse, C., et al. (2005). Transition from non-motile behaviour to directed migration during early PGC development in zebrafish. *Journal of Cell Science*, *118*, 4027–4038. https://doi.org/10.1242/jcs.02522.

Bohacek, J., & Mansuy, I. M. (2015). Molecular insights into transgenerational non-genetic inheritance of acquired behaviours. *Nature Reviews. Genetics*, *16*, 641–652. https://doi.org/10.1038/nrg3964.

Bontems, F., Stein, A., Marlow, F., Lyautey, J., Gupta, T., Mullins, M. C., et al. (2009). Bucky ball organizes germ plasm assembly in zebrafish. *Current Biology*, *19*, 414–422. https://doi.org/10.1016/j.cub.2009.01.038.

Bowles, J., Feng, C.-W., Miles, K., Ineson, J., Spiller, C., & Koopman, P. (2016). ALDH1A1 provides a source of meiosis-inducing retinoic acid in mouse fetal ovaries. *Nature Communications*, 7, 10845. https://doi.org/10.1038/ncomms10845.

Bowles, J., Feng, C.-W., Spiller, C., Davidson, T.-L., Jackson, A., & Koopman, P. (2010). FGF9 suppresses meiosis and promotes male germ cell fate in mice. *Developmental Cell*, *19*, 440–449. https://doi.org/10.1016/j.devcel.2010.08.010.

Bowles, J., Knight, D., Smith, C., Wilhelm, D., Richman, J., Mamiya, S., et al. (2006). Retinoid signaling determines germ cell fate in mice. *Science*, *312*, 596–600. https://doi.org/10.1126/science.1125691.

Brennecke, J., Aravin, A. A., Stark, A., Dus, M., Kellis, M., Sachidanandam, R., et al. (2007). Discrete small RNA-generating loci as master regulators of transposon activity in Drosophila. *Cell*, *128*, 1089–1103. https://doi.org/10.1016/j.cell.2007.01.043.

Brock, A., Chang, H., & Huang, S. (2009). Non-genetic heterogeneity—A mutation-independent driving force for the somatic evolution of tumours. *Nature Reviews. Genetics*, *10*, 336–342. https://doi.org/10.1038/nrg2556.

Buss, L. W. (1988). *The evolution of individuality*. Princeton: Princeton University Press. https://doi.org/10.1515/9781400858712.

Campbell, C. D., Chong, J. X., Malig, M., Ko, A., Dumont, B. L., Han, L., et al. (2012). Estimating the human mutation rate using autozygosity in a founder population. *Nature Genetics*, *44*, 1277–1281.

Cantú, A. V., Altshuler-Keylin, S., & Laird, D. J. (2016). Discrete somatic niches coordinate proliferation and migration of primordial germ cells via Wnt signaling. *The Journal of Cell Biology*, *214*, 215–229. https://doi.org/10.1083/jcb.201511061.

Cao, L., Shitara, H., Horii, T., Nagao, Y., Imai, H., Abe, K., et al. (2007). The mitochondrial bottleneck occurs without reduction of mtDNA content in female mouse germ cells. *Nature Genetics*, *39*, 386–390. https://doi.org/10.1038/ng1970.

Casper, A., & Van Doren, M. (2006). The control of sexual identity in the Drosophila germline. *Development*, *133*, 2783–2791.

Childs, A. J., Cowan, G., Kinnell, H. L., Anderson, R. A., & Saunders, P. T. K. (2011). Retinoic acid signalling and the control of meiotic entry in the human fetal gonad. *PLoS One*, *6*, e20249. https://doi.org/10.1371/journal.pone.0020249.

Chubb, J. R., Trcek, T., Shenoy, S. M., & Singer, R. H. (2006). Transcriptional pulsing of a developmental gene. *Current Biology*, *16*, 1018–1025. https://doi.org/10.1016/j.cub.2006.03.092.

Coffman, C. R., Strohm, R. C., Oakley, F. D., Yamada, Y., Przychodzin, D., & Boswell, R. E. (2002). Identification of X-linked genes required for migration and programmed cell death of Drosophila melanogaster germ cells. *Genetics*, *162*, 273.

Conrad, D. F., Keebler, J. E., DePristo, M. A., Lindsay, S. J., Zhang, Y., Casals, F., et al. (2011). Variation in genome-wide mutation rates within and between human families. *Nature Genetics*, *43*, 712–714.

Coucouvanis, E. C., Sherwood, S. W., Carswell-Crumpton, C., Spack, E. G., & Jones, P. P. (1993). Evidence that the mechanism of prenatal germ cell death in the mouse is apoptosis. *Experimental Cell Research, 209*, 238–247. https://doi.org/10.1006/excr.1993.1307.
Crow, J. F. (2000). The origins, patterns and implications of human spontaneous mutation. *Nature Reviews. Genetics, 1*, 40–47.
Dawson, E. P., Lanza, D. G., Webster, N. J., Benton, S. M., Suetake, I., & Heaney, J. D. (2018). Delayed male germ cell sex-specification permits transition into embryonal carcinoma cells with features of primed pluripotency. *Development, 145*. dev156612. https://doi.org/10.1242/dev.156612.
Delbruck, M. (1945). The burst size distribution in the growth of bacterial viruses (Bacteriophages). *Journal of Bacteriology, 50*, 131–135.
Diaz-Diaz, C., Fernandez de Manuel, L., Jimenez-Carretero, D., Montoya, M. C., Clavería, C., & Torres, M. (2017). Pluripotency surveillance by Myc-driven competitive elimination of differentiating cells. *Developmental Cell42*. https://doi.org/10.1016/j.devcel.2017.08.011. 585–599.e4.
Drake, J. W., Charlesworth, B., Charlesworth, D., & Crow, J. F. (1998). Rates of spontaneous mutation. *Genetics, 148*, 1667–1686.
Eddy, E. M. (1975). Germ plasm and the differentiation of germ cell line. *International Review of Cytology, 43*, 229–280.
Ellis, R. E. (2008). Sex determination in the Caenorhabditis elegans germ line. *Current Topics in Developmental Biology, 83*, 41–64. https://doi.org/10.1016/S0070-2153(08)00402-X.
Ephrussi, A., & Lehmann, R. (1992). Induction of germ cell formation by oskar. *Nature, 358*, 387–392. https://doi.org/10.1038/358387a0.
Evans, T., Wade, C. M., Chapman, F. A., Johnson, A. D., & Loose, M. (2014). Acquisition of germ plasm accelerates vertebrate evolution. *Science, 344*, 200–203. https://doi.org/10.1126/science.1249325.
Ewen, K. A., & Koopman, P. (2010). Mouse germ cell development: From specification to sex determination. *Molecular and Cellular Endocrinology, 323*, 76–93. https://doi.org/10.1016/j.mce.2009.12.013.
Ewen-Campen, B., Donoughe, S., Clarke, D. N., & Extavour, C. G. (2013). Germ cell specification requires zygotic mechanisms rather than germ plasm in a basally branching insect. *Current Biology, 23*(10), 835–842.
Ewen-Campen, B., Srouji, J. R., Schwager, E. E., & Extavour, C. G. (2012). Oskar predates the evolution of germ plasm in insects. *Current Biology, 22*, 2278–2283.
Extavour, C. G. (2003). Mechanisms of germ cell specification across the metazoans: Epigenesis and preformation. *Development, 130*, 5869–5884. https://doi.org/10.1242/dev.00804.
Falkenberg, M., Larsson, N.-G., & Gustafsson, C. M. (2007). DNA replication and transcription in mammalian mitochondria. *Annual Review of Biochemistry, 76*, 679–699. https://doi.org/10.1146/annurev.biochem.76.060305.152028.
Fernández-Pérez, D., Brieño-Enríquez, M. A., Isoler-Alcaraz, J., Larriba, E., & Del Mazo, J. (2018). MicroRNA dynamics at the onset of primordial germ and somatic cell sex differentiation during mouse embryonic gonad development. *RNA, 24*, 287–303. https://doi.org/10.1261/rna.062869.117.
Folmes, C. D. L., & Terzic, A. (2014). Metabolic determinants of embryonic development and stem cell fate. *Reproduction, Fertility, and Development, 27*, 82–88. https://doi.org/10.1071/RD14383.
Fujiwara, T., Dunn, N. R., & Hogan, B. L. (2001). Bone morphogenetic protein 4 in the extraembryonic mesoderm is required for allantois development and the localization and survival of primordial germ cells in the mouse. *PNAS, 98*, 13739–13744.
Fukuyama, M., Rougvie, A. E., & Rothman, J. H. (2006). *C. elegans* DAF-18/PTEN mediates nutrient-dependent arrest of cell cycle and growth in the germline. *Current Biology, 16*, 773–779. https://doi.org/10.1016/j.cub.2006.02.073.

Gallo, C. M., Wang, J. T., Motegi, F., & Seydoux, G. (2010). Cytoplasmic partitioning of P granule components is not required to specify the germline in *C. elegans*. *Science*, *330*, 1685–1689. https://doi.org/10.1126/science.1193697.

Gangaraju, V. K., Yin, H., Weiner, M. M., Wang, J., Huang, X. A., & Lin, H. (2010). Drosophila Piwi functions in Hsp90-mediated suppression of phenotypic variation. *Nature Genetics*, *43*, 153–158. https://doi.org/10.1038/ng.743.

Gardner, R. L., & Rossant, J. (1979). Investigation of the fate of 4-5 day post-coitum mouse inner cell mass cells by blastocyst injection. *Journal of Embryology and Experimental Morphology*, *52*, 141–152.

Gartner, A., Boag, P. R., & Blackwell, T. K. (2008). *Germline survival and apoptosis. In WormBook (Ed.), The* C. elegans *Research Community*. WormBook. https://doi.org/10.1895/wormbook.1.145.1. http://www.wormbook.org.

Gilboa, L., & Lehmann, R. (2006). Soma-germline interactions coordinate homeostasis and growth in the Drosophila gonad. *Nature*, *443*, 97–100. https://doi.org/10.1038/nature05068.

Ginsburg, M., & Eyal-Giladi, H. (1986). Temporal and spatial aspects of the gradual migration of primordial germ cells from the epiblast into the germinal crescent in the avian embryo. *Journal of Embryology and Experimental Morphology*, *95*, 53–71.

Ginsburg, M., & Eyal-Giladi, H. (1987). Primordial germ cells of the young chick blastoderm originate from the central zone of the area pellucida irrespective of the embryo-forming process. *Development*, *101*, 209–219.

Ginsburg, M., Snow, M. H., & McLaren, A. (1990). Primordial germ cells in the mouse embryo during gastrulation. *Development*, *110*, 521–528.

Glaser, R. L., Broman, K. W., Schulman, R. L., Eskenazi, B., Wyrobek, A. J., & Jabs, E. W. (2003). The paternal-age effect in Apert syndrome is due, in part, to the increased frequency of mutations in sperm. *American Journal of Human Genetics*, *73*, 939–947. https://doi.org/10.1086/378419.

Gomperts, M., Garcia-Castro, M., Wylie, C., & Heasman, J. (1994). Interactions between primordial germ cells play a role in their migration in mouse embryos. *Development*, *120*, 135–141.

Goriely, A., McVean, G. A. T., Röjmyr, M., Ingemarsson, B., & Wilkie, A. O. M. (2003). Evidence for selective advantage of pathogenic FGFR2 mutations in the male germ line. *Science*, *301*, 643–646. https://doi.org/10.1126/science.1085710.

Greenbaum, M. P., Iwamori, T., Buchold, G. M., & Matzuk, M. M. (2011). Germ cell intercellular bridges. *Cold Spring Harbor Perspectives in Biology*, *3*, a005850. a005850 https://doi.org/10.1101/cshperspect.a005850.

Grubb, B. J. (2006). Developmental biology, Eighth Edition. *Integrative and Comparative Biology*, *46*(5), 652–653. https://doi.org/10.1093/icb/icl011.

Grün, D., Lyubimova, A., Kester, L., Wiebrands, K., Basak, O., Sasaki, N., et al. (2015). Single-cell messenger RNA sequencing reveals rare intestinal cell types. *Nature*, *525*, 251.

Gumienny, T. L., Lambie, E., Hartwieg, E., Horvitz, H. R., & Hengartner, M. O. (1999). Genetic control of programmed cell death in the *Caenorhabditis elegans* hermaphrodite germline. *Development*, *126*, 1011–1022. https://doi.org/10.5167/uzh-1041.

Hackett, J. A., & Surani, M. A. (2012). DNA methylation dynamics during the mammalian life cycle. *Philosophical Transactions of the Royal Society of London. Series B, Biological Sciences368*. https://doi.org/10.1098/rstb.2011.0328. 20110328–20110328.

Hagemann-Jensen, M., Abdullayev, I., Sandberg, R., & Faridani, O. R. (2018). Small-seq for single-cell small-RNA sequencing. *Nature Protocols*, *13*, 2407–2424. https://doi.org/10.1038/s41596-018-0049-y.

Hajkova, P., Erhardt, S., Lane, N., Haaf, T., El-Maarri, O., Reik, W., et al. (2002). Epigenetic reprogramming in mouse primordial germ cells. *Mechanisms of Development*, *117*, 15–23.

Hancks, D. C., & Kazazian, H. H. (2012). Active human retrotransposons: Variation and disease. *Current Opinion in Genetics & Development*, *22*(3), 191–203.
Hay, B., Ackerman, L., Barbel, S., Jan, L. Y., & Jan, Y. N. (1988). Identification of a component of Drosophila polar granules. *Development*, *103*, 625–640.
Hayashi, Y., Hayashi, M., & Kobayashi, S. (2004). Nanos suppresses somatic cell fate in Drosophila germ line. *PNAS*, *101*, 10338. https://doi.org/10.1073/pnas.0401647101.
Heaney, J. D., Anderson, E. L., Michelson, M. V., Zechel, J. L., Conrad, P. A., Page, D. C., et al. (2012). Germ cell pluripotency, premature differentiation and susceptibility to testicular teratomas in mice. *Development*, *139*, 1577–1586. https://doi.org/10.1242/dev.076851.
Henderson, S. A., & Edwards, R. G. (1968). Chiasma frequency and maternal age in mammals. *Nature*, *218*, 22–28.
Heyer, B. S., MacAuley, A., Behrendtsen, O., & Werb, Z. (2000). Hypersensitivity to DNA damage leads to increased apoptosis during early mouse development. *Genes & Development*, *14*, 2072–2084.
Hikabe, O., Hamazaki, N., Nagamatsu, G., Obata, Y., Hirao, Y., Hamada, N., et al. (2016). Reconstitution in vitro of the entire cycle of the mouse female germ line. *Nature*, *539*, 299–303. https://doi.org/10.1038/nature20104.
Hill, P. W. S., Leitch, H. G., Requena, C. E., Sun, Z., Amouroux, R., Roman-Trufero, M., et al. (2018). Epigenetic reprogramming enables the transition from primordial germ cell to gonocyte. *Nature*, *555*, 392–396. https://doi.org/10.1038/nature25964.
Hird, S. N., Paulsen, J. E., & Strome, S. (1996). Segregation of germ granules in living Caenorhabditis elegans embryos: Cell-type-specific mechanisms for cytoplasmic localisation. *Development*, *122*, 1303–1312.
Hirshfield, A. N. (1992). Heterogeneity of cell populations that contribute to the formation of primordial follicles in rats. *Biology of Reproduction*, *47*, 466–472.
Huang, S. (2009). Non-genetic heterogeneity of cells in development: More than just noise. *Development*, *136*, 3853–3862. https://doi.org/10.1242/dev.035139.
Hubbard, E. J. A., Korta, D. Z., & Dalfo, D. (2013). Physiological control of germline development. *Advances in Experimental Medicine and Biology*, *757*, 101–131. https://doi.org/10.1007/978-1-4614-4015-4_5.
Hunt, P. A., & Hassold, T. J. (2008). Human female meiosis: What makes a good egg go bad? *Trends in Genetics*, *24*, 86–93.
Hurd, T. R., Herrmann, B., Sauerwald, J., Sanny, J., Grosch, M., & Lehmann, R. (2016). Long Oskar controls mitochondrial inheritance in *Drosophila melanogaster*. *Developmental Cell*, *39*(5), 560–571.
Ishikura, Y., Yabuta, Y., Ohta, H., Hayashi, K., Nakamura, T., Okamoto, I., et al. (2016). In-vitro derivation and propagation of spermatogonial stem cell activity from mouse pluripotent stem cells. *Cell Reports*, *17*, 2789–2804. https://doi.org/10.1016/j.celrep.2016.11.026.
Johnson, A. D., Richardson, E., Bachvarova, R. F., & Crother, B. I. (2011). Evolution of the germ line-soma relationship in vertebrate embryos. *Reproduction*, *141*, 291–300. https://doi.org/10.1530/REP-10-0474.
Jónsson, H., Sulem, P., Kehr, B., Kristmundsdottir, S., Zink, F., Hjartarson, E., et al. (2017). Parental in uence on human germline *de novo* mutations in 1,548 trios from Iceland. *Nature*, *549*, 519–522.
Kanatsu-Shinohara, M., Naoki, H., & Shinohara, T. (2016). Nonrandom germline transmission of mouse spermatogonial stem cells. *Developmental Cell*, *38*, 248–261. https://doi.org/10.1016/j.devcel.2016.07.011.
Kazazian, H. H., & Moran, J. V. (2017). Mobile DNA in health and disease. *The New England Journal of Medicine*, *377*(4), 361–370. http://doi:10.1056/NEJMra1510092.
Keightley, P. D., Ness, R. W., Halligan, D. L., & Haddrill, P. R. (2014). Estimation of the spontaneous mutation rate per nucleotide site in a Drosophila melanogaster full-sib family. *Genetics*, *196*, 313–320. https://doi.org/10.1534/genetics.113.158758.

Khrapko, K., Coller, H. A., Andre, P. C., Li, X. C., Hanekamp, J. S., & Thilly, W. G. (1997). Mitochondrial mutational spectra in human cells and tissues. *PNAS*, *94*, 13798–13803.

Kimble, J. E., & White, J. G. (1981). On the control of germ cell development in *Caenorhabditis elegans*. *Developmental Biology*, *81*, 208–219.

Kohler, S. W., Provost, G. S., Fieck, A., Kretz, P. L., Bullock, W. O., Sorge, J. A., et al. (1991). Spectra of spontaneous and mutagen-induced mutations in the lacI gene in transgenic mice. *Proceedings of the National Academy of Sciences of the United States of America*, *88*, 7958–7962.

Kong, A., Frigge, M. L., Masson, G., Besenbacher, S., Sulem, P., Magnusson, G., et al. (2012). Rate of *de novo* mutations and the importance of father's age to disease risk. *Nature*, *488*, 471–475.

Köprunner, M., Thisse, C., Thisse, B., & Raz, E. (2001). A zebrafish nanos-related gene is essential for the development of primordial germ cells. *Genes & Development*, *15*, 2877–2885. https://doi.org/10.1101/gad.212401.

Koubova, J., Menke, D. B., Zhou, Q., Capel, B., Griswold, M. D., & Page, D. C. (2006). Retinoic acid regulates sex-specific timing of meiotic initiation in mice. *Proceedings of the National Academy of Sciences of the United States of America*, *103*, 2474–2479. https://doi.org/10.1073/pnas.0510813103.

Kuramochi-Miyagawa, S., Kimura, T., Ijiri, T. W., Isobe, T., Asada, N., Fujita, Y., et al. (2004). Mili, a mammalian member of piwi family gene, is essential for spermatogenesis. *Development*, *131*, 839–849. https://doi.org/10.1242/dev.00973.

Laird, D. J. (2018). How to lose your inheritance. *Nature*, *562*, 497–498. https://doi.org/10.1038/d41586-018-06849-5.

Laird, D. J., Altshuler-Keylin, S., Kissner, M. D., Zhou, X., & Anderson, K. V. (2011). Ror2 enhances polarity and directional migration of primordial germ cells. *PLoS Genetics*, 7, e1002428. https://doi.org/10.1371/journal.pgen.1002428.

Laird, D. J., Chang, W.-T., Weissman, I. L., & Lauzon, R. J. (2005). Identification of a novel gene involved in asexual organogenesis in the budding ascidian Botryllus schlosseri. *Developmental Dynamics*, *234*, 997–1005. https://doi.org/10.1002/dvdy.20583.

Lawson, K. A., Dunn, N. R., Roelen, B. A., Zeinstra, L. M., Davis, A. M., Wright, C. V., et al. (1999). Bmp4 is required for the generation of primordial germ cells in the mouse embryo. *Genes & Development*, *13*, 424–436.

Le Bouffant, R., Souquet, B., Duval, N., Duquenne, C., Hervé, R., Frydman, N., et al. (2011). Msx1 and Msx2 promote meiosis initiation. *Development*, *138*, 5393–5402. https://doi.org/10.1242/dev.068452.

Lee, H., Kim, S., Park, T., Rengaraj, D., Park, K., Lee, H., et al. (2013). Compensatory proliferation of endogenous chicken primordial germ cells after elimination by busulfan treatment. *Stem Cell Research & Therapy*, *4*, 136.

Lei, L., & Spradling, A. C. (2013). Mouse primordial germ cells produce cysts that partially fragment prior to meiosis. *Development*, *140*, 2075–2081. https://doi.org/10.1242/dev.093864.

Lei, L., & Spradling, A. C. (2016). Mouse oocytes differentiate through organelle enrichment from sister cyst germ cells. *Science*, *352*, 95–99. https://doi.org/10.1126/science.aad2156.

Levsky, J. M., & Singer, R. H. (2003). Fluorescence in situ hybridization: Past, present and future. *Journal of Cell Science*, *116*, 2833–2838. https://doi.org/10.1242/jcs.00633.

Longley, M. J., Nguyen, D., Kunkel, T. A., & Copeland, W. C. (2001). The fidelity of human DNA polymerase gamma with and without exonucleolytic proofreading and the p55 accessory subunit. *The Journal of Biological Chemistry*, *276*, 38555–38562. https://doi.org/10.1074/jbc.M105230200.

Lynch, M. (2010). Evolution of the mutation rate. *Trends in Genetics*, *26*, 345–352. https://doi.org/10.1016/j.tig.2010.05.003.

Maatouk, D. M., Kellam, L. D., Mann, M. R. W., Lei, H., Li, E., Bartolomei, M. S., et al. (2006). DNA methylation is a primary mechanism for silencing postmigratory primordial germ cell genes in both germ cell and somatic cell lineages. *Development*, *133*, 3411–3418. https://doi.org/10.1242/dev.02500.

Madsen, C. S., Ghivizzani, S. C., & Hauswirth, W. W. (1993). In vivo and in vitro evidence for slipped mispairing in mammalian mitochondria. *PNAS*, *90*, 7671–7675.

Mahowald, A. P. (1962). Fine structure of pole cells and polar granules in *Drosophila melanogaster*. *The Journal of Experimental Zoology*, *151*, 201–215.

Mahowald, A. P. (2001). Assembly of the *Drosophila* germ plasm. *International Review of Cytology*, *203*, 187–213.

Malki, S., van der Heijden, G. W., O'Donnell, K. A., Martin, S. L., & Bortvin, A. (2014). A role for retrotransposon LINE-1 in fetal oocyte attrition in mice. *Developmental Cell*, *29*, 521–533. https://doi.org/10.1016/j.devcel.2014.04.027.

Marlow, F. (2015). Primordial germ cell specification and migration. *F1000Res*, *16*, 4. https://doi.org/10.12688/f1000research.6995.1.

Matson, C. K., Murphy, M. W., Griswold, M. D., Yoshida, S., Bardwell, V. J., & Zarkower, D. (2010). The mammalian Doublesex homolog DMRT1 is a transcriptional gatekeeper that controls the mitosis versus meiosis decision in male germ cells. *Developmental Cell*, *19*, 612–624. https://doi.org/10.1016/j.devcel.2010.09.010.

McDole, K., Guignard, L., Amat, F., Berger, A., Malandain, G., Royer, L. A., et al. (2018). In toto imaging and reconstruction of post-implantation mouse development at the single-cell level. *Cell175*. https://doi.org/10.1016/j.cell.2018.09.031. 859–876.e33.

McGee, E. A. (2000). The regulation of apoptosis in preantral ovarian follicles. *Biological Signals and Receptors*, *9*, 81–86. https://doi.org/10.1159/000014626.

McGranahan, N., & Swanton, C. (2017). Clonal heterogeneity and tumor evolution: Past, present, and the future. *Cell*, *168*, 613–628. https://doi.org/10.1016/j.cell.2017.01.018.

McIver, S. C., Roman, S. D., Nixon, B., & McLaughlin, E. A. (2012). miRNA and mammalian male germ cells. *Human Reproduction Update*, *18*, 44–59. https://doi.org/10.1093/humupd/dmr041.

McLaren, A. (2003). Primordial germ cells in the mouse. *Developmental Biology*, *262*, 1–15.

McLaren, A., & Southee, D. (1997). Entry of mouse embryonic germ cells into meiosis. *Developmental Biology*, *187*, 107–113. https://doi.org/10.1006/dbio.1997.8584.

Menke, D. B., Koubova, J., & Page, D. C. (2003). Sexual differentiation of germ cells in XX mouse gonads occurs in an anterior-to-posterior wave. *Developmental Biology*, *262*, 303–312.

Michaelson, J. J., Shi, Y., Gujral, M., Zheng, H., Malhotra, D., Jin, X., et al. (2012). Whole-genome sequencing in autism identifies hot spots for de novo germline mutation. *Cell*, *151*, 1431–1442.

Milholland, B., Dong, X., Zhang, L., Hao, X., Suh, Y., & Vijg, J. (2017). Differences between germline and somatic mutation rates in humans and mice. *Nature Communications*, *8*, 15183. https://doi.org/10.1038/ncomms15183.

Mishra, P., & Chan, D. C. (2014). Mitochondrial dynamics and inheritance during cell division, development and disease. *Nature*, *15*, 634–646. https://doi.org/10.1038/nrm3877.

Molyneaux, K. A., Stallock, J., Schaible, K., & Wylie, C. (2001). Time-lapse analysis of living mouse germ cell migration. *Developmental Biology*, *240*, 488–498. https://doi.org/10.1006/dbio.2001.0436.

Morita-Fujimura, Y., Tokitake, Y., & Matsui, Y. (2009). Heterogeneity of mouse primordial germ cells reflecting the distinct status of their differentiation, proliferation and apoptosis can be classified by the expression of cell surface proteins integrin α6 and c-Kit. *Development, Growth & Differentiation*, *51*, 567–583. https://doi.org/10.1111/j.1440-169X.2009.01119.x.

Muller-Sieburg, C. E., Sieburg, H. B., Bernitz, J. M., & Cattarossi, G. (2012). Stem cell heterogeneity: Implications for aging and regenerative medicine. *Blood, 119*, 3900–3907. https://doi.org/10.1182/blood-2011-12-376749.

Nguyen, D. H., & Laird, D. J. (2019). Apoptosis in the fetal testis eliminates developmentally defective germ cell clones. *bioRxiv*. https://doi.org/10.1101/601013.

Nieuwkoop, P. D. (1947). Experimental observations on the origin and determination of the germ cells, and on the development of the lateral plates and germ ridges in the urodeles. *Archives Néerlandaises de Zoologie, 8*(1), 1–205.

Nieuwkoop, P. D., & Sutasurya, L. A. (1979). Primordial germ cells in the chordates: Embryogenesis and phylogenesis. In *Volume 7 of developmental and cell biology series*. Cambridge University Press. ISBN: 0521223032. CUP Archive.

Nieuwkoop, P. D., & Sutasurya, L. A. (1981). Primordial germ cells in the invertebrates: From epigenesis to preformation. In *Volume 10 of developmental and cell biology series*. CUP Archive. ISBN: 0521221897.

Nikolic, A., Volarevic, V., Armstrong, L., Lako, M., & Stojkovic, M. (2016). Primordial germ cells: Current knowledge and perspectives. *Stem Cells International, 2016*, 1741072–1741078. https://doi.org/10.1155/2016/1741072.

Nishikawa, K., & Kinjo, A. R. (2018). Mechanism of evolution by genetic assimilation: Equivalence and independence of genetic mutation and epigenetic modulation in phenotypic expression. *Biophysical Reviews, 10*, 667–676. https://doi.org/10.1007/s12551-018-0403-x.

Ohinata, Y., Payer, B., O'Carroll, D., Ancelin, K., Ono, Y., Sano, M., et al. (2005). Blimp1 is a critical determinant of the germ cell lineage in mice. *Nature, 436*, 207–213. https://doi.org/10.1038/nature03813.

Ohno, M. (2018). Spontaneous de novo germline mutations in humans and mice: Rates, spectra, causes and consequences. *Genes & Genetic Systems, 94*, 13–22. https://doi.org/10.1266/ggs.18-00015.

Ohkubo, Y., Shirayoshi, Y., & Nakatsuji, N. (1996). Autonomous regulation of proliferation and growth arrest in mouse primordial germ cells studied by mixed and clonal cultures. *Experimental Cell Research, 222*, 291–297. https://doi.org/10.1006/excr.1996.0037.

Ohta, K., Yamamoto, M., Lin, Y., Hogg, N., Akiyama, H., Behringer, R. R., et al. (2012). Male differentiation of germ cells induced by embryonic age-specific Sertoli cells in mice. *Biology of Reproduction, 86*, 112. https://doi.org/10.1095/biolreprod.111.095943.

Percharde, M., Wong, P., & Ramalho-Santos, M. (2017). Global hypertranscription in the mouse embryonic germline. *Cell Reports, 19*, 1987–1996. https://doi.org/10.1016/j.celrep.2017.05.036.

Pfeifer, S. P. (2017). Direct estimate of the spontaneous germ line mutation rate in African green monkeys. *Evolution, 71*, 2858–2870. https://doi.org/10.1111/evo.13383.

Pinz, K. G., Shibutani, S., & Bogenhagen, D. F. (1995). Action of mitochondrial DNA polymerase gamma at sites of base loss or oxidative damage. *The Journal of Biological Chemistry, 270*, 9202–9206.

Polani, P. E., & Crolla, J. A. (1991). A test of the production line hypothesis of mammalian oogenesis. *Human Genetics, 88*, 64–70.

Rahbari, R., Wuster, A., Lindsay, S. J., Hardwick, R. J., Alexandrov, L. B., Al Turki, S., et al. (2016). Timing, rates and spectra of human germline mutation. *Nature Genetics, 48*, 126–133.

Raj, A., Peskin, C. S., Tranchina, D., Vargas, D. Y., & Tyagi, S. (2006). Stochastic mRNA synthesis in mammalian cells. *PLoS Biology, 4*, e309. https://doi.org/10.1371/journal.pbio.0040309.

Raj, A., & van Oudenaarden, A. (2008). Nature, nurture, or chance: Stochastic gene expression and its consequences. *Cell, 135*, 216–226. https://doi.org/10.1016/j.cell.2008.09.050.

Ramasamy, S., Wang, H., Quach, H. N. B., & Sampath, K. (2006). Zebrafish Staufen1 and Staufen2 are required for the survival and migration of primordial germ cells. *Developmental Biology*, *292*, 393–406. https://doi.org/10.1016/j.ydbio.2006.01.014.

Raz, E. (2002). Primordial germ cell development in zebrafish. *Seminars in Cell & Developmental Biology*, *13*, 489–495.

Raz, E., & Reichman-Fried, M. (2006). Attraction rules: Germ cell migration in zebrafish. *Current Opinion in Genetics & Development*, *16*, 355–359. https://doi.org/10.1016/j.gde.2006.06.007.

Reichman-Fried, M., Minina, S., & Raz, E. (2004). Autonomous modes of behavior in primordial germ cell migration. *Developmental Cell*, *6*, 589–596.

Reizel, Y., Itzkovitz, S., Adar, R., Elbaz, J., Jinich, A., Chapal-Ilani, N., et al. (2012). Cell lineage analysis of the mammalian female germline. *PLoS Genetics*, *8*, e1002477. https://doi.org/10.1371/journal.pgen.1002477.

Reznik, B., Cincotta, S. A., Jaszczak, R. G., Mateo, L. J., Shen, J., Cao, M., et al. (2019). Heterogeneity of transposon expression and activation of the repressive network in human fetal germ cells. *Development*, *146*. dev171157.

Richardson, S. R., Gerdes, P., Gerhardt, D. J., Sanchez-Luque, F. J., Bodea, G.-O., Muñoz-Lopez, M., et al. (2017). Heritable L1 retrotransposition in the mouse primordial germline and early embryo. *Genome Research*, *27*(8), 1395–1405. http://doi:10.1101/gr.219022.116.

Richardson, B. E., & Lehmann, R. (2010). Mechanisms guiding primordial germ cell migration: Strategies from different organisms. *Nature*, *11*, 37–49. https://doi.org/10.1038/nrm2815.

Roach, J. C., Glusman, G., Smit, A. F. A., Huff, C. D., Hubley, R., Shannon, P. T., et al. (2010). Analysis of genetic inheritance in a family quartet by whole-genome sequencing. *Science*, *328*, 636–639.

Rohrschneider, M. R., & Nance, J. (2013). The union of somatic gonad precursors and primordial germ cells during *Caenorhabditis elegans* embryogenesis. *Developmental Biology*, *379*, 139–151. https://doi.org/10.1016/j.ydbio.2013.03.019.

Rucker, E. B., Dierisseau, P., Wagner, K.-U., Garrett, L., Wynshaw-Boris, A., Flaws, J. A., et al. (2000). Bcl-x and Bax regulate mouse primordial germ cell survival and apoptosis during embryogenesis. *Molecular Endocrinology*, *14*, 1038–1052. https://doi.org/10.1210/mend.14.7.0465.

Runyan, C., Gu, Y., Shoemaker, A., Looijenga, L., & Wylie, C. (2008). The distribution and behavior of extragonadal primordial germ cells in Bax mutant mice suggest a novel origin for sacrococcygeal germ cell tumors. *The International Journal of Developmental Biology*, *52*(4), 333–344. https://doi.org/10.1387/ijdb.072486cr.

Runyan, C., Schaible, K., Molyneaux, K., Wang, Z., Levin, L., & Wylie, C. (2006). Steel factor controls midline cell death of primordial germ cells and is essential for their normal proliferation and migration. *Development*, *133*, 4861–4869. https://doi.org/10.1242/dev.02688.

Rutherford, S. L., & Lindquist, S. (1998). Hsp90 as a capacitor for morphological evolution. *Nature*, *396*, 336–342. https://doi.org/10.1038/24550.

Saitou, M., Payer, B., O'Carroll, D., Ohinata, Y., & Surani, M. A. (2005). Blimp1 and the emergence of the germ line during development in the mouse. *Cell Cycle*, *4*, 1736–1740. https://doi.org/10.4161/cc.4.12.2209.

Sano, H., Renault, A. D., & Lehmann, R. (2005). Control of lateral migration and germ cell elimination by the Drosophila melanogaster lipid phosphate phosphatases Wunen and Wunen 2. *The Journal of Cell Biology*, *171*, 675. https://doi.org/10.1083/jcb.200506038.

Santos, A. C., & Lehmann, R. (2004). Germ cell specification and migration in Drosophila and beyond. *Current Biology*, *14*, R578–R589. https://doi.org/10.1016/j.cub.2004.07.018.

Sarikaya, D. P., & Extavour, C. G. (2015). The Hippo pathway regulates homeostatic growth of stem cell niche precursors in the Drosophila ovary. *PLoS Genetics*, *11*, e1004962.

Schaner, C. E., Deshpande, G., Schedl, P. D., & Kelly, W. G. (2003). A conserved chromatin architecture marks and maintains the restricted germ cell lineage in worms and flies. *Developmental Cell*, *5*, 747–757. https://doi.org/10.1016/S1534-5807(03)00327-7.

Schroeder, T. (2010). Hematopoietic stem cell heterogeneity: Subtypes, not unpredictable behavior. *Cell Stem Cell*, *6*, 203–207. https://doi.org/10.1016/j.stem.2010.02.006.

Ségurel, L., Wyman, M. J., & Przeworski, M. (2014). Determinants of mutation rate variation in the human germline. *Annual Review of Genomics and Human Genetics*, *15*, 47–70.

Seisenberger, S., Andrews, S., Krueger, F., Arand, J., Walter, J., Santos, F., et al. (2012). The dynamics of genome-wide DNA methylation reprogramming in mouse primordial germ cells. *Molecular Cell*, *48*, 849–862. https://doi.org/10.1016/j.molcel.2012.11.001.

Seki, Y., Hayashi, K., Itoh, K., Mizugaki, M., Saitou, M., & Matsui, Y. (2005). Extensive and orderly reprogramming of genome-wide chromatin modifications associated with specification and early development of germ cells in mice. *Developmental Biology*, *278*, 440–458. https://doi.org/10.1016/j.ydbio.2004.11.025.

Seki, Y., Yamaji, M., Yabuta, Y., Sano, M., Shigeta, M., Matsui, Y., et al. (2007). Cellular dynamics associated with the genome-wide epigenetic reprogramming in migrating primordial germ cells in mice. *Development*, *134*, 2627–2638. https://doi.org/10.1242/dev.005611.

Seoane, J. (2017). Cancer: Division hierarchy leads to cell heterogeneity. *Nature*, *549*, 164–166. https://doi.org/10.1038/nature23546.

Seydoux, G., & Braun, R. E. (2006). Pathway to totipotency: Lessons from germ cells. *Cell*, *127*, 891–904. https://doi.org/10.1016/j.cell.2006.11.016.

Slaidina, M., & Lehmann, R. (2017). Quantitative differences in a single maternal factor determine survival probabilities among drosophila germ cells. *Current Biology*, *27*, 291–297. https://doi.org/10.1016/j.cub.2016.11.048.

Smeds, L., Qvarnstrom, A., & Ellegren, H. (2016). Direct estimate of the rate of germline mutation in a bird. *Genome Research*, *26*, 1211–1218. https://doi.org/10.1101/gr.204669.116.

Smith, J. L., Wilson, J. E., & Macdonald, P. M. (1992). Overexpression of oskar directs ectopic activation of nanos and presumptive pole cell formation in *Drosophila* embryos. *Cell*, *70*, 849–859.

Snee, M. J., & Macdonald, P. M. (2004). Live imaging of nuage and polar granules: Evidence against a precursor-product relationship and a novel role for Oskar in stabilization of polar granule components. *Journal of Cell Science*, *117*, 2109–2120. https://doi.org/10.1242/jcs.01059.

Snee, M. J., & Macdonald, P. M. (2009). Dynamic organization and plasticity of sponge bodies. *Developmental Dynamics*, *238*, 918–930. https://doi.org/10.1002/dvdy.21914.

Snow, M. H. L. (1977). Gastrulation in the mouse: Growth and regionalization of the epiblast. *Journal of Embryology and Experimental Morphology*, *42*, 293–303.

Solter, D. (2016). Preformation versus epigenesis in early mammalian development. *Current Topics in Developmental Biology*, *117*, 377–391. https://doi.org/10.1016/bs.ctdb.2015.11.006.

Soper, S. F. C., van der Heijden, G. W., Hardiman, T. C., Goodheart, M., Martin, S. L., de Boer, P., et al. (2008). Mouse Maelstrom, a component of nuage, is essential for spermatogenesis and transposon repression in meiosis. *Developmental Cell*, *15*, 285–297. https://doi.org/10.1016/j.devcel.2008.05.015.

Soriano, P., & Jaenisch, R. (1986). Retroviruses as probes for mammalian development: Allocation of cells to the somatic and germ cell lineages. *Cell*, *46*(1), 19–29.

Specchia, V., Piacentini, L., Tritto, P., Fanti, L., D'Alessandro, R., Palumbo, G., et al. (2010). Hsp90 prevents phenotypic variation by suppressing the mutagenic activity of transposons. *Nature*, *463*, 662–665. https://doi.org/10.1038/nature08739.

Spencer, S. L., Gaudet, S., Albeck, J. G., Burke, J. M., & Sorger, P. K. (2009). Non-genetic origins of cell-to-cell variability in TRAIL-induced apoptosis. *Nature*, *459*, 428–432. https://doi.org/10.1038/nature08012.

Spiller, C., Burnet, G., & Bowles, J. (2017). Regulation of fetal male germ cell development by members of the TGFβ superfamily. *Stem Cell Research*, *24*, 174–180. https://doi.org/10.1016/j.scr.2017.07.016.

Spiller, C., Koopman, P., & Bowles, J. (2017). Sex determination in the mammalian germline. *Annual Review of Genetics*, *51*, 265–285. https://doi.org/10.1146/annurev-genet-120215-035449.

Spudich, J. L., & Koshland, D. E. J. (1976). Non-genetic individuality: Chance in the single cell. *Nature*, *262*, 467–471.

Stallock, J., Molyneaux, K., Schaible, K., Knudson, C. M., & Wylie, C. (2003). The pro-apoptotic gene Bax is required for the death of ectopic primordial germ cells during their migration in the mouse embryo. *Development*, *130*, 6589–6597. https://doi.org/10.1242/dev.00898.

Stepanik, V., Dunipace, L., Bae, Y.-K., Macabenta, F., Sun, J., Trisnadi, N., et al. (2016). The migrations of Drosophila muscle founders and primordial germ cells are interdependent. *Development*, *143*, 3206–3215. https://doi.org/10.1242/dev.134346.

Stewart, J. B., & Chinnery, P. F. (2015). The dynamics of mitochondrial DNA heteroplasmy: Implications for human health and disease. *Nature Reviews. Genetics*, *16*, 530–542. https://doi.org/10.1038/nrg3966.

Strome, S., & Lehmann, R. (2007). Germ versus soma decisions: Lessons from flies and worms. *Science*, *316*, 392–393. https://doi.org/10.1126/science.1140846.

Strome, S., & Wood, W. B. (1983). Generation of asymmetry and segregation of germ-line granules in early C. elegans embryos. *Cell*, *35*, 15–25.

Sulston, J. E., Schierenberg, E., White, J. G., & Thomson, J. N. (1983). The embryonic cell lineage of the nematode Caenorhabditis elegans. *Developmental Biology*, *100*, 64–119.

Suter, D. M., Molina, N., Gatfield, D., Schneider, K., Schibler, U., & Naef, F. (2011). Mammalian genes are transcribed with widely different bursting kinetics. *Science*, *332*, 472–474. https://doi.org/10.1126/science.1198817.

Suzuki, A., & Saga, Y. (2008). Nanos2 suppresses meiosis and promotes male germ cell differentiation. *Genes & Development*, *22*, 430–435. https://doi.org/10.1101/gad.1612708.

Tam, P. P., & Snow, M. H. (1981). Proliferation and migration of primordial germ cells during compensatory growth in mouse embryos. *Journal of Embryology and Experimental Morphology*, *64*, 133–147.

Tam, P. P., & Zhou, S. X. (1996). The allocation of epiblast cells to ectodermal and germ-line lineages is influenced by the position of the cells in the gastrulating mouse embryo. *Developmental Biology*, *178*(1), 124–132.

Teixeira, F. K., Okuniewska, M., Malone, C. D., Coux, R.-X., Rio, D. C., & Lehmann, R. (2017). piRNA-mediated regulation of transposon alternative splicing in the soma and germ line. *Nature*, *552*, 268–272. https://doi.org/10.1038/nature25018.

Uchimura, A., Higuchi, M., Minakuchi, Y., Ohno, M., Toyoda, A., Fujiyama, A., et al. (2015). Germline mutation rates and the long-term phenotypic effects of mutation accumulation in wild-type laboratory mice and mutator mice. *Genome Research*, *25*, 1125–1134.

Ueno, H., Turnbull, B. B., & Weissman, I. L. (2009). Two-step oligoclonal development of male germ cells. *PNAS*, *106*, 175–180. https://doi.org/10.1073/pnas.0810325105.

Underwood, E. M., Caulton, J. H., Allis, C. D., & Mahowald, A. P. (1980). Developmental fate of pole cells in Drosophila melanogaster. *Developmental Biology*, 77, 303–314. https://doi.org/10.1016/0012-1606(80)90476-5.

Vasiliauskaitė, L., Berrens, R. V., Ivanova, I., Carrieri, C., Reik, W., Enright, A. J., et al. (2018). Defective germline reprogramming rewires the spermatogonial transcriptome. *Nature Structural & Molecular Biology*, *25*, 394–404. https://doi.org/10.1038/s41594-018-0058-0.

Wai, T., Teoli, D., & Shoubridge, E. A. (2008). The mitochondrial DNA genetic bottleneck results from replication of a subpopulation of genomes. *Nature Genetics*, *40*, 1484–1488. https://doi.org/10.1038/ng.258.

Walter, C. A., Intano, G. W., McCarrey, J. R., McMahan, C. A., & Walter, R. B. (1998). Mutation frequency declines during spermatogenesis in young mice but increases in old mice. *PNAS*, *95*, 10015–10019.

Wang, R. A., Nakane, P. K., & Koji, T. (1998). Autonomous cell death of mouse male germ cells during fetal and postnatal period. *Biology of Reproduction*, *58*, 1250–1256.

Wang, N., Zheng, J., Chen, Z., Liu, Y., Dura, B., Kwak, M., et al. (2019). Single-cell microRNA-mRNA co-sequencing reveals non-genetic heterogeneity and mechanisms of microRNA regulation. *Nature Communications*, *10*, 1–12. https://doi.org/10.1038/s41467-018-07981-6.

Weidinger, G., Stebler, J., Slanchev, K., Dumstrei, K., Wise, C., Lovell-Badge, R., et al. (2003). Dead end, a novel vertebrate germ plasm component, is required for zebrafish primordial germ cell migration and survival. *Current Biology*, *13*, 1429–1434. https://doi.org/10.1016/S0960-9822(03)00537-2.

Weissman, I. L. (2015). Stem cells are units of natural selection for tissue formation, for germline development, and in cancer development. *PNAS*, *112*, 8922–8928. https://doi.org/10.1073/pnas.1505464112.

West, J. A., Viswanathan, S. R., Yabuuchi, A., Cunniff, K., Takeuchi, A., Park, I.-H., et al. (2009). A role for Lin28 in primordial germ-cell development and germ-cell malignancy. *Nature*, *460*, 909–913. https://doi.org/10.1038/nature08210.

Western, P. S., Miles, D. C., van den Bergen, J. A., Burton, M., & Sinclair, A. H. (2008). Dynamic regulation of mitotic arrest in fetal male germ cells. *Stem Cells*, *26*, 339–347. https://doi.org/10.1634/stemcells.2007-0622.

Whittle, C. A., & Extavour, C. G. (2016). Refuting the hypothesis that the acquisition of germ plasm accelerates animal evolution. *Nature Communications*, 7, 12637. https://doi.org/10.1038/ncomms12637.

Wolke, U., Jezuit, E. A., & Priess, J. R. (2007). Actin-dependent cytoplasmic streaming in C. elegans oogenesis. *Development*, *134*, 2227–2236. https://doi.org/10.1242/dev.004952.

Wong, W. S., Solomon, B. D., Bodian, D. L., Kothiyal, P., Eley, G., Huddleston, K. C., et al. (2016). New observations on maternal age effect on germline *de novo* mutations. *Nature Communications*, 7, 10486.

Yabuta, Y., Kurimoto, K., Ohinata, Y., Seki, Y., & Saitou, M. (2006). Gene expression dynamics during germline specification in mice identified by quantitative single-cell gene expression profiling. *Biology of Reproduction*, *75*, 705–716.

Yamada, Y., Davis, K. D., & Coffman, C. R. (2008). Programmed cell death of primordial germ cells in Drosophila is regulated by p53 and the Outsiders monocarboxylate transporter. *Development*, *135*, 207–216. https://doi.org/10.1242/dev.010389.

Ying, Y., Liu, X. M., Marble, A., Lawson, K. A., & Zhao, G. Q. (2000). Requirement of Bmp8b for the generation of primordial germ cells in the mouse. *Molecular Endocrinology*, *14*, 1053–1063.

Young, R. A. (2011). Control of the embryonic stem cell state. *Cell*, *144*, 940–954. https://doi.org/10.1016/j.cell.2011.01.032.

Youngren, K. K., Coveney, D., Peng, X., Bhattacharya, C., Schmidt, L. S., Nickerson, M. L., et al. (2005). The *Ter* mutation in the dead end gene causes germ cell loss and testicular germ cell tumours. *Nature*, *435*, 360–364. https://doi.org/10.1038/nature03595.

Zhang, J., Zhang, M., Acampora, D., Vojtek, M., Yuan, D., Simeone, A., et al. (2018). OTX2 restricts entry to the mouse germline. *Nature*, *562*, 595–599. https://doi.org/10.1038/s41586-018-0581-5.

Zheng, C.-J., Luebeck, E. G., Byers, B., & Moolgavkar, S. H. (2005). On the number of founding germ cells in humans. *Theoretical Biology & Medical Modelling*, *2*, 32. https://doi.org/10.1186/1742-4682-2-32.

CHAPTER SIX

Germline stem cell homeostasis

Jonathan O. Nelson[†], Cuie Chen[†], Yukiko M. Yamashita*
Life Sciences Institute, Department of Cell and Developmental Biology, Howard Hughes Medical Institute, University of Michigan, Ann Arbor, MI, United States
*Corresponding author: e-mail address: yukikomy@umich.edu

Contents

1. Introduction 204
2. Specifying and maintaining germline stem cells 205
   2.1 The niche signaling that maintains germline stem cells 205
   2.2 Cell adhesion maintains the niche architecture 208
   2.3 The mechanisms that restrict the effective range of the niche signaling 209
   2.4 Cell intrinsic factors required for GSC self-renewal 210
   2.5 Somatic control of germline differentiation 212
3. Asymmetric stem cell division 213
   3.1 The mechanisms of asymmetric stem cell division 214
   3.2 Asymmetric segregation of cellular components 216
4. Transit-amplifying divisions to reduce the burden on stem cells 218
5. Dedifferentiation maintains GSC populations beyond the lifespan of individual cells 220
   5.1 Dedifferentiation creates GSCs to replace lost GSCs 221
   5.2 The mechanism of dedifferentiation 222
   5.3 Are dedifferentiated GSCs equal to native GSCs? 224
6. Cell death in GSC homeostasis 225
   6.1 SG death induced by somatic CC apoptosis shifts tissue homeostasis during starvation 225
   6.2 SG death as a mechanism to protect germline genome integrity 226
   6.3 Differential sensitivity to cell death among distinct cell populations 227
7. The disruption of germline homeostasis during aging 228
8. Maintaining the germline over successive generations 229
9. Conclusion 233
Acknowledgments 233
References 233

[†] These authors equally contributed to this work.

Current Topics in Developmental Biology, Volume 135
ISSN 0070-2153
https://doi.org/10.1016/bs.ctdb.2019.04.006

Abstract

In many species, germline stem cells (GSCs) function to sustain gametogenesis throughout the life of organismal life span. As the source of gametes, the only cell type that can pass the genetic information to the next generation, GSCs play a fundamental role in maximizing the quantity of gametes that animals produce, while ensuring their highest quality. GSCs are maintained by the signals from their niches, and germ cells that exited the niche undergo differentiation to generate functional gametes. GSC population is sustained by a multitude of mechanisms such as asymmetric stem cell divisions and dedifferentiation of partially differentiated germ cells. In this review, we summarize the mechanisms that maintain GSC homeostasis to ensure life-long production of functional gametes.

1. Introduction

Germ cells are the only cell type that is passed to the next generation. Whereas essentially all somatic cells are mortal unless transformed, the germline has continued their lineage, and all living organisms on the planet can be tracked back to their parents' parents' parents' parents'... germ cells, eventually back to the gonad of ancestral species, and to the origin of animals. Evolutionarily speaking, passage of germline to the next generation is the whole and sole purpose of life, and accordingly, it is probably not an exaggeration to state that understanding germ cells is understanding life. Nevertheless, we have little understanding how the germline achieves this remarkable feat of immortality. The secret(s) to germline immortality must lie somewhere between germ cell formation during embryogenesis and terminal differentiation of eggs and sperms, all of which processes are reasonably well documented to date. This likely means that we must be seeing those secrets without understanding them.

In addition to infinity of genome transmission through germline, evolution must have selected strategies that maximize the number of gametes that successfully produce the next generation of the gametes. Many species utilize the stem cell system, germline stem cells (GSCs), to sustain gamete production throughout their life. GSCs continue to proliferate throughout reproductive life of organisms to generate cells that undergo differentiation, while self-renewing themselves. The studies in past decades have revealed the mechanisms that regulate GSC identify, proliferation and maintenance. GSCs have many common features with somatic stem cells in the mechanisms of self-renewal and differentiation.

In this review, we summarize recent progress in understanding how GSCs are maintained and how GSCs sustain continuous gamete production throughout the life of animals. We will also discuss how these processes of GSC homeostasis may be possibly tied to germline immortality. As the stem cells in mammalian spermatogenesis (spermatogonial stem cells (SSCs)) is covered in a separate chapter (chapter "Heterogeneous, dynamic, and stochastic nature of mammalian spermatogenic stem cells" by Yoshida), we will mostly focus on the knowledge from invertebrate model systems, mostly *Drosophila*, while referring to knowledge in mammalian SSCs when relevant.

2. Specifying and maintaining germline stem cells

In most species studied to date, germ cells are set aside early on during embryogenesis as primordial germ cells (PGCs), either by maternally deposited fate determinants as in *Drosophila* (Williamson & Lehmann, 1996), *C. elegans* (Wang & Seydoux, 2013), *Xenopus* (Kloc et al., 2001) and zebrafish (Raz, 2003) or by induction as in mammals (Saitou & Yamaji, 2012; Tang, Kobayashi, Irie, Dietmann, & Surani, 2016). These PGCs migrate within the embryos to find somatic gonadal cells to form functional gonads. In gonads, PGCs may become GSCs, which continue to produce gametes throughout the reproductive life span of organisms in many systems including invertebrate as well as zebrafish and medaka fish, although mammalian females apparently lack adult GSC population (Draper, McCallum, & Moens, 2007; Lei & Spradling, 2013; Nakamura, Kobayashi, Nishimura, Higashijima, & Tanaka, 2010; Tanaka, 2016).

2.1 The niche signaling that maintains germline stem cells

After cells acquire the identity of germ cells as PGCs, they will form a gonad by interacting with somatic support cells. In many species, a subset of PGCs become GSCs to continue to produce gametes throughout the reproductive life span of organisms. GSCs reside in a specialized microenvironment, niches, to maintain their identity. The essence of the stem cell niche function is to create signaling environment that allows stem cell self-renewal and production of daughter cells that are committed to differentiation. By making self-renewal dependent on niche-derived extrinsic factors, the niche mechanism also safeguards against overproliferation of stem cells. Accordingly, the niche component cells are mostly post-mitotic, although niche cells can be induced to proliferate under somewhat non-physiological conditions

(Greenspan & Matunis, 2018; Voog et al., 2014). Model organisms such as *Drosophila* and *C. elegans* have been instrumental in studying the niche mechanisms that sustain GSCs owing to their relatively simple anatomy (Lehmann, 2012; Spradling, Fuller, Braun, & Yoshida, 2011). In contrast to these "closed niches" in *Drosophila* and *C. elegans*, mouse GSCs (spermatogonial stem cells, SSCs) do not appear to stay in a defined space and move around within the seminiferous tubules (Yoshida, Sukeno, & Nabeshima, 2007), while requiring signaling ligands such as GDNF from the Sertoli cells, which is received by the receptor GFRα1 of SSCs (Meng et al., 2000; Spradling et al., 2011).

The *Drosophila* female GSC niche is organized by post-mitotic terminal filament (TF) cells and cap cells, supporting 2–3 GSCs per ovariole (Fig. 1A) (Decotto & Spradling, 2005; Kirilly, Wang, & Xie, 2011; Morris & Spradling, 2011; Xie & Spradling, 2000). GSC identity depends on BMP (bone morphogenic protein) ligands Decapentaplegic (Dpp) and Glass bottom boat (Gbb), which are produced and secreted from cap cells (Chen & McKearin, 2003). Dpp/Gbb ligands in turn activate Thickveins (Tkv) and Saxophone (Sax) receptors in GSCs to specify GSC identity. Additionally, JAK-STAT signaling in the niche cells promote the production of Dpp to maintain GSCs (Decotto & Spradling, 2005; Lopez-Onieva, Fernandez-Minan, & Gonzalez-Reyes, 2008; Wang, Li, & Cai, 2008). The *Drosophila* male GSC niche is organized by a cluster of post-mitotic somatic hub cells, as well as cyst stem cells (CySCs) (Fig. 1B). GSCs are attached to the hub to receive the niche ligands Upd (Kiger, Jones, Schulz, Rogers, & Fuller, 2001; Tulina & Matunis, 2001) and Dpp/Gbb (Kawase, Wong, Ding, & Xie, 2004; Schulz et al., 2004; Shivdasani & Ingham, 2003). CySCs, which are also attached to the hub, encapsulate GSCs to help create the GSC niche environment together with hub cells (Leatherman & Dinardo, 2008, 2010).

A number of other signaling pathways have been identified that regulate GSC behavior and/or niche formation, including Wnt signaling, Hedgehog (Hh) signaling and Notch signaling, revealing the complexity of the niche signaling. Wnt signaling in escort cells, somatic cells that surround GSCs and differentiating germ cells, is required to restrict the BMP signaling outside the niche, thereby promoting differentiation of germ cells in the *Drosophila* ovary (Luo, Wang, Fan, Liu, & Cai, 2015; Mottier-Pavie, Palacios, Eliazer, Scoggin, & Buszczak, 2016; Upadhyay et al., 2016; Wang et al., 2015). Hh is required in somatic cyst stem cells (CySC) for their maintenance in the *Drosophila* testis (Amoyel, Sanny, Burel, & Bach, 2013; Amoyel, Simons, & Bach, 2014; Lv et al., 2016; Michel, Kupinski, Raabe, & Bokel, 2012; Zhang, Lv, Jiang, Zhang, & Zhao, 2013). Hh likely regulates CySC

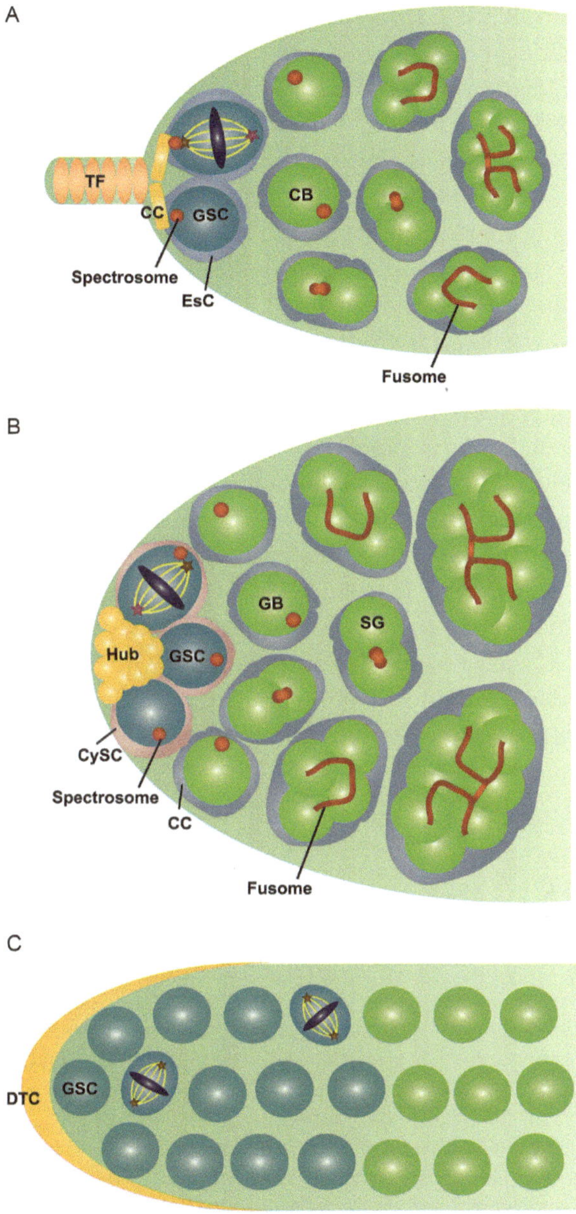

Fig. 1 Anatomy of GSC niches (A) *Drosophila* female GSC niche. The niche contains three types of somatic cells: terminal filament cells, cap cells and escort cells. The GSC divides asymmetrically to yield one GSC that remains in the niche and one cystoblast (CB) that exits the niche. The CB initiates differentiation and undergoes four rounds of division with incomplete cytokinesis to form an interconnected 16-cell cyst. *(Continued)*

maintenance through regulation of niche competition (Amoyel et al., 2014). In the *Drosophila* ovary, Hh is produced by cap cells, received by escort cells, where it activates the production of BMP ligands to sustain GSC self-renewal (Rojas-Rios, Guerrero, & Gonzalez-Reyes, 2012). Also in the *Drosophila* ovary, Notch signaling regulates the niche size, where GSCs produce ligands, Serrate and Delta, whereas the cap cells receive these ligands to activate the pathway to ultimately regulate the expression of BMP pathway components (Song, Call, Kirilly, & Xie, 2007; Ward et al., 2006).

In *C. elegans*, GSCs are maintained as a pool in the distal end of the gonad. Their niche is comprised of a single somatic cell, the distal tip cell (DTC), which extends long processes to surround the mitotically-active GSC region (Fig. 1C) (Byrd, Knobel, Affeldt, Crittenden, & Kimble, 2014; Crittenden, Leonhard, Byrd, & Kimble, 2006). Like many other niche component cells, DTCs are also post-mitotic and support GSCs for the entire reproductive life span of *C. elegans*. Germ cells initiate differentiation and transition to meiosis gradually as they move away from the DTC toward the proximal end of the germline. The maintenance of GSC identity is governed by Notch signaling, where DTC produces Notch ligand Lag-2 and GSCs express the Notch receptor GLP-1(Austin & Kimble, 1987; Crittenden, Troemel, Evans, & Kimble, 1994; Henderson, Gao, Lambie, & Kimble, 1994) (see also Kimble, 2011, for review).

2.2 Cell adhesion maintains the niche architecture

To remain in the niche, GSCs are physically attached to the niche component cells. *Drosophila* male and female GSCs attach to the hub cells and cap

Fig. 1—Cont'd Only one cell within the cyst eventually becomes oocyte, while the others become nurse cells to assist oocyte maturation. The spectrosome orients the spindle of the GSC and fusomes regulate cystocytes division and oocyte determination. The daughter centrosome (brown star) is inherited by the GSC upon division. (B) *Drosophila* male GSC niche. The niche contains somatic hub cells and CySCs. GSCs divide asymmetrically to produce one self-renewed GSC and one differentiating daughter named gonialblast (GB). The GB differentiates and undergoes four rounds of division to generate interconnected 16-cell spermatogonia (SGs). The SGs mature into spermatocytes (SCs) and subsequently give rise to sperms. CySCs also divide asymmetrically to give rise to one CySC and one cyst cell (CC). A pair of CySCs encapsulates each GSC, and a pair of CCs encapsulates each differentiating germ cell. The mother centrosome (purple star) orients the spindle of the GSC and is inherited by the GSC upon division. (C) *C. elegans* GSC niche. The niche contains a single distal tip cell (DTC). GSCs divide without consistent spindle orientation. As cells move away from the DTC, they gradually differentiate and enter meiosis.

cells, respectively, via E-cadherin-mediated adherens junctions (Song, Zhu, Doan, & Xie, 2002; Yamashita, Jones, & Fuller, 2003). Formation of adherens junctions is further supported by underlying actin cytoskeletons and associated signaling, involving profilin, Rap-GEF/Rap signaling, and the transmembrane receptor tyrosine phosphatase Leukocyte-antigen-related-like (Lar) (Shields, Spence, Yamashita, Davies, & Fuller, 2014; Srinivasan, Mahowald, & Fuller, 2012; Wang et al., 2006). Adhesion to the niche is a critical aspect of GSC identity, since it is a prerequisite to receive the niche-derived self-renewal ligands. Accordingly, loss of (or weakened) adherens junctions leads to GSC loss due to differentiation (Song & Xie, 2002; Voog, D'Alterio, & Jones, 2008). In addition, the level of the adherens junction between GSC and the niche cells determines the "competitiveness" of GSCs for niche occupancy, causing GSCs with weaker adhesion to leave the niche and differentiate. On the contrary, stronger adhesion can cause non-stem cells to gain access to the niche, outcompeting existing GSCs (Jin et al., 2008; Lim et al., 2015).

In addition to GSC-niche anchorage, cell adhesion is critical for supporting the overall niche architecture. For example, in the absence of *SOCS36E* in the *Drosophila* testis, increased adhesion of CySCs to the hub due to upregulation of βPS-integrin, results in CySCs dominating the niche outcompeting the GSCs (Issigonis et al., 2009). Also, the hub is anchored to the apical tip of the testis via integrin-mediated adhesion (Tanentzapf, Devenport, Godt, & Brown, 2007).

2.3 The mechanisms that restrict the effective range of the niche signaling

Whereas GSCs are maintained by the niche, their daughters that initiate differentiation are juxtaposed to GSCs upon completion of mitosis. Considering that some self-renewal ligands can diffuse far from the source, defining the boundary of the niche between GSC and its juxtaposed daughter requires elaborate mechanisms: the niche must provide sufficient self-renewal ligands for the GSCs while depriving them from the GSCs' differentiating daughters. A few mechanisms operate to limit the diffusion of the self-renewal ligands. For example, in the *Drosophila* ovary, extracellular matrix (ECM) components surrounding the GSC niche, such as the glypican molecule Dally, restricts the diffusion of Dpp, such that its concentration is maintained high within the niche space, and low outside the niche (Guo & Wang, 2009; Hayashi, Kobayashi, & Nakato, 2009). In addition, type IV collagens such as Viking and Dcg-1 have also been shown to restrict the diffusion of Dpp, thereby forming an appropriate gradient of Dpp (Wang, Harris, Bayston, & Ashe, 2008). As in the females, the ECM plays a critical role in concentrating niche-secreted ligands within the niche, where Dally-like protein (Dlp) concentrate BMP ligands to maintain GSCs within the niche (Hayashi et al., 2009). In addition, Magu, a Dally binding protein that regulates Dpp signaling (Vuilleumier et al., 2010), is specifically expressed in the hub cells to promote GSC self-renewal (Zheng, Wang, Vargas, & DiNardo, 2011).

Another mechanism to limit the diffusion of the niche ligands is via targeted delivery/reception of the ligands. *Drosophila* male GSCs form thin cellular protrusions, microtubule-based (MT)-nanotubes, which extend into the invagination of hub cells (Inaba, Buszczak, & Yamashita, 2015). This creates a specialized surface, where Dpp secreted from the hub and Tkv produced in the GSCs engage in productive signal transduction. This mechanism allows the hub cells to avoid "broadcasting" the niche ligand Dpp, and effectively excludes differentiating germ cells from engaging in the niche signaling. The trafficking of Tkv receptor molecules from the cell body of GSCs into MT-nanotubes relies on intraflagellar transport. It remains unknown how the secretion of Dpp ligand may be limited toward the surface of MT-nanotubes. In the *Drosophila* ovary, cap cells extend cytonemes, actin-rich filopodia (Kornberg & Roy, 2014; Ramirez-Weber & Kornberg, 1999), to deliver Hedgehog (Hh) ligands specifically to adjacent escort cells (Rojas-Rios et al., 2012). Such refinement for niche signals may achieve spatial and temporal control with precision.

2.4 Cell intrinsic factors required for GSC self-renewal

Niche signaling ultimately regulates the gene expression of cell intrinsic factors within GSCs to specify their identity. Decades of work revealed many genes involved in GSC self-renewal, which revealed several common themes of GSC identity.

First, the immediate downstream effect of "self-renewal signaling" is often repression of differentiation. Directly downstream of the niche BMP signaling in the *Drosophila* male and female GSCs is the repression of Bam (Chen & McKearin, 2003; Song et al., 2004), the master regulator of differentiation (McKearin & Spradling, 1990; Ohlstein & McKearin, 1997). Bam interacts with benign gonial cell neoplasm (Bgcn) to promote differentiation (Ohlstein, Lavoie, Vef, Gateff, & McKearin, 2000). In parallel with BMP-mediated transcriptional repression of Bam, a Nanos

(Nos)-Pumillio translational repressor complex is required for GSC maintenance (Chen & McKearin, 2005; Forbes & Lehmann, 1998; Gilboa & Lehmann, 2004; Lin & Spradling, 1997; Tang et al., 2016; Wang & Lin, 2004). Nos-Pumillio complex and Bam-Bgcn complex antagonize each other in specifying GSC identity and promoting differentiation (Harris, Pargett, Sutcliffe, Umulis, & Ashe, 2011; Li, Minor, Park, McKearin, & Maines, 2009). Mei-P26 functions to prevent Bam expression and thus maintains GSCs (Li, Maines, Tastan, McKearin, & Buszczak, 2012).

Second, as is the case with any cell types, GSC fate is determined by a unique gene expression profile, which is supported by epigenetic regulation. Numerous epigenetic regulators have been identified to date, some of which regulating the core niche signaling components. GSC intrinsic epigenetic regulators include NURF (nucleosome remodeling factor) (Ables & Drummond-Barbosa, 2010; Cherry & Matunis, 2010), ISWI (Xi & Xie, 2005), SWI/SNF (He et al., 2014), Lid (H3K4me3-specific histone demethylase) (Tarayrah, Li, Gan, & Chen, 2015), histone H3 methyltransferase Su(var)3–9 (Xing & Li, 2015), histone H3 K9 trimethylase Eggless/SetDB1 (Clough, Tedeschi, & Hazelrigg, 2014; Rangan et al., 2011; Wang et al., 2011), PRC2 component E(z) (Eun et al., 2017), histone acetyl transferase Enok (Xin et al., 2013), histone acetyltransferase compex (McCarthy, Deiulio, Martin, Upadhyay, & Rangan, 2018). Other epigenetic regulators function in somatic lineage to regulate GSC self-renewal/ differentiation: dUTX (H3K27me3-specific histone demethylase) (Tarayrah, Herz, Shilatifard, & Chen, 2013), E(z) (Eun, Shi, Cui, Zhao, & Chen, 2014), histone deubiquitylase Scrawny (Buszczak, Paterno, & Spradling, 2009), histone demethylase Lsd1 (Eliazer et al., 2014; Eliazer, Shalaby, & Buszczak, 2011). Many of these epigenetic regulators function through the regulation of the major niche signaling such as JAK-STAT and BMP. Understanding the chicken-or-egg question of how niche signaling and epigenetic regulation are intertwined to achieve GSC homeostasis awaits future investigations.

In the *C. elegans* gonad, the downstream effectors of the major niche signaling, i.e., Notch signaling, are two PUF (Pumillio/FBF) proteins, FBF-1 and FBF-2. These FBF-1/FBF-2 proteins function redundantly to generally repress translation of mRNAs that promote differentiation (Crittenden et al., 2002; Kalchhauser, Farley, Pauli, Ryder, & Ciosk, 2011; Kershner & Kimble, 2010; Prasad et al., 2016), similar to the function of *Drosophila* Pumillio as described above. Another PUF protein, PUF-8, functions to maintain GSCs by promoting GSC mitosis and ensuring differentiation (Ariz, Mainpal, & Subramaniam, 2009; Mainpal, Priti, & Subramaniam, 2011; Racher & Hansen, 2012; Subramaniam & Seydoux, 2003). Two additional critical downstream targets of Notch signaling, LST-1 and SYGL-1, form a complex with FBF-1/FBF-2 to promote GSC self-renewal (Brenner & Schedl, 2016; Kershner, Shin, Hansen, & Kimble, 2014; Shin et al., 2017). Detailed examination of nascent transcripts of these genes revealed that Notch-dependent activation of transcription occurs in a steep gradient, being confined to the distal region of the gonad in close proximity to the DTC cell body, while germ cells contacting DTC protrusions at the far (proximal) side of the gonad rarely actively transcribe Notch target genes (Lee, Sorensen, Lynch, & Kimble, 2016). Likely reflecting differential activation of Notch signaling in distal vs. proximal pools of GSCs, these pools also exhibit distinct proliferation rates (Cinquin, Crittenden, Morgan, & Kimble, 2010).

2.5 Somatic control of germline differentiation

The gonad is a complex tissue where germ cells and supporting somatic cells intermingle to regulate each other for successful gametogenesis. Not only GSC self-renewal but also the differentiation of germ cells is supported by somatic cells.

In the *Drosophila* ovary, escort cells surround differentiating germ cells in germaria to support their differentiation (Fig. 1A) (Decotto & Spradling, 2005; Kirilly et al., 2011; Morris & Spradling, 2011). Similarly, in the *Drosophila* testis, a pair of somatic cyst cells (CCs) encapsulate differentiating SGs and then spermatocytes (SCs) to support their differentiation (Zoller & Schulz, 2012). Encapsulation is mediated by EGF signaling: CCs express EGF receptor (EGFR), whereas germ cells express the ligand Spitz to mediate the encapsulation of SGs by CCs via regulation of Rac and Rho (Banisch, Maimon, Dadosh, & Gilboa, 2017; Sarkar et al., 2007; Schulz, Wood, Jones, Tazuke, & Fuller, 2002). The encapsulation of germ cells by somatic cells (CCs and ECs) is critical for promoting the differentiation of SGs. When encapsulation of SG cysts by CCs is compromised, either due to the lack of Egfr signaling or absence of CCs, SGs fail to differentiate beyond the 4-cell SG stage (Fairchild, Smendziuk, & Tanentzapf, 2015; Kiger, White-Cooper, & Fuller, 2000; Lim & Fuller, 2012; Schulz et al., 2002; Tran, Brenner, & DiNardo, 2000). It was shown that CCs form a permeability barrier to isolate germ cells through formation of septate junctions between two CCs that together encapsulate SG cysts (Fairchild et al., 2015).

This permeability barrier likely promotes differentiation, at least in part by isolating SGs from niche signaling, as the disruption of the permeability barrier extends the range of BMP signals from the niche (Fairchild et al., 2015). Sertoli cells in mammalian testis provide similar somatic control (Hai et al., 2014; Spradling et al., 2011). Sertoli cells, like CCs in the *Drosophila* testis, form a permeability barrier (blood-testis barrier) to support spermatogenesis (Stanton, 2016), implying that the general logic behind spermatogonial differentiation is conserved between *Drosophila* and mammals.

3. Asymmetric stem cell division

Asymmetric divisions, which generate one stem cell and one differentiating cells, are widely observed in many stem cell systems, including *Drosophila* male and female GSCs. Generally, asymmetric stem cell divisions are considered to be a mechanism that balances self-renewal and differentiation. Although asymmetric stem cell division indeed results in balanced self-renewal and differentiation, it is not essential for a few reasons. First, GSC identity is predominantly determined by factors from the niche. Accordingly, even when GSCs divide symmetrically, the availability of the niche space sets the upper limit of the GSC numbers and thus the homeostasis of GSC numbers will not be drastically perturbed. Second, GSCs can be recreated from partially differentiated germ cells (cystocytes in the female germline and SGs in the male germline, see below) (Brawley & Matunis, 2004; Kai & Spradling, 2004), therefore loss of GSCs due to symmetric differentiation can easily be compensated. Moreover, there are many stem cell systems that do not employ asymmetric divisions. For example, mouse SSCs are stochastically maintained at a population level without any evidence of asymmetric divisions (see Chapter 7, by Yoshida) (Klein, Nakagawa, Ichikawa, Yoshida, & Simons, 2010; Nakagawa, Sharma, Nabeshima, Braun, & Yoshida, 2010). Similarly, GSCs in the *C. elegans* gonad are maintained at a population level, with the range of DTC extension predominantly determining the size of GSC population. Germ cells within the "proliferation zone" defined by DTC extension divide with no apparent asymmetries and gradually lose their proliferative potential and Notch-dependent transcriptional activation of targets as they move farther away from the distal end of the gonad (Crittenden et al., 2006; Lee et al., 2016; Rosu & Cohen-Fix, 2017). Therefore, the asymmetric stem cell division is not a requirement to maintain stem cell population in homeostasis. However, extensive studies revealed elaborate mechanisms of asymmetric divisions of *Drosophila* GSCs. We will discuss how these asymmetries might be important for the long-term function of GSCs, not necessarily limited to creating the fate asymmetry of self-renewal vs. differentiation.

3.1 The mechanisms of asymmetric stem cell division

In both *Drosophila* male and female GSCs, stem cells divide asymmetrically by orienting their mitotic spindle in the context of niche geometry such that only one daughter cell maintains intimate contact with the niche and retains stem cell identity, whereas the other daughter is displaced outside of the niche and initiates differentiation.

In *Drosophila* male GSCs, the mitotic spindle is oriented perpendicular to the hub by stereotypical centrosome positioning during interphase (Yamashita et al., 2003; Yamashita, Mahowald, Perlin, & Fuller, 2007) (Fig. 2A). The mother centrosome consistently positions near the hub-GSC junction and the daughter centrosome migrates to the distal side. Adherens junctions formed between the hub and GSCs, which help anchor GSCs to the hub as described above, also serves as a landmark to polarize GSCs toward the hub cells (Inaba, Yuan, Salzmann, Fuller, & Yamashita, 2010). Adenomatous polyposis coli (APC) 2 protein and the polarity protein Bazooka (Baz) are recruited to the hub-GSC junction in an E-cadherin dependent manner, to correctly position the centrosome to the hub-GSC junction (Fig. 2B) (Inaba, Venkei, & Yamashita, 2015; Inaba et al., 2010; Yamashita et al., 2003). Moreover, the niche ligand Upd and its receptor Dome in GSCs are directly involved in centrosome and spindle orientation in GSCs via physical interaction of Dome with the microtubule-binding protein Eb1 (Fig. 2A) (Chen et al., 2018). These results indicate that the niche can directly orient stem cell division to ensure an asymmetric outcome, obligatorily combining stem cell self-renewal and asymmetric division.

In the *Drosophila* female GSCs, the spindle is anchored to the spectrosome, a germline-specific organelle that is always located at the apical side of the GSCs (Deng & Lin, 1997). In comparison with female GSC, male GSCs position the spectrosome randomly during interphase (Yamashita et al., 2003), revealing the distinct mechanisms for orienting the GSC spindle in male vs. female (the centrosome vs. spectrosome, respectively). However, female GSCs also exhibit cell cycle-dependent centrosome orientation toward the cap cells (Lu et al., 2012; Salzmann et al., 2014). Also, in male GSCs, in the absence of the centrosome, the spectrosome becomes

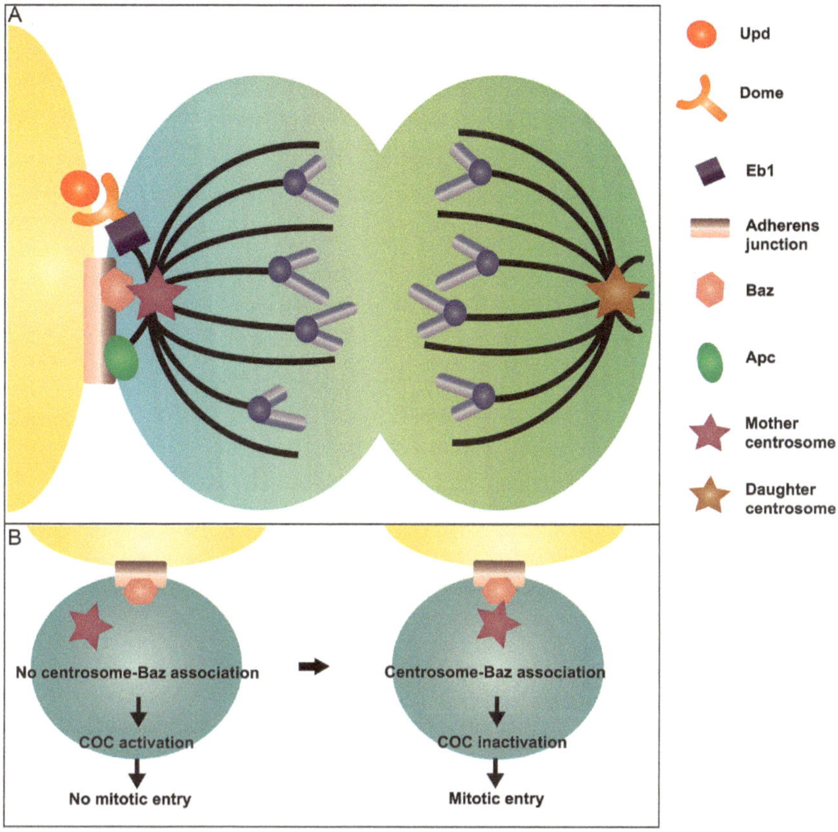

Fig. 2 Asymmetric division of the *Drosophila* male GSC. (A) Various mechanisms of asymmetric GSC division. The male GSC spindle orientation is set up by stereotypical centrosome positioning, where the mother centrosome is always anchored to the hub and GSC interface. Apc2 and Baz regulate GSC divisions via connecting the adherens junction to the centrosome via astral MTs. Also, the niche ligand Upd and its receptor Dome directly instruct spindle orientation via receptor-Eb1 interaction. (B) The centrosome orientation checkpoint (COC) and Baz function in COC. The COC is a mechanism in the *Drosophila* male GSCs to monitor centrosome orientation in interphase and blocks mitotic entry if centrosomes are not oriented correctly. Baz is a critical regulator for the COC, and Baz-centrosome association is recognized by the COC as correct centrosome orientation.

consistently positioned near the hub cells and anchors the spindle pole, leading to correct spindle orientation (Yuan, Chiang, Cheng, Salzmann, & Yamashita, 2012), suggesting that the spectrosome and the centrosome systems are two parallel mechanisms to ensure correct spindle orientation.

GSC spindle orientation is further backed up by a checkpoint mechanism: the *Drosophila* male GSCs have a "centrosome orientation checkpoint"

(COC), which actively monitors correct centrosome orientation before mitotic entry (Cheng et al., 2008). The COC arrests GSCs in G2 phase of the cell cycle upon detecting misoriented centrosomes (Cheng et al., 2008; Inaba et al., 2010; Venkei & Yamashita, 2015). Baz, which forms a specialized platform at the hub-GSC junction, to which the centrosome closely associates prior to mitotic entry, is likely sensed by GSCs as correct centrosome orientation, triggering mitotic entry (Fig. 2B) (Inaba, Venkei, et al., 2015). Studies in the past decade identified several components of COCs: Cnn, a major pericentrosomal material is required for COC, possibly by recruiting components of COCs to the centrosome (Inaba et al., 2010; Venkei & Yamashita, 2015). Par-1, a kinase, that is also involved in COC partly by phosphorylating Baz (Inaba, Venkei, et al., 2015). Additionally, Par-1 functions to sequester Cyclin A to the spectrosome such that Cyclin A cannot promote mitotic entry unless the COC is cleared to allow mitotic entry (Yuan et al., 2012). A checkpoint-like activity was observed in female GSCs as well, where GSCs arrest in mitotic prophase upon sensing incorrect centrosome/spindle orientation (Lu et al., 2012).

3.2 Asymmetric segregation of cellular components

In the examples detailed above, GSC identity is mostly defined by the niche signaling, and asymmetric segregation of cell-intrinsic fate determinants, as has been observed in *Drosophila* neuroblasts and the *C. elegans* zygote (Gonczy, 2008; Knoblich, 2008), has not been reported. Given that ectopic activation of niche signaling is often sufficient to induce GSC tumors, it is likely that the GSC fate is predominantly determined by the niche signaling. Indeed, GSCs' immediate daughters, GBs, are observed to frequently "crawl back" into the niche to assume GSC identity (Sheng & Matunis, 2011), suggesting that there are no dominating cell-intrinsic fate determinants that make a GB a GB (or a GSC a GSC). Despite this, asymmetric segregation of many cellular components has been observed during GSC divisions. These asymmetries do not appear to dominantly determine the fate of germ cells per se (self-renew or differentiate), but likely contribute to certain aspects of asymmetric GSC divisions.

One of the best-established asymmetries with important functional roles is centrosome inheritance. The centrosome is the major microtubule organizing center (MTOC) in animal cells and plays key roles for cell division and polarity. Two centrosomes within a cell are inherently asymmetric, very much like DNA strands, due to semi-conservative nature of their duplication (Nigg & Stearns, 2011). A centrosome contains two centrioles, one mother centriole and one daughter centriole, where the mother was the template for the daughter production in the previous cell cycle. When this centrosome undergoes replication, each of its two centrioles becomes a template for new daughter centrioles. Now each cell has two centrosomes: one with the oldest centriole and a new centriole, and the other with the first-time-mother centriole and a new centriole. The former is called the mother centrosome and the latter the daughter centrosome.

Several stem cells have been observed to segregate their mother vs. daughter centrosomes non-randomly during asymmetric cell divisions. For instance, *Drosophila* male GSCs, mouse radial glial progenitor cells and embryonic stem cells induced to divide asymmetrically inherit the mother centrosome consistently (Habib et al., 2013; Wang et al., 2009; Yamashita et al., 2007), whereas *Drosophila* female GSCs and neuroblasts obtain the daughter centrosome (Conduit & Raff, 2010; Januschke, Llamazares, Reina, & Gonzalez, 2011; Salzmann et al., 2014). Although these studies have demonstrated the importance of mother-daughter centrosome asymmetry in spindle orientation, whether it plays additional roles such as contributing to asymmetric fates remains an open question. However, as the organizer of spindles, the very cell division apparatus, centrosome asymmetry may serve as a platform for asymmetric segregation of other cellular components, which in turn contribute to some aspects of fate asymmetry. For example, a pool of pSmad1, a transcription factor downstream of BMP signaling, destined for degradation is associated with the spindle pole and segregated asymmetrically to only one of two daughter cells in cultured cells (Fuentealba, Eivers, Geissert, Taelman, & De Robertis, 2008). In another example, both *Drosophila* male and female GSCs have been shown to inherit the midbody ring in a stereotypical manner, correlating with the mode of centrosome inheritance (Salzmann et al., 2014). The midbody ring is a remnant of the contractile ring formed during cytokinesis, and its correlation with stem cell fate has been documented in stem cells and cancer cell lines (Ettinger et al., 2011; Kuo et al., 2011), pointing to a possibility it may carry information that contributes to cell fate.

In addition, *Drosophila* male GSCs segregate sister chromatids of X and Y chromosomes non-randomly (Yadlapalli & Yamashita, 2013). Centrosome asymmetry is essential for such biased chromosome segregations, supporting the idea that centrosome asymmetry can host a plethora of asymmetries. Intriguingly, it has been shown that old vs. new histone H3 segregates asymmetrically during *Drosophila* male GSC divisions, where older histone is preferentially retained in the GSCs (Tran, Lim, Xie, & Chen, 2012), leading to the idea that epigenetic memories are asymmetrically inherited during GSC divisions. Whether asymmetric histone segregation requires centrosome asymmetry remains unknown.

"Age factors" have been observed to be associated with the centrosome. For example, large protein aggregates of misfolded proteins termed aggrosomes are associated with the spindle pole/centrosome and inherited asymmetrically to one of the daughter cells in culture cells as well as in *Drosophila* neuroblasts (Rujano et al., 2006). It has been proposed that asymmetric inheritance of aging factors may help maintain stem cells to stay "young" (Moore, Pilz, Arauzo-Bravo, Barral, & Jessberger, 2015). By the extension of this idea, GSCs may utilize a similar mechanism to rejuvenate—and in this case, the implication is not just rejuvenation of any cells, but it is rejuvenation of germline, the cells that transmit the genome to the next generation.

Considering that germ cells are "non-essential" for organismal survival, germ cells would have the luxury of pushing this asymmetry to the extreme, creating one entirely rejuvenated cell and one fatally bad cell. The system utilizes the resource to create two cells to generate just one (but very good) cell. Depending on how many rounds of such asymmetric divisions are required to rejuvenate a cell, somatic cells might not be able to afford such luxury, whereas germ cells might be able to do it. The fact that many stem cells undergo asymmetric divisions, even when it is not absolutely required to maintain the stem cell population and balance self-renewal and differentiation, may be explained by this kind of rejuvenation, independently of the need for fate asymmetry. In this regard, other stem cell population that do not undergo stereotypical asymmetric divisions such as mammalian spermatogonial stem cells and *C. elegans* GSCs might be dividing asymmetrically once in a while to rejuvenate themselves.

4. Transit-amplifying divisions to reduce the burden on stem cells

Although the mechanisms described above, such as asymmetric stem cell division and niche-mediated stem cell maintenance, play a central role for maintaining tissue homeostasis, these features demand cell division in order to replace lost differentiated cells. Asymmetric stem cell division alone cannot support tissue homeostasis beyond the replicative lifespan of individual stem cells, since stem cells are subject to disruptions such as DNA damage and oxidative stress, and the accumulation of these disruptions is thought to be a key component to tissue degeneration and aging (Schultz & Sinclair, 2016). However, many individual stem cells live shorter than the organismal lifespan, and thus mechanisms are required to protect the stem cell population as a whole in the face of the limited life span of these individual stem cells. This limited lifespan can become particularly problematic in tissues with high turnover, such as the male germline, where rapid stem cell division can cause individual stem cells to quickly decline (Schultz & Sinclair, 2016; Wallenfang, Nayak, & DiNardo, 2006).

Tissues often protect their stem cell populations by employing transit-amplifying divisions and/or quiescence to minimize the burden of cell division on stem cells and protect them from exhaustion. Transit-amplifying divisions magnify the number of differentiated cells produced from a single stem cell division, and is employed by broad range of tissues (Beumer & Clevers, 2016; Rangel-Huerta & Maldonado, 2017; Watt, 2001), including *Drosophila* male GSCs and mouse SSCs to reduce the number of GSC divisions needed to yield a large number of differentiated cells (de Rooij, 2017; Fuller, 1998). Once a *Drosophila* male GSC creates 1 differentiating daughter, the gonialblast (GB), it undergoes 4 rounds of mitotic divisions to yield 16 spermatogonia, prior to committing to terminal differentiation as spermatocytes (Fuller, 1998). This amplification allows each GSC to produce 64 sperm cells from a single division due to 4 transit-amplifying divisions followed by meiosis, instead of 4 sperm cells as would be the case with no transit-amplifying divisions, leading to a 16-fold reduction in the number of GSC divisions needed to produce the same number of sperm. However, the *C. elegans* hermaphrodite germline does not appear to have any transit-amplifying divisions after differentiation from GSCs, possibly due to the need to rapidly replace meiotic germ cells in response to environmental and developmental conditions that completely eliminate these cells (Fox & Schedl, 2015).

The ability to produce multiple differentiated cells from a single stem cell division affords some tissues to maintain homeostasis without requiring all stem cells to be mitotically active. Hematopoietic stem cells have a subset of their population that is quiescent, creating a reserve store of stem cells that is maintained until needed (Nakamura-Ishizu, Takizawa, & Suda, 2014). The quiescent hematopoietic stem cell population becomes active after tissue disruption or when the active stem cell population is depleted, restoring tissue damage that is beyond the ability to be repaired by the active stem cell population alone (Wilson et al., 2008). Furthermore, tissues with little turnover, such as muscle, can have their entire stem cell population be quiescent, since there is no regular need for differentiation of new cells (Schultz, 1976). This quiescence is relieved in response to tissue injury, allowing for quick production of new cells to replace the damaged tissue without overstressing the replicative limits of individual stem cells (Cornelison, Filla, Stanley, Rapraeger, & Olwin, 2001). Stem cell quiescence is critical for these tissues to preserve homeostasis so that they only proliferate and differentiate in response to tissue insult or injury, and loss of quiescence in these cells disrupts tissue homeostasis over time (Cheung et al., 2012; Hausburg et al., 2015; Zhang et al., 2006). Limiting the number of cell divisions by periods of quiescence prevents the stem cells from reaching their individual replicative limit and gives the ability to maintain tissue homeostasis throughout an organism's lifespan. While quiescence can be a beneficial mechanism to protect stem cell populations, the high demand for gametes makes it difficult for GSCs to have extended periods outside the cell cycle. Some GSC populations can become quiescent in response to environmental conditions, such as *C. elegans* GSCs during starvation (Angelo & Van Gilst, 2009; Seidel & Kimble, 2015), but neither male nor female *Drosophila* GSCs have been found to exit the cell cycle, even during starvation (Drummond-Barbosa & Spradling, 2001; Yang & Yamashita, 2015), suggesting that other mechanisms maintain these stem cell populations throughout an animal's lifespan.

5. Dedifferentiation maintains GSC populations beyond the lifespan of individual cells

Although stem cell quiescence and transit-amplifying divisions slow down the exhaustion of stem cells, these mechanisms cannot maintain stem cells forever, and additional tissue insults may occur that deplete stem cells beyond the capacity of these mechanisms. *Drosophila* male GSCs have an estimated 14-day half-life during normal mating conditions, however, the overall GSC population is more stable than expected from the 14 days of half-life (Wallenfang et al., 2006), indicating that there are robust mechanisms to replace lost GSCs during aging to maintain homeostasis. Symmetric cell division is one mechanism to replace lost stem cells and occurs in *Drosophila* female GSCs (Xie & Spradling, 2000). Interestingly, however, symmetric division is uncommonly observed in *Drosophila* male GSCs. Instead, the excess of differentiating SG afforded by transit-amplifying divisions allows for SG dedifferentiation to replace lost GSCs (Fig. 3A).

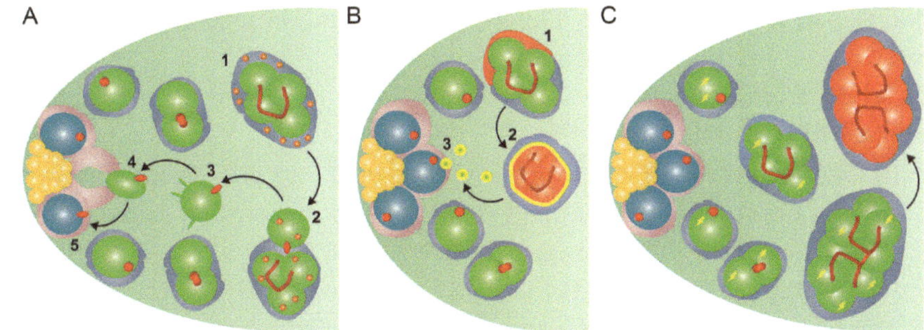

Fig. 3 Death of transit-amplifying cells in germline homeostasis and integrity. (A) SG dedifferentiation replenishes lost GSCs. (1) The earliest known feature of SG dedifferentiation is a pulse of JNK signaling (orange) in CCs. (2) This signal is shortly followed by a pulse of JNK in SGs that initiate dedifferentiation and break their cytoplasmic bridges with the rest of the SG cyst, leaving a remnant ring canal, and escape the encapsulating CCs. (3) Dedifferentiating SGs distant from the niche migrate to the GSC niche, frequently forming filopodia-like protrusions. (4) Upon reaching the niche, dedifferentiating cells are re-encapsulated by CySCs. (5) Establishment in the niche completes GSC dedifferentiation and fate is maintained by niche signaling. (B) Phagocytosis of dead SGs by CCs may maintain GSC homeostasis during starvation. (1) During protein starvation, individual CCs from primarily 4-cell cysts initiate caspase-dependent cell death (red). (2) After CC death, the SG cyst is entirely engulfed in a phagolysosome (yellow) by the remaining CC or a nearby CCs. (3) SG autophagy within the phagolysosome is proposed to provide recycled metabolites for the GSCs to maintain GSC population throughout protein starvation. (C) SG connectivity increases sensitivity to DNA damage. Single cell GBs that experience DNA damage (yellow thunderbolts) only enter cell death when they suffer large amounts of damage. SG-cysts share DNA damage signals across their cytoplasmic bridges, so that more connected cysts (8- and 16-cell SGs) initiate cell death of the entire cysts (red cells) even when the damage of an individual SG within the cyst is below the level that would normally trigger death of an individual cell.

5.1 Dedifferentiation creates GSCs to replace lost GSCs

Dedifferentiation is a process whereby partially differentiated cells revert to a stem cell identity. Dedifferentiation of *Drosophila* germ cells is observed both in the ovary and testes, where differentiating cystocytes or SGs, respectively, regain GSC identity after induced GSCs loss, either by transient expression of *bam* or transient inactivation of STAT (Brawley & Matunis, 2004; Kai & Spradling, 2004; Sheng, Brawley, & Matunis, 2009). The differentiating cells (cystocytes and SGs) that are interconnected with cytoplasmic bridges break apart during dedifferentiation and re-occupy the stem cell niche as a single cell (Brawley & Matunis, 2004; Kai & Spradling, 2004). Spermatocytes are believed to be incapable of dedifferentiation, because testes that lost all spermatogonia yet maintained spermatocytes are unable to recover GSCs

(Brawley & Matunis, 2004). This is not surprising considering these cells have entered premeiotic S phase.

While dedifferentiation was first identified in the *Drosophila* germlines in response to genetic ablation of GSCs, dedifferentiation also occurs during aging without external perturbation (Cheng et al., 2008). Also, dedifferentiation was observed when animals recover from protein starvation, which causes reduction in GSC number (Herrera & Bach, 2018; McLeod, Wang, Wong, & Jones, 2010; Yang & Yamashita, 2015), indicating that dedifferentiation restores GSC homeostasis after environmental perturbations. Lineage tracing of 4- and 8-cell SG during normal aging indicated that at least 40% of the GSCs in animals at 50 days of age are from dedifferentiated cells (Cheng et al., 2008), revealing that dedifferentiation normally maintains GSC homeostasis under otherwise unperturbed conditions. Overexpression of *bam* in 4- and 8-cell SG inhibits dedifferentiation of these cells, causing a significant reduction in GSC population after 45 days of mating, but not when males are reared separately from females (Herrera & Bach, 2018), suggesting the dedifferentiation of these cells becomes essential when GSCs are under constant demand of proliferation. However, these experiments only linage-trace and inhibit dedifferentiation in 4- and 8-cell SGs, and precludes the assessment of the contribution of dedifferentiation from GBs and 2-cell SGs. Indeed, approximately 7% of all GBs produced during live observation of testes cultured in normal conditions crawl back to occupy the niche position and dedifferentiate, and 75% of all dedifferentiated cells observed after GSC ablation came from GBs or 2-cell SG (Sheng & Matunis, 2011). Therefore, it is plausible that early stage germ cells more readily dedifferentiate, perhaps due to proximity to the hub being a determinant of dedifferentiation. If this is the case, the GSC pool may be predominantly maintained through dedifferentiation of GBs and 2-cell SGs when the tissue turnover is modest. The robust dedifferentiation activity throughout adulthood provides the germline with a mechanism to maintain homeostasis, even after successive periods of starvation and recovery (Herrera & Bach, 2018).

5.2 The mechanism of dedifferentiation

The critical contribution of dedifferentiation to germline homeostasis suggests that regulation of this process is key for germline maintenance, however, much remains unknown about the genetic and environmental factors that influence dedifferentiation. Since JAK-STAT signaling from the hub is critical for GSC identity, it is unsurprising that this signaling pathway is also necessary for dedifferentiation (Sheng et al., 2009). Shortly after recovery from GSC elimination by *bam* mis-expression, SGs immediately near to the hub express STAT92E, an indication of JAK-STAT signaling activity and GSC identity in these cells. JAK-STAT inhibition reduces dedifferentiation after germline *bam* mis-expression, indicating JAK-STAT signaling is required for GSC dedifferentiation (Sheng et al., 2009). However, given the necessity of JAK-STAT signaling to maintain GSC identity, and that there is no change in *upd* expression in the hub when dedifferentiation is induced (Sheng et al., 2009), it remains unclear whether this pathway is required for SG dedifferentiation per se or simply for maintaining GSC identity after dedifferentiation is complete. The contribution of another key niche signaling factor, BMP signaling, in dedifferentiation is yet to be tested.

The process of initiating SGs to dedifferentiate into GSCs requires more than simply the changing of cell fate. Since SGs are interconnected by cytoplasmic bridges during differentiation, this connection must break apart to create individual SGs that can dedifferentiate into GSCs (Brawley & Matunis, 2004). However, the mechanisms by which cytoplasmic bridges are lost and stabilized ring canals between SGs are resolved are not well understood. It has been shown that during GSC divisions, cytokinesis is delayed to form a temporary bridge with GBs, and is resumed to complete abscission and GSC division (Lenhart & DiNardo, 2015). Perhaps similar mechanisms to resume the completion of cytokinesis are employed in dedifferentiating SGs. In addition to breaking the cytoplasmic bridges during differentiation, SGs must break free from the CCs that encapsulate the SG cyst. CCs briefly express the JNK target *puc* soon after starvation conditions are relieved, and blocking JNK signaling prevents dedifferentiation in recovery from starvation, even though this pathway does not normally contribute to maintain the GSC population (Herrera & Bach, 2018). Since *puc* was also shown to be activated in CCs when SG cysts undergo starvation-induced cell death (Yang & Yamashita, 2015), it is possible that CC death triggered by JNK signaling during starvation might be key for allowing SGs to release from encapsulation and dedifferentiate. Once dedifferentiating SGs break free from their cysts and encapsulating CCs, they must find their way to the niche to complete dedifferentiation. Small actin protrusions were observed in SGs when dedifferentiation was induced by GSC ablation (Sheng et al., 2009), and SGs were observed to be migrating to the hub during dedifferentiation (Sheng & Matunis, 2011). It is not documented whether these protrusions are oriented in a particular direction (e.g., toward hub cells), but given their destination (hub), it is likely that dedifferentiating SGs are guided by chemotaxic signals. The identification of such chemotaxic signals would be of particular interest to understand the process of GSC dedifferentiation. Migration is likely also necessary during dedifferentiation for SGs to navigate through CySCs, which surround the hub cells upon GSC elimination (Sheng et al., 2009). In addition, activation of JNK signaling was observed in germ cells upon induction of dediffererntiation (Herrera & Bach, 2018). JNK signaling is critical for cell migration in developing *Drosophila* tissues (Hou, Goldstein, & Perrimon, 1997; Riesgo-Escovar & Hafen, 1997), and activation of this pathway in SGs may promote SG migration to the hub for dedifferentiation. Finally, in order to engraft in the niche, dedifferentiated SGs must establish interaction with the niche supporting cells, such as hub cells. Expression of the aminopeptidase *Slamdance* (*sda*), which is required for processing and maturation of E-cadherin, in hub cells directly influences dedifferentiation activity (Lim et al., 2015). *Sda* mutants have reduced frequency of dedifferentiated GSCs during both aging and after genetic GSC ablation by *bam* mis-expression, and *sda* overexpression in the hub increases the number of dedifferentiated GSCs in testes, causing dedifferentiation in animals as young as 1-day old (Lim et al., 2015), indicating that *sda* expression is sufficient to induce dedifferentiation. These findings suggest that *sda* may promote the establishment of dedifferentiated GSCs in the niche. Interestingly, *sda* overexpression does not increase the overall number of GSCs (Lim et al., 2015), suggesting that dedifferentiated GSCs outcompete native GSCs at the hub, similar to the female GSC niche (Jin et al., 2008). However, E-cadherin overexpression only partially suppresses *sda* mutant dedifferentiation defects (Lim et al., 2015), indicating that *sda* may have activity required for dedifferentiation that is independent of E-cadherin. Together these studies indicate that dedifferentiation requires a complex network of signaling and reorganization of cellular architecture to allow conversion of differentiating cells into stem cells.

5.3 Are dedifferentiated GSCs equal to native GSCs?

Although dedifferentiation is a key aspect of germline homeostasis, the GSCs generated by dedifferentiation are not identical to native GSCs produced during development. GSCs derived from dedifferentiated SGs are defective in two unique features of GSC biology. Dedifferentiated GSCs that accumulate during aging have misoriented centrosome, causing a lower division rate compared to those with oriented centrosomes (Cheng et al., 2008), due to activation of the centrosome orientation checkpoint (Inaba et al., 2010). Additionally, dedifferentiated GSCs do not have the non-random sister chromatid segregation that is observed in normal GSCs (Yadlapalli & Yamashita, 2013). However, the function of non-random sister chromatid segregation in GSCs remains unknown, thus it is unclear how this difference influences the activity of dedifferentiated GSCs. Males that have prolonged GSC turnover from cyclical starvation and refeeding over a 45-day life span do have reduced sperm production (Herrera & Bach, 2018), however, it is unclear if this reduction is due to a defective ability in dedifferentiated GSCs or another effect of constant environmental perturbation throughout life. Better characterization of the functional consequences of dedifferentiation will improve our understanding of how this process affects the fertility of aged animals.

6. Cell death in GSC homeostasis

Not only is the production of cells important for tissue homeostasis, but cell death plays a key role in maintaining tissue homeostasis, and accordingly, regulation of cell death in a cell-type specific manner is of critical importance. Distinct types and modes of cell death have been reported in *Drosophila* testis, illuminating the complexity of regulating cell death in tissue homeostasis.

6.1 SG death induced by somatic CC apoptosis shifts tissue homeostasis during starvation

Tissue homeostasis needs to not only maintain the constant flow of cell production in a steady state, but also adjust the flow of cell production based on various environmental inputs. During protein starvation, there is an overall reduction in germ cells, including reduced numbers of both GSCs and SGs (McLeod et al., 2010; Yang & Yamashita, 2015). Once germ cell numbers are reduced though, they can maintain homeostasis for an extended period of starvation and expand to normal numbers when relieved from starvation (McLeod et al., 2010; Yang & Yamashita, 2015). The source of GSC loss is unclear, but the lack of increased TUNEL staining in starving GSCs suggests the reduction is due to excess differentiation and not GSC death (McLeod et al., 2010). The SG reduction is primarily due to 4-cell SG death, causing a substantial decrease in the number of 4-, 8-, and 16-cell SGs (Yang & Yamashita, 2015). The SG death is independent of caspase activity in the germline (Yacobi-Sharon, Namdar, & Arama, 2013), but dependent on apoptosis of somatic CCs that encapsulate SGs (Yang & Yamashita, 2015). SGs cannot survive without encapsulation by CCs as described above, thus apoptosis of CCs leads to SG death as well. Surprisingly, blocking CC apoptosis disrupts GSC maintenance and slows down the GSC cell cycle, which prevents germline recovery after starvation is relieved (Yang & Yamashita, 2015). A likely reason for SG/CC death during protein starvation is to spare nutrients for GSCs by eliminating SGs, and to recycle nutrients from dead SGs to GSCs. Surviving CCs and CySCs near dying SG/CCs during protein starvation are highly positive for lysosomes (Chiang, Yang, & Yamashita, 2017; Yang & Yamashita, 2015), suggesting that they take up the material from the nearby dying cells. The endosomal component *spict* likely participates in this process of recycling nutrients. Spict is expressed in dying CCs and then trafficked into neighboring surviving CCs, and is required for GSC maintenance during starvation (Chiang et al., 2017). Together the coordinated death and uptake of materials of dead SGs suggests that the sacrifice of differentiating germ cells is important for maintaining germline homeostasis during protein starvation (Fig. 3B), and understanding the factors that regulate this process is important for uncovering the mechanism that can maintain the germline during environmental disruptions.

6.2 SG death as a mechanism to protect germline genome integrity

In contrast to the death of early SG cysts during protein starvation, which is dependent on the death of somatic CCs, late SGs undergo cell death independent of CC apoptosis or nutrients conditions. Instead, these late SGs likely die due to increased sensitivity to DNA damage. It was found that SGs are extremely sensitive to X-ray irradiation, and the sensitivity dramatically increases as the number of SGs in the cyst increases (e.g., 16-cell SGs are considerably more sensitive to X-ray irradiation than 2-cell SGs) (Lu & Yamashita, 2017). This sensitivity of 16-cell SG to irradiation depends on germ cell connectivity (Lu & Yamashita, 2017). Because of this connectivity, even when only a subset of germ cells within a cyst contains detectable DNA damage, the entire cyst of SGs underwent cell death, likely indicating that the signal that triggers cell death is shared among SGs within the cyst. This sharing of cell death signal effectively increases the sensitivity of germ cells to DNA damage (Fig. 3C), and this mechanism might be utilized to protect the integrity of the germline genome (Lu, Jensen, Lei, & Yamashita, 2017). It is of note that germ cell connectivity is broadly conserved during evolution, suggesting that sharing of cell death signals may be a general mechanism to increase the sensitivity to DNA damage, thus to protect the integrity of the germline genome (Lu et al., 2017).

6.3 Differential sensitivity to cell death among distinct cell populations

The priming of SGs for death upon environmental disruptions appears to be a key feature to maintain germline homeostasis, but GSCs themselves are particularly resistant to environmental insults. While about 20% of SGs in normal testes are detected as dying by TUNEL signal, TUNEL positive cells are rarely found in the "stem cell" region near the hub (Hasan, Hetie, & Matunis, 2015). Even when animals are exposed to environmental challenges that induce SG death, such as a high dose of X-ray irradiation or protein starvation, TUNEL positive cells are still infrequently found adjacent to the hub (Hasan et al., 2015). This resistance of GSCs to cell death is partially caused by expression of *inhibitor of apoptosis 1* (*DIAP1*), since inhibiting *DIAP1* expression in GSCs increases cell death during normal conditions (Hasan et al., 2015). *DIAP1* expression in GSCs is directly influenced by JAK-STAT signaling, and inhibiting JAK-STAT increases GSC death after irradiation (Hasan et al., 2015), suggesting that hub signaling protects GSCs from normal cell death activity. While GSCs do not appear to die in response to these environmental perturbations, the number of GSCs in the testis is reduced during starvation (McLeod et al., 2010; Yang & Yamashita, 2015), suggesting that damaged GSCs may differentiate rather than die in response to cellular perturbations, similar to an observation in mammalian stem cells (Inomata et al., 2009). These damaged differentiating cells may then undergo cell death as late stage SGs, and their cellular contents can be recycled to maintain the undamaged GSCs.

Hub cells are also usually resistant to cell death. There is no cell loss in the hub when exposed to 40G irradiation, but GSCs do die from this exposure (Volin, Zohar-Fux, Gonen, Porat-Kuperstein, & Toledano, 2018). Multiple miRNAs that target the pro-apoptotic genes *hid* and *grim* protect hub cells by preventing the activation of cell death pathways, and inhibition of the miRNA pathway by *dcr1* knockdown causes progressive loss of hub cells during aging and hub sensitivity to irradiation (Volin et al., 2018). Since hub cells are quiescent, there is likely little cost to allowing genomic perturbations to accumulate in these cells, and their necessity to maintain GSC populations provides an advantage to protecting them from death. These multiple cell-type-specific cell death responses to environmental insults appear to enact a cellular quality control at the late spermatogonial stage to prevent the transmission of damaged genomes to the next generation, while preventing excessive loss of the GSC population. A small amount of GSC loss by differentiation or occasional cell death can be tolerated in these instances because dedifferentiation will restore the lost GSCs and hub signals establish cell-death protection in these new GSCs, maintaining the homeostasis after tissue damage.

7. The disruption of germline homeostasis during aging

Thus far, we have discussed the robust ability of tissues to maintain themselves for a very long time, however, all animals and tissues eventually age and suffer tissue degeneration. The number of GSCs in the testis is marginally reduced in aged males (Wallenfang et al., 2006), indicating that germline homeostasis is disrupted during aging. Although the loss of tissue homeostasis during aging impacts nearly all animals and tissues, it is unclear if the effects of aging are due to an accumulation of inevitable insults over time, or genetic programs that limit the replicative capacity of tissues, perhaps as a mechanism to restrict tumor formation.

One change in germline signaling activity that occurs during aging is the progressive reduction of *upd* expression in hub cells, causing decreased GSC maintenance in older animals (Boyle, Wong, Rocha, & Jones, 2007; Toledano, D'Alterio, Czech, Levine, & Jones, 2012). The reduction in *upd* over time is due to increasing expression of the *let-7* miRNA, which inhibits the expression of the *upd* mRNA-binding protein Imp (Toledano et al., 2012). Imp protects *upd* mRNA from degradation, and its inhibition causes decreased *upd* expression and JAK-STAT signaling in the hub (Toledano et al., 2012). This pathway means that the activation of *let-7* expression in essence acts as an "aging switch" that disrupts GSC maintenance, and blocking this switch by expression of a *let-7*-resistant *Imp* prevents the reduction of GSCs in aged animals (Toledano et al., 2012). However, it remains unclear how *let-7* expression is regulated, and whether its expression is a programmed molecular clock that limits the replicative lifespan of the testis, or if it is responding to perturbations that have accumulated in the germline over time during aging.

A reduction in germ cell production during aging is not only influenced by reduced hub signaling, but also a slowing of GSC cell cycle. The *string* (*stg*) gene is specifically expressed in both GSCs and CySCs, and is required for the maintenance of both of these cell types (Inaba, Yuan, & Yamashita, 2011). Reduced *stg* expression in 10- to 20-day-old GSCs, but not CySCs, causes a reduced GSC mitotic index (Inaba et al., 2011). Interestingly, the reduced expression appears consistent across all GSCs, suggesting that the reduction is not the result of dedifferentiated spermatogonia with low *stg* expression, but in fact coordinated reduction of expression across all cells. However, similar to the case of reduced Upd expression during aging, it remains unclear whether reduced *stg* expression is programmed or in response to age-related, inevitable decline in certain aspects of GSC characteristics.

In addition to these age-related decline in signaling and cell cycle progression, the germline accumulates natural perturbations during aging that impact tissue homeostasis. The tandem repeats of the ribosomal DNA (rDNA) locus are susceptible to copy number loss due to intra-chromatid exchange between rDNA copies (Fig. 4A) (Srivastava & Schlessinger, 1991), and the number of rDNA copies in GSCs is progressively reduced during aging (Lu, Nelson, Watase, Warsinger-Pepe, & Yamashita, 2018). rDNA copy insufficiency can lead to cell death (Ganley & Kobayashi, 2014), potentially disrupting tissue homeostasis. Furthermore, the defects associated with dedifferentiated GSCs, such as misoriented centrosomes and reduced cell proliferation (Cheng et al., 2008), also likely drive age-related decline in tissue homeostasis. These stochastic detriments to GSCs during aging likely contribute to the disruption of homeostasis in the germline of aged animals. For this reason, it will be critical to understand how these perturbations influence the genetic factors that are altered during aging to truly dissect the contributions of inevitable vs. programmed influences on aging.

8. Maintaining the germline over successive generations

The fact that the germline is the only cell lineage to pass its genome from generation to generation makes germ cells essentially immortal. However, given the limited life span of individual cells and the gradual decline in germ cell functionality during aging, it is remarkable that germ cells can establish a new life and a new germline therein in each successive generation. While it is unclear how the germline is able to in essence reset its age at each generation, it likely occurs through a combination of the following mechanisms: (1) the ability to protect the germline genome, (2) the ability to select against the damaged genomes, and (3) the ability to rejuvenate the effects of individual aging in each generation. Perhaps the most well

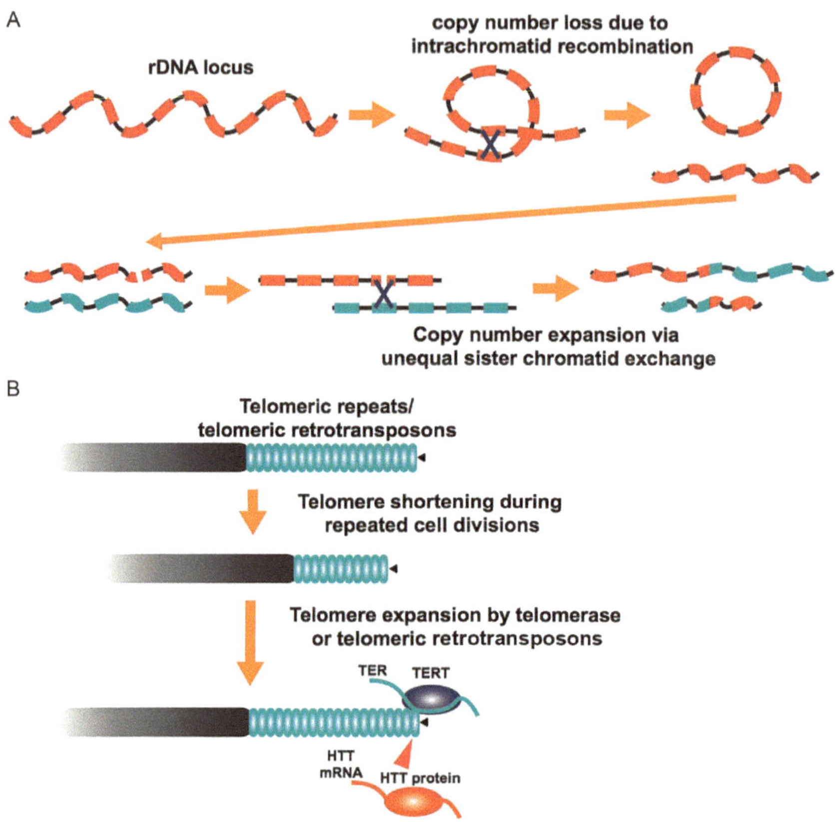

Fig. 4 Instable genomic regions are replenished in the germline to maintain the genome over successive generations. (A) Copy number dynamics of rDNA repeats. Tandem duplications of the rDNA cistron (red) containing the 5.8S, 18S, and 28S rRNA genes comprise the rDNA locus. Intra-chromatid recombination between distant rDNA repeats causes circularization and excision of the intervening rDNA copies, leading to reduction of the rDNA copy number at the chromosomal locus. Double-stranded DNA breaks in an rDNA copy can be repaired by homologous recombination with a sister chromatid (blue). If recombination occurs between misaligned rDNA copies, this produces unequal sister chromatid exchange, creating one chromatid with an expanded rDNA locus while the other sister has reduced rDNA copies. (B) Telomeres are composed of repeats of a short (~6 bp) sequence in most eukaryotes, or Het-A, TAHRE, and TART (HTT) retrotransposable elements in *Drosophila*. Telomere repeats are shortened by every cell cycle due to the end-replication problem (inability of DNA polymerase to replicate the very 3′ end of DNA). Telomeres are extended by a telomerase complex composed of a telomerase RNA (TER) and a telomerase reverse-transcriptase protein (TERT) in most of eukaryotes. Alternatively, in *Drosophila,* HTT retrotransposable elements use their own mRNA as a template to insert a new copy of themselves into telomere ends to expand telomeric repeats.

understood among these is the piRNA pathway, which protects the germline from the propagation of transposable elements (reviewed in Toth, Pezic, Stuwe, & Webster, 2016). As described above, the germline exhibits extremely high sensitivity to DNA damages owing to germ cell connectivity (Lu et al., 2017), and this mechanism may serve to select against mutations, and thus to maintain the integrity of germline genome that will be inherited. It is of note that this mechanism of maintaining genomic integrity, i.e., discarding anything that is subpar, is a wasteful method, and likely can be afforded only by germ cells, due to their inessential contribution to the viability of an individual organism.

No matter how robust the mechanisms are that protect the genome of the germline from mutagenic effects, it is unlikely that the germline can prevent the inevitable disruption of inherently instable regions of the genome, such as rDNA repeats and telomeres (Fig. 4A and B). The most well-established example of this genomic rejuvenation is the activity of telomere extension by telomerase (reviewed in Blackburn & Collins, 2011) (Fig. 4B). In the absence of telomerase, the telomeres at chromosome ends progressively deteriorate, and telomerase activity in the germline and embryonic stem cells prevents the accumulated trans-generational elimination of telomeres (Blasco et al., 1997; Herrera et al., 1999). Sperm or oocytes with severely reduced telomeres have a reduced capacity for fertilization and have aberrant first zygotic divisions (Liu, Blasco, Trimarchi, & Keefe, 2002). Furthermore, to prevent the inheritance of deteriorated telomeres, germ cells with short telomeres are removed during meiosis in mammalian spermatogenesis (Hemann et al., 2001).

Drosophila lack the telomerase enzyme that normally extends telomeres in most eukaryotes (Greider & Blackburn, 1985), and instead telomeres are maintained by multicopy retrotransposons that duplicate at chromosome ends (Levis, Ganesan, Houtchens, Tolar, & Sheen, 1993) (Fig. 4B), making it difficult to eliminate telomere maintenance to dissect its activity in the *Drosophila* germline. The observation of telomere formation at the new ends of broken chromosomes in the male germline suggests the ability for telomere elongation, however, it is unclear if this telomere formation includes the addition of new telomeric retrotransposons, or is simply the binding of telomere cap proteins to the chromosome end (Ahmad & Golic, 1998; Titen & Golic, 2010). Like most other transposable elements, the telomeric retrotransposons are transcriptionally repressed by heterochromatin established by the piRNA pathway in the ovary (Radion et al., 2018; Shpiz et al., 2011), and inhibition of piRNA factors *spn-E* and *aubergine* leads to retroelement attachment to broken chromosome ends (Savitsky, Kwon, Georgiev, Kalmykova, & Gvozdev, 2006). It is unclear if this piRNA-mediated silencing of these retrotransposons is relieved during any stage of gametogenesis to provide telomere elongation in either the male or female germline. While it appears that telomere extension is possible under these specific conditions, how telomeres are normally maintained in the *Drosophila* germline, and the activity and regulation of the telomere retrotransposons, remains undetermined, and the potential consequences of disrupting these retrotransposons should be investigated to uncover the activity of *Drosophila* telomere maintenance.

In addition to telomere extension, the rDNA locus also appears to experience rejuvenation in the *Drosophila* male germline. rDNA copies that are lost from the germline during aging or in response to dietary excess cause offspring to inherit reduced rDNA copies (Aldrich & Maggert, 2015; Lu et al., 2018), yet the sons of aged animals that have diminished rDNA have the capacity to expand the rDNA locus in their GSCs to its original size (Lu et al., 2018). This ability of the germline to rapidly expand the rDNA locus has also been observed in animals with large deletions of the rDNA locus (Atwood, 1969; Ritossa, 1968). The ability to rapidly expand the rDNA locus in the germline has been proposed to be due to homologous recombination between misaligned rDNA copies on sister chromatids, causing unequal sister chromatid exchange (USCE) (Fig. 4A) (Tartof, 1974). Mutants with defective homologous recombination-mediated double-strand break repair have a reduced capacity to expand rDNA (Hawley, Marcus, Cameron, Schwartz, & Zitron, 1985; Hawley & Tartof, 1983; Lu et al., 2018), suggesting this mechanism does contribute to rDNA copy recovery. However, direct evidence of USCE at rDNA loci is yet to be observed, and a potential source for double-strand breaks that would prompt homologous recombination-mediated repair is unknown. While animals that inherit reduced rDNA or large rDNA deletions from aged fathers can recover the lost copies, this mechanism does not seem to occur in response to dietary effects, as the lost rDNA from dietary excess is maintained for several generations (Aldrich & Maggert, 2015). These differences suggest that mechanisms may exist to distinguish changes in rDNA from nutritional influences or normal aging, and the ability to restore rDNA may differ depending on the source of its loss. Nevertheless, the existence of robust mechanisms to restore the effects of aging in the parental germline in GSCs of young animals indicates that germline immortality can be supported by features that counteract the effects of aging.

9. Conclusion

We reviewed the recent progress in our understanding of GSC homeostasis. As the source of life-long production of gametes, multitude of mechanisms ensures GSC maintenance such as asymmetric stem cell division and dedifferentiation. Moreover, the mechanisms that govern GSC homeostasis and the differentiation of their progeny may be tied to the mechanisms that maintain genome integrity not only through the life time but also from a generation to another. It will be of particular interest for the future studies to expand our knowledge toward understanding how the immortality of the germline may be achieved.

Acknowledgments

We apologize our colleagues whose work we were unable to cite due to space limitations. We thank the Yamashita laboratory for discussions. The research in the Yamashita laboratory is supported by Howard Hughes Medical Institute.

References

Ables, E. T., & Drummond-Barbosa, D. (2010). The steroid hormone ecdysone functions with intrinsic chromatin remodeling factors to control female germline stem cells in Drosophila. *Cell Stem Cell*, *7*, 581–592.

Ahmad, K., & Golic, K. G. (1998). The transmission of fragmented chromosomes in Drosophila melanogaster. *Genetics*, *148*, 775–792.

Aldrich, J. C., & Maggert, K. A. (2015). Transgenerational inheritance of diet-induced genome rearrangements in Drosophila. *PLoS Genetics*, *11*, e1005148.

Amoyel, M., Sanny, J., Burel, M., & Bach, E. A. (2013). Hedgehog is required for CySC self-renewal but does not contribute to the GSC niche in the Drosophila testis. *Development*, *140*, 56–65.

Amoyel, M., Simons, B. D., & Bach, E. A. (2014). Neutral competition of stem cells is skewed by proliferative changes downstream of Hh and Hpo. *The EMBO Journal*, *33*, 2295–2313.

Angelo, G., & Van Gilst, M. R. (2009). Starvation protects germline stem cells and extends reproductive longevity in C. elegans. *Science*, *326*, 954–958.

Ariz, M., Mainpal, R., & Subramaniam, K. (2009). C. elegans RNA-binding proteins PUF-8 and MEX-3 function redundantly to promote germline stem cell mitosis. *Developmental Biology*, *326*, 295–304.

Atwood, K. C. (1969). Some aspects of the bobbed problem in Drosophila. *Genetics*, *61*, 319–327. Suppl.

Austin, J., & Kimble, J. (1987). Glp-1 is required in the germ line for regulation of the decision between mitosis and meiosis in C. elegans. *Cell*, *51*, 589–599.

Banisch, T. U., Maimon, I., Dadosh, T., & Gilboa, L. (2017). Escort cells generate a dynamic compartment for germline stem cell differentiation via combined Stat and Erk signalling. *Development*, *144*, 1937–1947.

Beumer, J., & Clevers, H. (2016). Regulation and plasticity of intestinal stem cells during homeostasis and regeneration. *Development*, *143*, 3639–3649.

Blackburn, E. H., & Collins, K. (2011). Telomerase: An RNP enzyme synthesizes DNA. *Cold Spring Harbor Perspectives in Biology*, *3*, a003558.

Blasco, M. A., Lee, H. W., Hande, M. P., Samper, E., Lansdorp, P. M., DePinho, R. A., et al. (1997). Telomere shortening and tumor formation by mouse cells lacking telomerase RNA. *Cell*, *91*, 25–34.

Boyle, M., Wong, C., Rocha, M., & Jones, D. L. (2007). Decline in self-renewal factors contributes to aging of the stem cell niche in the Drosophila testis. *Cell Stem Cell*, *1*, 470–478.

Brawley, C., & Matunis, E. (2004). Regeneration of male germline stem cells by spermatogonial dedifferentiation in vivo. *Science*, *304*, 1331–1334.

Brenner, J. L., & Schedl, T. (2016). Germline stem cell differentiation entails regional control of cell fate regulator GLD-1 in Caenorhabditis elegans. *Genetics*, *202*, 1085–1103.

Buszczak, M., Paterno, S., & Spradling, A. C. (2009). Drosophila stem cells share a common requirement for the histone H2B ubiquitin protease scrawny. *Science*, *323*, 248–251.

Byrd, D. T., Knobel, K., Affeldt, K., Crittenden, S. L., & Kimble, J. (2014). A DTC niche plexus surrounds the germline stem cell pool in Caenorhabditis elegans. *PLoS One*, *9*, e88372.

Chen, C., Cummings, R., Mordovanakis, A., Hunt, A. J., Mayer, M., Sept, D., et al. (2018). Cytokine receptor-Eb1 interaction couples cell polarity and fate during asymmetric cell division. *eLife*, 7, e33685.

Chen, D., & McKearin, D. (2003). Dpp signaling silences bam transcription directly to establish asymmetric divisions of germline stem cells. *Current Biology*, *13*, 1786–1791.

Chen, D., & McKearin, D. (2005). Gene circuitry controlling a stem cell niche. *Current Biology*, *15*, 179–184.

Cheng, J., Turkel, N., Hemati, N., Fuller, M. T., Hunt, A. J., & Yamashita, Y. M. (2008). Centrosome misorientation reduces stem cell division during ageing. *Nature*, *456*, 599–604.

Cherry, C. M., & Matunis, E. L. (2010). Epigenetic regulation of stem cell maintenance in the Drosophila testis via the nucleosome-remodeling factor NURF. *Cell Stem Cell*, *6*, 557–567.

Cheung, T. H., Quach, N. L., Charville, G. W., Liu, L., Park, L., Edalati, A., et al. (2012). Maintenance of muscle stem-cell quiescence by microRNA-489. *Nature*, *482*, 524–528.

Chiang, A. C., Yang, H., & Yamashita, Y. M. (2017). Spict, a cyst cell-specific gene, regulates starvation-induced spermatogonial cell death in the Drosophila testis. *Scientific Reports*, 7, 40245.

Cinquin, O., Crittenden, S. L., Morgan, D. E., & Kimble, J. (2010). Progression from a stem cell-like state to early differentiation in the C. elegans germ line. *Proceedings of the National Academy of Sciences of the United States of America*, *107*, 2048–2053.

Clough, E., Tedeschi, T., & Hazelrigg, T. (2014). Epigenetic regulation of oogenesis and germ stem cell maintenance by the Drosophila histone methyltransferase Eggless/dSetDB1. *Developmental Biology*, *388*, 181–191.

Conduit, P. T., & Raff, J. W. (2010). Cnn dynamics drive centrosome size asymmetry to ensure daughter centriole retention in drosophila neuroblasts. *Current Biology*, *20*, 2187–2192.

Cornelison, D. D., Filla, M. S., Stanley, H. M., Rapraeger, A. C., & Olwin, B. B. (2001). Syndecan-3 and syndecan-4 specifically mark skeletal muscle satellite cells and are implicated in satellite cell maintenance and muscle regeneration. *Developmental Biology*, *239*, 79–94.

Crittenden, S. L., Bernstein, D. S., Bachorik, J. L., Thompson, B. E., Gallegos, M., Petcherski, A. G., et al. (2002). A conserved RNA-binding protein controls germline stem cells in Caenorhabditis elegans. *Nature*, *417*, 660–663.

Crittenden, S. L., Leonhard, K. A., Byrd, D. T., & Kimble, J. (2006). Cellular analyses of the mitotic region in the Caenorhabditis elegans adult germ line. *Molecular Biology of the Cell, 17*, 3051–3061.

Crittenden, S. L., Troemel, E. R., Evans, T. C., & Kimble, J. (1994). GLP-1 is localized to the mitotic region of the C. elegans germ line. *Development, 120*, 2901–2911.

de Rooij, D. G. (2017). The nature and dynamics of spermatogonial stem cells. *Development, 144*, 3022.

Decotto, E., & Spradling, A. C. (2005). The Drosophila ovarian and testis stem cell niches: Similar somatic stem cells and signals. *Developmental Cell, 9*, 501–510.

Deng, W., & Lin, H. (1997). Spectrosomes and fusomes anchor mitotic spindles during asymmetric germ cell divisions and facilitate the formation of a polarized microtubule array for oocyte specification in Drosophila. *Developmental Biology, 189*, 79–94.

Draper, B. W., McCallum, C. M., & Moens, C. B. (2007). Nanos1 is required to maintain oocyte production in adult zebrafish. *Developmental Biology, 305*, 589–598.

Drummond-Barbosa, D., & Spradling, A. C. (2001). Stem cells and their progeny respond to nutritional changes during Drosophila oogenesis. *Developmental Biology, 231*, 265–278.

Eliazer, S., Palacios, V., Wang, Z., Kollipara, R. K., Kittler, R., & Buszczak, M. (2014). Lsd1 restricts the number of germline stem cells by regulating multiple targets in escort cells. *PLoS Genetics, 10*, e1004200.

Eliazer, S., Shalaby, N. A., & Buszczak, M. (2011). Loss of lysine-specific demethylase 1 nonautonomously causes stem cell tumors in the Drosophila ovary. *Proceedings of the National Academy of Sciences of the United States of America, 108*, 7064–7069.

Ettinger, A. W., Wilsch-Brauninger, M., Marzesco, A. M., Bickle, M., Lohmann, A., Maliga, Z., et al. (2011). Proliferating versus differentiating stem and cancer cells exhibit distinct midbody-release behaviour. *Nature Communications, 2*, 503.

Eun, S. H., Feng, L., Cedeno-Rosario, L., Gan, Q., Wei, G., Cui, K., et al. (2017). Polycomb group gene E(z) is required for Spermatogonial dedifferentiation in Drosophila adult testis. *Journal of Molecular Biology, 429*, 2030–2041.

Eun, S. H., Shi, Z., Cui, K., Zhao, K., & Chen, X. (2014). A non-cell autonomous role of E(z) to prevent germ cells from turning on a somatic cell marker. *Science, 343*, 1513–1516.

Fairchild, M. J., Smendziuk, C. M., & Tanentzapf, G. (2015). A somatic permeability barrier around the germline is essential for Drosophila spermatogenesis. *Development, 142*, 268–281.

Forbes, A., & Lehmann, R. (1998). Nanos and Pumilio have critical roles in the development and function of Drosophila germline stem cells. *Development, 125*, 679–690.

Fox, P. M., & Schedl, T. (2015). Analysis of germline stem cell differentiation following loss of GLP-1 notch activity in Caenorhabditis elegans. *Genetics, 201*, 167–184.

Fuentealba, L. C., Eivers, E., Geissert, D., Taelman, V., & De Robertis, E. M. (2008). Asymmetric mitosis: Unequal segregation of proteins destined for degradation. *Proceedings of the National Academy of Sciences of the United States of America, 105*, 7732–7737.

Fuller, M. T. (1998). Genetic control of cell proliferation and differentiation in Drosophila spermatogenesis. *Seminars in Cell & Developmental Biology, 9*, 433–444.

Ganley, A. R., & Kobayashi, T. (2014). Ribosomal DNA and cellular senescence: New evidence supporting the connection between rDNA and aging. *FEMS Yeast Research, 14*, 49–59.

Gilboa, L., & Lehmann, R. (2004). Repression of primordial germ cell differentiation parallels germ line stem cell maintenance. *Current Biology, 14*, 981–986.

Gonczy, P. (2008). Mechanisms of asymmetric cell division: Flies and worms pave the way. *Nature Reviews. Molecular Cell Biology, 9*, 355–366.

Greenspan, L. J., & Matunis, E. L. (2018). Retinoblastoma intrinsically regulates niche cell quiescence, identity, and niche number in the adult Drosophila testis. *Cell Reports*, *24*. 3466-3476.e3468.

Greider, C. W., & Blackburn, E. H. (1985). Identification of a specific telomere terminal transferase activity in Tetrahymena extracts. *Cell*, *43*, 405–413.

Guo, Z., & Wang, Z. (2009). The glypican Dally is required in the niche for the maintenance of germline stem cells and short-range BMP signaling in the Drosophila ovary. *Development*, *136*, 3627–3635.

Habib, S. J., Chen, B. C., Tsai, F. C., Anastassiadis, K., Meyer, T., Betzig, E., et al. (2013). A localized Wnt signal orients asymmetric stem cell division in vitro. *Science*, *339*, 1445–1448.

Hai, Y., Hou, J., Liu, Y., Liu, Y., Yang, H., Li, Z., et al. (2014). The roles and regulation of Sertoli cells in fate determinations of spermatogonial stem cells and spermatogenesis. *Seminars in Cell & Developmental Biology*, *29*, 66–75.

Harris, R. E., Pargett, M., Sutcliffe, C., Umulis, D., & Ashe, H. L. (2011). Brat promotes stem cell differentiation via control of a bistable switch that restricts BMP signaling. *Developmental Cell*, *20*, 72–83.

Hasan, S., Hetie, P., & Matunis, E. L. (2015). Niche signaling promotes stem cell survival in the Drosophila testis via the JAK-STAT target DIAP1. *Developmental Biology*, *404*, 27–39.

Hausburg, M. A., Doles, J. D., Clement, S. L., Cadwallader, A. B., Hall, M. N., Blackshear, P. J., et al. (2015). Post-transcriptional regulation of satellite cell quiescence by TTP-mediated mRNA decay. *eLife*, *4*, e03390.

Hawley, R. S., Marcus, C. H., Cameron, M. L., Schwartz, R. L., & Zitron, A. E. (1985). Repair-defect mutations inhibit rDNA magnification in Drosophila and discriminate between meiotic and premeiotic magnification. *Proceedings of the National Academy of Sciences of the United States of America*, *82*, 8095–8099.

Hawley, R. S., & Tartof, K. D. (1983). The effect of mei-41 on rDNA redundancy in Drosophila melanogaster. *Genetics*, *104*, 63–80.

Hayashi, Y., Kobayashi, S., & Nakato, H. (2009). Drosophila glypicans regulate the germline stem cell niche. *The Journal of Cell Biology*, *187*, 473–480.

He, J., Xuan, T., Xin, T., An, H., Wang, J., Zhao, G., et al. (2014). Evidence for chromatin-remodeling complex PBAP-controlled maintenance of the Drosophila ovarian germline stem cells. *PLoS One*, *9*, e103473.

Hemann, M. T., Rudolph, K. L., Strong, M. A., DePinho, R. A., Chin, L., & Greider, C. W. (2001). Telomere dysfunction triggers developmentally regulated germ cell apoptosis. *Molecular Biology of the Cell*, *12*, 2023–2030.

Henderson, S. T., Gao, D., Lambie, E. J., & Kimble, J. (1994). Lag-2 may encode a signaling ligand for the GLP-1 and LIN-12 receptors of C. elegans. *Development*, *120*, 2913–2924.

Herrera, S. C., & Bach, E. A. (2018). JNK signaling triggers spermatogonial dedifferentiation during chronic stress to maintain the germline stem cell pool in the Drosophila testis. *eLife*, 7, e1005815.

Herrera, E., Samper, E., Martin-Caballero, J., Flores, J. M., Lee, H. W., & Blasco, M. A. (1999). Disease states associated with telomerase deficiency appear earlier in mice with short telomeres. *The EMBO Journal*, *18*, 2950–2960.

Hou, X. S., Goldstein, E. S., & Perrimon, N. (1997). Drosophila Jun relays the Jun amino-terminal kinase signal transduction pathway to the decapentaplegic signal transduction pathway in regulating epithelial cell sheet movement. *Genes & Development*, *11*, 1728–1737.

Inaba, M., Buszczak, M., & Yamashita, Y. M. (2015). Nanotubes mediate niche-stem-cell signalling in the Drosophila testis. *Nature*, *523*, 329–332.

Inaba, M., Venkei, Z. G., & Yamashita, Y. M. (2015). The polarity protein Baz forms a platform for the centrosome orientation during asymmetric stem cell division in the Drosophila male germline. *eLife*, *4*, e04960.
Inaba, M., Yuan, H., Salzmann, V., Fuller, M. T., & Yamashita, Y. M. (2010). E-cadherin is required for centrosome and spindle orientation in Drosophila male germline stem cells. *PLoS One*, *5*, e12473.
Inaba, M., Yuan, H., & Yamashita, Y. M. (2011). String (Cdc25) regulates stem cell maintenance, proliferation and aging in Drosophila testis. *Development*, *138*, 5079–5086.
Inomata, K., Aoto, T., Binh, N. T., Okamoto, N., Tanimura, S., Wakayama, T., et al. (2009). Genotoxic stress abrogates renewal of melanocyte stem cells by triggering their differentiation. *Cell*, *137*, 1088–1099.
Issigonis, M., Tulina, N., de Cuevas, M., Brawley, C., Sandler, L., & Matunis, E. (2009). JAK-STAT signal inhibition regulates competition in the Drosophila testis stem cell niche. *Science*, *326*, 153–156.
Januschke, J., Llamazares, S., Reina, J., & Gonzalez, C. (2011). Drosophila neuroblasts retain the daughter centrosome. *Nature Communications*, *2*, 243.
Jin, Z., Kirilly, D., Weng, C., Kawase, E., Song, X., Smith, S., et al. (2008). Differentiation-defective stem cells outcompete normal stem cells for niche occupancy in the Drosophila ovary. *Cell Stem Cell*, *2*, 39–49.
Kai, T., & Spradling, A. (2004). Differentiating germ cells can revert into functional stem cells in Drosophila melanogaster ovaries. *Nature*, *428*, 564–569.
Kalchhauser, I., Farley, B. M., Pauli, S., Ryder, S. P., & Ciosk, R. (2011). FBF represses the Cip/Kip cell-cycle inhibitor CKI-2 to promote self-renewal of germline stem cells in C. elegans. *The EMBO Journal*, *30*, 3823–3829.
Kawase, E., Wong, M. D., Ding, B. C., & Xie, T. (2004). Gbb/Bmp signaling is essential for maintaining germline stem cells and for repressing bam transcription in the Drosophila testis. *Development*, *131*, 1365–1375.
Kershner, A. M., & Kimble, J. (2010). Genome-wide analysis of mRNA targets for Caenorhabditis elegans FBF, a conserved stem cell regulator. *Proceedings of the National Academy of Sciences of the United States of America*, *107*, 3936–3941.
Kershner, A. M., Shin, H., Hansen, T. J., & Kimble, J. (2014). Discovery of two GLP-1/Notch target genes that account for the role of GLP-1/Notch signaling in stem cell maintenance. *Proceedings of the National Academy of Sciences of the United States of America*, *111*, 3739–3744.
Kiger, A. A., Jones, D. L., Schulz, C., Rogers, M. B., & Fuller, M. T. (2001). Stem cell self-renewal specified by JAK-STAT activation in response to a support cell cue. *Science*, *294*, 2542–2545.
Kiger, A. A., White-Cooper, H., & Fuller, M. T. (2000). Somatic support cells restrict germline stem cell self-renewal and promote differentiation. *Nature*, *407*, 750–754.
Kimble, J. (2011). Molecular regulation of the mitosis/meiosis decision in multicellular organisms. *Cold Spring Harbor Perspectives in Biology*, *3*, a002683.
Kirilly, D., Wang, S., & Xie, T. (2011). Self-maintained escort cells form a germline stem cell differentiation niche. *Development*, *138*, 5087–5097.
Klein, A. M., Nakagawa, T., Ichikawa, R., Yoshida, S., & Simons, B. D. (2010). Mouse germ line stem cells undergo rapid and stochastic turnover. *Cell Stem Cell*, *7*, 214–224.
Kloc, M., Bilinski, S., Chan, A. P., Allen, L. H., Zearfoss, N. R., & Etkin, L. D. (2001). RNA localization and germ cell determination in Xenopus. *International Review of Cytology*, *203*, 63–91.
Knoblich, J. A. (2008). Mechanisms of asymmetric stem cell division. *Cell*, *132*, 583–597.
Kornberg, T. B., & Roy, S. (2014). Cytonemes as specialized signaling filopodia. *Development*, *141*, 729–736.

Kuo, T. C., Chen, C. T., Baron, D., Onder, T. T., Loewer, S., Almeida, S., et al. (2011). Midbody accumulation through evasion of autophagy contributes to cellular reprogramming and tumorigenicity. *Nature Cell Biology, 13*, 1214–1223.

Leatherman, J. L., & Dinardo, S. (2008). Zfh-1 controls somatic stem cell self-renewal in the Drosophila testis and nonautonomously influences germline stem cell self-renewal. *Cell Stem Cell, 3*, 44–54.

Leatherman, J. L., & Dinardo, S. (2010). Germline self-renewal requires cyst stem cells and stat regulates niche adhesion in Drosophila testes. *Nature Cell Biology, 12*, 806–811.

Lee, C., Sorensen, E. B., Lynch, T. R., & Kimble, J. (2016). C. elegans GLP-1/Notch activates transcription in a probability gradient across the germline stem cell pool. *eLife, 5*, e18370.

Lehmann, R. (2012). Germline stem cells: Origin and destiny. *Cell Stem Cell, 10*, 729–739.

Lei, L., & Spradling, A. C. (2013). Female mice lack adult germ-line stem cells but sustain oogenesis using stable primordial follicles. *Proceedings of the National Academy of Sciences of the United States of America, 110*, 8585–8590.

Lenhart, K. F., & DiNardo, S. (2015). Somatic cell encystment promotes abscission in germline stem cells following a regulated block in cytokinesis. *Developmental Cell, 34*, 192–205.

Levis, R. W., Ganesan, R., Houtchens, K., Tolar, L. A., & Sheen, F. M. (1993). Transposons in place of telomeric repeats at a Drosophila telomere. *Cell, 75*, 1083–1093.

Li, Y., Maines, J. Z., Tastan, O. Y., McKearin, D. M., & Buszczak, M. (2012). Mei-P26 regulates the maintenance of ovarian germline stem cells by promoting BMP signaling. *Development, 139*, 1547–1556.

Li, Y., Minor, N. T., Park, J. K., McKearin, D. M., & Maines, J. Z. (2009). Bam and Bgcn antagonize Nanos-dependent germ-line stem cell maintenance. *Proceedings of the National Academy of Sciences of the United States of America, 106*, 9304–9309.

Lim, J. G., & Fuller, M. T. (2012). Somatic cell lineage is required for differentiation and not maintenance of germline stem cells in Drosophila testes. *Proceedings of the National Academy of Sciences of the United States of America, 109*, 18477–18481.

Lim, C., Gandhi, S., Biniossek, M. L., Feng, L., Schilling, O., Urban, S., et al. (2015). An aminopeptidase in the Drosophila testicular niche acts in germline stem cell maintenance and Spermatogonial dedifferentiation. *Cell Reports, 13*, 315–325.

Lin, H., & Spradling, A. C. (1997). A novel group of pumilio mutations affects the asymmetric division of germline stem cells in the Drosophila ovary. *Development, 124*, 2463–2476.

Liu, L., Blasco, M., Trimarchi, J., & Keefe, D. (2002). An essential role for functional telomeres in mouse germ cells during fertilization and early development. *Developmental Biology, 249*, 74–84.

Lopez-Onieva, L., Fernandez-Minan, A., & Gonzalez-Reyes, A. (2008). Jak/Stat signalling in niche support cells regulates dpp transcription to control germline stem cell maintenance in the Drosophila ovary. *Development, 135*, 533–540.

Lu, W., Casanueva, M. O., Mahowald, A. P., Kato, M., Lauterbach, D., & Ferguson, E. L. (2012). Niche-associated activation of rac promotes the asymmetric division of Drosophila female germline stem cells. *PLoS Biology, 10*, e1001357.

Lu, K., Jensen, L., Lei, L., & Yamashita, Y. M. (2017). Stay connected: A germ cell strategy. *Trends in Genetics, 33*, 971–978.

Lu, K. L., Nelson, J. O., Watase, G. J., Warsinger-Pepe, N., & Yamashita, Y. M. (2018). Transgenerational dynamics of rDNA copy number in Drosophila male germline stem cells. *eLife, 7*, e32421.

Lu, K. L., & Yamashita, Y. M. (2017). Germ cell connectivity enhances cell death in response to DNA damage in the Drosophila testis. *eLife, 6*, e27960.

Luo, L., Wang, H., Fan, C., Liu, S., & Cai, Y. (2015). Wnt ligands regulate Tkv expression to constrain Dpp activity in the Drosophila ovarian stem cell niche. *The Journal of Cell Biology, 209*, 595–608.
Lv, X., Pan, C., Zhang, Z., Xia, Y., Chen, H., Zhang, S., et al. (2016). SUMO regulates somatic cyst stem cell maintenance and directly targets the Hedgehog pathway in adult Drosophila testis. *Development, 143*, 1655–1662.
Mainpal, R., Priti, A., & Subramaniam, K. (2011). PUF-8 suppresses the somatic transcription factor PAL-1 expression in C. elegans germline stem cells. *Developmental Biology, 360*, 195–207.
McCarthy, A., Deiulio, A., Martin, E. T., Upadhyay, M., & Rangan, P. (2018). Tip60 complex promotes expression of a differentiation factor to regulate germline differentiation in female Drosophila. *Molecular Biology of the Cell, 29*, 2933–2945. https://doi.org/10.109/mbc.E18-06-0385.
McKearin, D. M., & Spradling, A. C. (1990). Bag-of-marbles: A Drosophila gene required to initiate both male and female gametogenesis. *Genes & Development, 4*, 2242–2251.
McLeod, C. J., Wang, L., Wong, C., & Jones, D. L. (2010). Stem cell dynamics in response to nutrient availability. *Current Biology, 20*, 2100–2105.
Meng, X., Lindahl, M., Hyvonen, M. E., Parvinen, M., de Rooij, D. G., Hess, M. W., et al. (2000). Regulation of cell fate decision of undifferentiated spermatogonia by GDNF. *Science, 287*, 1489–1493.
Michel, M., Kupinski, A. P., Raabe, I., & Bokel, C. (2012). Hh signalling is essential for somatic stem cell maintenance in the Drosophila testis niche. *Development, 139*, 2663–2669.
Moore, D. L., Pilz, G. A., Arauzo-Bravo, M. J., Barral, Y., & Jessberger, S. (2015). A mechanism for the segregation of age in mammalian neural stem cells. *Science, 349*, 1334–1338.
Morris, L. X., & Spradling, A. C. (2011). Long-term live imaging provides new insight into stem cell regulation and germline-soma coordination in the Drosophila ovary. *Development, 138*, 2207–2215.
Mottier-Pavie, V. I., Palacios, V., Eliazer, S., Scoggin, S., & Buszczak, M. (2016). The Wnt pathway limits BMP signaling outside of the germline stem cell niche in Drosophila ovaries. *Developmental Biology, 417*, 50–62.
Nakagawa, T., Sharma, M., Nabeshima, Y., Braun, R. E., & Yoshida, S. (2010). Functional hierarchy and reversibility within the murine spermatogenic stem cell compartment. *Science, 328*, 62–67.
Nakamura, S., Kobayashi, K., Nishimura, T., Higashijima, S., & Tanaka, M. (2010). Identification of germline stem cells in the ovary of the teleost medaka. *Science, 328*, 1561–1563.
Nakamura-Ishizu, A., Takizawa, H., & Suda, T. (2014). The analysis, roles and regulation of quiescence in hematopoietic stem cells. *Development, 141*, 4656–4666.
Nigg, E. A., & Stearns, T. (2011). The centrosome cycle: Centriole biogenesis, duplication and inherent asymmetries. *Nature Cell Biology, 13*, 1154.
Ohlstein, B., Lavoie, C. A., Vef, O., Gateff, E., & McKearin, D. M. (2000). The Drosophila cystoblast differentiation factor, benign gonial cell neoplasm, is related to DExH-box proteins and interacts genetically with bag-of-marbles. *Genetics, 155*, 1809–1819.
Ohlstein, B., & McKearin, D. (1997). Ectopic expression of the Drosophila Bam protein eliminates oogenic germline stem cells. *Development, 124*, 3651–3662.
Prasad, A., Porter, D. F., Kroll-Conner, P. L., Mohanty, I., Ryan, A. R., Crittenden, S. L., et al. (2016). The PUF binding landscape in metazoan germ cells. *RNA, 22*, 1026–1043.
Racher, H., & Hansen, D. (2012). PUF-8, a Pumilio homolog, inhibits the proliferative fate in the Caenorhabditis elegans germline. *G3 (Bethesda), 2*, 1197–1205.

Radion, E., Morgunova, V., Ryazansky, S., Akulenko, N., Lavrov, S., Abramov, Y., et al. (2018). Key role of piRNAs in telomeric chromatin maintenance and telomere nuclear positioning in Drosophila germline. *Epigenetics & Chromatin*, *11*, 40.

Ramirez-Weber, F. A., & Kornberg, T. B. (1999). Cytonemes: Cellular processes that project to the principal signaling center in Drosophila imaginal discs. *Cell*, *97*, 599–607.

Rangan, P., Malone, C. D., Navarro, C., Newbold, S. P., Hayes, P. S., Sachidanandam, R., et al. (2011). piRNA production requires heterochromatin formation in Drosophila. *Current Biology*, *21*, 1373–1379.

Rangel-Huerta, E., & Maldonado, E. (2017). Transit-amplifying cells in the fast lane from stem cells towards differentiation. *Stem Cells International*, *2017*, 7602951.

Raz, E. (2003). Primordial germ-cell development: The zebrafish perspective. *Nature Reviews. Genetics*, *4*, 690–700.

Riesgo-Escovar, J. R., & Hafen, E. (1997). Drosophila Jun kinase regulates expression of decapentaplegic via the ETS-domain protein Aop and the AP-1 transcription factor DJun during dorsal closure. *Genes & Development*, *11*, 1717–1727.

Ritossa, F. M. (1968). Unstable redundancy of genes for ribosomal RNA. *Proceedings of the National Academy of Sciences of the United States of America*, *60*, 509–516.

Rojas-Rios, P., Guerrero, I., & Gonzalez-Reyes, A. (2012). Cytoneme-mediated delivery of hedgehog regulates the expression of bone morphogenetic proteins to maintain germline stem cells in Drosophila. *PLoS Biology*, *10*, e1001298.

Rosu, S., & Cohen-Fix, O. (2017). Live-imaging analysis of germ cell proliferation in the C. elegans adult supports a stochastic model for stem cell proliferation. *Developmental Biology*, *423*, 93–100.

Rujano, M. A., Bosveld, F., Salomons, F. A., Dijk, F., van Waarde, M. A., van der Want, J. J., et al. (2006). Polarised asymmetric inheritance of accumulated protein damage in higher eukaryotes. *PLoS Biology*, *4*, e417.

Saitou, M., & Yamaji, M. (2012). Primordial germ cells in mice. *Cold Spring Harbor Perspectives in Biology*, *4*, a008375.

Salzmann, V., Chen, C., Chiang, C. Y., Tiyaboonchai, A., Mayer, M., & Yamashita, Y. M. (2014). Centrosome-dependent asymmetric inheritance of the midbody ring in Drosophila germline stem cell division. *Molecular Biology of the Cell*, *25*, 267–275.

Sarkar, A., Parikh, N., Hearn, S. A., Fuller, M. T., Tazuke, S. I., & Schulz, C. (2007). Antagonistic roles of Rac and Rho in organizing the germ cell microenvironment. *Current Biology*, *17*, 1253–1258.

Savitsky, M., Kwon, D., Georgiev, P., Kalmykova, A., & Gvozdev, V. (2006). Telomere elongation is under the control of the RNAi-based mechanism in the Drosophila germline. *Genes & Development*, *20*, 345–354.

Schultz, E. (1976). Fine structure of satellite cells in growing skeletal muscle. *The American Journal of Anatomy*, *147*, 49–70.

Schultz, M. B., & Sinclair, D. A. (2016). When stem cells grow old: Phenotypes and mechanisms of stem cell aging. *Development*, *143*, 3–14.

Schulz, C., Kiger, A. A., Tazuke, S. I., Yamashita, Y. M., Pantalena-Filho, L. C., Jones, D. L., et al. (2004). A misexpression screen reveals effects of bag-of-marbles and TGF beta class signaling on the Drosophila male germ-line stem cell lineage. *Genetics*, *167*, 707–723.

Schulz, C., Wood, C. G., Jones, D. L., Tazuke, S. I., & Fuller, M. T. (2002). Signaling from germ cells mediated by the rhomboid homolog stet organizes encapsulation by somatic support cells. *Development*, *129*, 4523–4534.

Seidel, H. S., & Kimble, J. (2015). Cell-cycle quiescence maintains Caenorhabditis elegans germline stem cells independent of GLP-1/Notch. *eLife*, *4*, e10832.

Sheng, X. R., Brawley, C. M., & Matunis, E. L. (2009). Dedifferentiating spermatogonia outcompete somatic stem cells for niche occupancy in the Drosophila testis. *Cell Stem Cell*, *5*, 191–203.

Sheng, X. R., & Matunis, E. (2011). Live imaging of the Drosophila spermatogonial stem cell niche reveals novel mechanisms regulating germline stem cell output. *Development*, *138*, 3367–3376.

Shields, A. R., Spence, A. C., Yamashita, Y. M., Davies, E. L., & Fuller, M. T. (2014). The actin-binding protein profilin is required for germline stem cell maintenance and germ cell enclosure by somatic cyst cells. *Development*, *141*, 73–82.

Shin, H., Haupt, K. A., Kershner, A. M., Kroll-Conner, P., Wickens, M., & Kimble, J. (2017). SYGL-1 and LST-1 link niche signaling to PUF RNA repression for stem cell maintenance in Caenorhabditis elegans. *PLoS Genetics*, *13*, e1007121.

Shivdasani, A. A., & Ingham, P. W. (2003). Regulation of stem cell maintenance and transit amplifying cell proliferation by tgf-Beta signaling in Drosophila spermatogenesis. *Current Biology*, *13*, 2065–2072.

Shpiz, S., Olovnikov, I., Sergeeva, A., Lavrov, S., Abramov, Y., Savitsky, M., et al. (2011). Mechanism of the piRNA-mediated silencing of Drosophila telomeric retrotransposons. *Nucleic Acids Research*, *39*, 8703–8711.

Song, X., Call, G. B., Kirilly, D., & Xie, T. (2007). Notch signaling controls germline stem cell niche formation in the Drosophila ovary. *Development*, *134*, 1071–1080.

Song, X., Wong, M. D., Kawase, E., Xi, R., Ding, B. C., McCarthy, J. J., et al. (2004). Bmp signals from niche cells directly repress transcription of a differentiation-promoting gene, bag of marbles, in germline stem cells in the Drosophila ovary. *Development*, *131*, 1353–1364.

Song, X., & Xie, T. (2002). DE-cadherin-mediated cell adhesion is essential for maintaining somatic stem cells in the Drosophila ovary. *Proceedings of the National Academy of Sciences of the United States of America*, *99*, 14813–14818.

Song, X., Zhu, C. H., Doan, C., & Xie, T. (2002). Germline stem cells anchored by adherens junctions in the Drosophila ovary niches. *Science*, *296*, 1855–1857.

Spradling, A., Fuller, M. T., Braun, R. E., & Yoshida, S. (2011). Germline stem cells. *Cold Spring Harbor Perspectives in Biology*, *3*, a002642.

Srinivasan, S., Mahowald, A. P., & Fuller, M. T. (2012). The receptor tyrosine phosphatase Lar regulates adhesion between Drosophila male germline stem cells and the niche. *Development*, *139*, 1381–1390.

Srivastava, A. K., & Schlessinger, D. (1991). Structure and organization of ribosomal DNA. *Biochimie*, *73*, 631–638.

Stanton, P. G. (2016). Regulation of the blood-testis barrier. *Seminars in Cell & Developmental Biology*, *59*, 166–173.

Subramaniam, K., & Seydoux, G. (2003). Dedifferentiation of primary spermatocytes into germ cell tumors in C. elegans lacking the pumilio-like protein PUF-8. *Current Biology*, *13*, 134–139.

Tanaka, M. (2016). Germline stem cells are critical for sexual fate decision of germ cells. *Bioessays*, *38*, 1227–1233.

Tanentzapf, G., Devenport, D., Godt, D., & Brown, N. H. (2007). Integrin-dependent anchoring of a stem-cell niche. *Nature Cell Biology*, *9*, 1413–1418.

Tang, W. W., Kobayashi, T., Irie, N., Dietmann, S., & Surani, M. A. (2016). Specification and epigenetic programming of the human germ line. *Nature Reviews. Genetics*, *17*, 585–600.

Tarayrah, L., Herz, H. M., Shilatifard, A., & Chen, X. (2013). Histone demethylase dUTX antagonizes JAK-STAT signaling to maintain proper gene expression and architecture of the Drosophila testis niche. *Development*, *140*, 1014–1023.

Tarayrah, L., Li, Y., Gan, Q., & Chen, X. (2015). Epigenetic regulator lid maintains germline stem cells through regulating JAK-STAT signaling pathway activity. *Biology Open*, *4*, 1518–1527.
Tartof, K. D. (1974). Unequal mitotic sister chromatin exchange as the mechanism of ribosomal RNA gene magnification. *Proceedings of the National Academy of Sciences of the United States of America*, *71*, 1272–1276.
Titen, S. W., & Golic, K. G. (2010). Healing of euchromatic chromosome breaks by efficient de novo telomere addition in Drosophila melanogaster. *Genetics*, *184*, 309–312.
Toledano, H., D'Alterio, C., Czech, B., Levine, E., & Jones, D. L. (2012). The let-7-Imp axis regulates ageing of the Drosophila testis stem-cell niche. *Nature*, *485*, 605–610.
Toth, K. F., Pezic, D., Stuwe, E., & Webster, A. (2016). The piRNA pathway guards the germline genome against transposable elements. *Advances in Experimental Medicine and Biology*, *886*, 51–77.
Tran, J., Brenner, T. J., & DiNardo, S. (2000). Somatic control over the germline stem cell lineage during Drosophila spermatogenesis. *Nature*, *407*, 754–757.
Tran, V., Lim, C., Xie, J., & Chen, X. (2012). Asymmetric division of Drosophila male germline stem cell shows asymmetric histone distribution. *Science*, *338*, 679–682.
Tulina, N., & Matunis, E. (2001). Control of stem cell self-renewal in Drosophila spermatogenesis by JAK-STAT signaling. *Science*, *294*, 2546–2549.
Upadhyay, M., Martino Cortez, Y., Wong-Deyrup, S., Tavares, L., Schowalter, S., Flora, P., et al. (2016). Transposon dysregulation modulates dWnt4 signaling to control germline stem cell differentiation in Drosophila. *PLoS Genetics*, *12*, e1005918.
Venkei, Z. G., & Yamashita, Y. M. (2015). The centrosome orientation checkpoint is germline stem cell specific and operates prior to the spindle assembly checkpoint in Drosophila testis. *Development*, *142*, 62–69.
Volin, M., Zohar-Fux, M., Gonen, O., Porat-Kuperstein, L., & Toledano, H. (2018). microRNAs selectively protect hub cells of the germline stem cell niche from apoptosis. *The Journal of Cell Biology*, *217*, 3829–3838.
Voog, J., D'Alterio, C., & Jones, D. L. (2008). Multipotent somatic stem cells contribute to the stem cell niche in the Drosophila testis. *Nature*, *454*, 1132–1136.
Voog, J., Sandall, S. L., Hime, G. R., Resende, L. P., Loza-Coll, M., Aslanian, A., et al. (2014). Escargot restricts niche cell to stem cell conversion in the Drosophila testis. *Cell Reports*, *7*, 722–734.
Vuilleumier, R., Springhorn, A., Patterson, L., Koidl, S., Hammerschmidt, M., Affolter, M., et al. (2010). Control of Dpp morphogen signalling by a secreted feedback regulator. *Nature Cell Biology*, *12*, 611–617.
Wallenfang, M. R., Nayak, R., & DiNardo, S. (2006). Dynamics of the male germline stem cell population during aging of Drosophila melanogaster. *Aging Cell*, *5*, 297–304.
Wang, S., Gao, Y., Song, X., Ma, X., Zhu, X., Mao, Y., et al. (2015). Wnt signaling-mediated redox regulation maintains the germ line stem cell differentiation niche. *eLife*, *4*, e08174.
Wang, X., Harris, R. E., Bayston, L. J., & Ashe, H. L. (2008). Type IV collagens regulate BMP signalling in Drosophila. *Nature*, *455*, 72–77.
Wang, L., Li, Z., & Cai, Y. (2008). The JAK/STAT pathway positively regulates DPP signaling in the Drosophila germline stem cell niche. *The Journal of Cell Biology*, *180*, 721–728.
Wang, Z., & Lin, H. (2004). Nanos maintains germline stem cell self-renewal by preventing differentiation. *Science*, *303*, 2016–2019.

Wang, X., Pan, L., Wang, S., Zhou, J., McDowell, W., Park, J., et al. (2011). Histone H3K9 trimethylase eggless controls germline stem cell maintenance and differentiation. *PLoS Genetics*, *7*, e1002426.

Wang, J. T., & Seydoux, G. (2013). Germ cell specification. *Advances in Experimental Medicine and Biology*, *757*, 17–39.

Wang, H., Singh, S. R., Zheng, Z., Oh, S. W., Chen, X., Edwards, K., et al. (2006). Rap-GEF signaling controls stem cell anchoring to their niche through regulating DE-cadherin-mediated cell adhesion in the Drosophila testis. *Developmental Cell*, *10*, 117–126.

Wang, X., Tsai, J. W., Imai, J. H., Lian, W. N., Vallee, R. B., & Shi, S. H. (2009). Asymmetric centrosome inheritance maintains neural progenitors in the neocortex. *Nature*, *461*, 947–955.

Ward, E. J., Shcherbata, H. R., Reynolds, S. H., Fischer, K. A., Hatfield, S. D., & Ruohola-Baker, H. (2006). Stem cells signal to the niche through the Notch pathway in the Drosophila ovary. *Current Biology*, *16*, 2352–2358.

Watt, F. M. (2001). Stem cell fate and patterning in mammalian epidermis. *Current Opinion in Genetics & Development*, *11*, 410–417.

Williamson, A., & Lehmann, R. (1996). Germ cell development in Drosophila. *Annual Review of Cell and Developmental Biology*, *12*, 365–391.

Wilson, A., Laurenti, E., Oser, G., van der Wath, R. C., Blanco-Bose, W., Jaworski, M., et al. (2008). Hematopoietic stem cells reversibly switch from dormancy to self-renewal during homeostasis and repair. *Cell*, *135*, 1118–1129.

Xi, R., & Xie, T. (2005). Stem cell self-renewal controlled by chromatin remodeling factors. *Science*, *310*, 1487–1489.

Xie, T., & Spradling, A. C. (2000). A niche maintaining germ line stem cells in the Drosophila ovary. *Science*, *290*, 328–330.

Xin, T., Xuan, T., Tan, J., Li, M., Zhao, G., & Li, M. (2013). The Drosophila putative histone acetyltransferase Enok maintains female germline stem cells through regulating Bruno and the niche. *Developmental Biology*, *384*, 1–12.

Xing, Y., & Li, W. X. (2015). Heterochromatin components in germline stem cell maintenance. *Scientific Reports*, *5*, 17463.

Yacobi-Sharon, K., Namdar, Y., & Arama, E. (2013). Alternative germ cell death pathway in Drosophila involves HtrA2/Omi, lysosomes, and a caspase-9 counterpart. *Developmental Cell*, *25*, 29–42.

Yadlapalli, S., & Yamashita, Y. M. (2013). Chromosome-specific nonrandom sister chromatid segregation during stem-cell division. *Nature*, *498*, 251–254.

Yamashita, Y. M., Jones, D. L., & Fuller, M. T. (2003). Orientation of asymmetric stem cell division by the APC tumor suppressor and centrosome. *Science*, *301*, 1547–1550.

Yamashita, Y. M., Mahowald, A. P., Perlin, J. R., & Fuller, M. T. (2007). Asymmetric inheritance of mother versus daughter centrosome in stem cell division. *Science*, *315*, 518–521.

Yang, H., & Yamashita, Y. M. (2015). The regulated elimination of transit-amplifying cells preserves tissue homeostasis during protein starvation in Drosophila testis. *Development*, *142*, 1756–1766.

Yoshida, S., Sukeno, M., & Nabeshima, Y. (2007). A vasculature-associated niche for undifferentiated spermatogonia in the mouse testis. *Science*, *317*, 1722–1726.

Yuan, H., Chiang, C. Y., Cheng, J., Salzmann, V., & Yamashita, Y. M. (2012). Regulation of cyclin A localization downstream of Par-1 function is critical for the centrosome orientation checkpoint in Drosophila male germline stem cells. *Developmental Biology*, *361*, 57–67.

Zhang, J., Grindley, J. C., Yin, T., Jayasinghe, S., He, X. C., Ross, J. T., et al. (2006). PTEN maintains haematopoietic stem cells and acts in lineage choice and leukaemia prevention. *Nature*, *441*, 518–522.
Zhang, Z., Lv, X., Jiang, J., Zhang, L., & Zhao, Y. (2013). Dual roles of Hh signaling in the regulation of somatic stem cell self-renewal and germline stem cell maintenance in Drosophila testis. *Cell Research*, *23*, 573–576.
Zheng, Q., Wang, Y., Vargas, E., & DiNardo, S. (2011). Magu is required for germline stem cell self-renewal through BMP signaling in the Drosophila testis. *Developmental Biology*, *357*, 202–210.
Zoller, R., & Schulz, C. (2012). The Drosophila cyst stem cell lineage: Partners behind the scenes? *Spermatogenesis*, *2*, 145–157.

CHAPTER SEVEN

Heterogeneous, dynamic, and stochastic nature of mammalian spermatogenic stem cells

Shosei Yoshida[a,b,*]

[a]Division of Germ Cell Biology, National Institute for Basic Biology, National Institutes of Natural Sciences, Okazaki, Aichi, Japan

[b]Department of Basic Biology, School of Life Science, SOKENDAI (Graduate University for Advanced Studies), Okazaki, Aichi, Japan

*Corresponding author: e-mail address: shosei@nibb.ac.jp

Contents

1. Introduction 246
   1.1 Key questions about SSCs 246
   1.2 The stem-cell system in *Drosophila* spermatogenesis 247
2. Overview of mouse spermatogenesis in adult testes 248
   2.1 Spermatogenic cell types 248
   2.2 Seminiferous tubules 251
   2.3 The seminiferous epithelial cycle and spermatogenic wave 251
   2.4 Interstitial compartment 254
3. Advantages and available techniques in the study of SSCs 255
   3.1 Advantageous anatomy of seminiferous tubules 255
   3.2 Transplantation 255
   3.3 In vitro SSC culture in 2D 256
   3.4 Ex vivo organ culture in 3D 256
   3.5 Intravital live imaging 257
   3.6 In vivo pulse-labeling fate analysis 257
   3.7 Single-cell "omics" study 257
   3.8 In vitro gametogenesis from pluripotent cells 257
   3.9 Statistics, mathematics, and theory 258
4. Identity and dynamics of SSCs 258
   4.1 The "A_s model" 258
   4.2 Heterogeneous and hierarchical transcriptional states within A_{undiff} population 260
   4.3 Extension of the A_s model 262
   4.4 Clonal fate variability: Population asymmetry and neutral competition 262
   4.5 Syncytial fragmentation: Interchange between topologically different states 264
   4.6 Context dependent interchange between different transcriptional states 266

Current Topics in Developmental Biology, Volume 135
ISSN 0070-2153
https://doi.org/10.1016/bs.ctdb.2019.04.008

4.7 Dynamic heterogeneity: A paradigm of stem-cell homeostasis? 267
4.8 Future questions about the SSC identity 268
5. Microenvironmental regulation of SSCs 268
5.1 Scattered distribution of A_{undiff} in the basal compartment 268
5.2 Biased distribution toward interstitium and blood vessels 270
5.3 Motion of spermatogonia 270
5.4 Spatiotemporal distribution of extracellular signaling factors in an open (facultative) niche microenvironment 270
5.5 Heterogeneous susceptibility to extracellular signals 274
5.6 Density regulation through the mechanism of mitogen competition 275
5.7 Next questions 276
6. Conclusion 277
Acknowledgments 278
References 278

Abstract

Mammalian testes produce a huge number of sperm over a long period. This process, essential for the continuity of life, depends on the delicate balance of self-renewal and differentiation of resident *stem cells*, termed spermatogenic (spermatogonial) stem cells or SSCs. SSCs have motivated many researchers to query their identity, behavior, and regulation in the tissue microenvironment. The study of SSCs has a long and prominent history: Taking advantage of the unique organization of the seminiferous tubules, and the accompanying seminiferous epithelial *cycle* and spermatogenic *wave*, intricate concepts of SSC dynamics were established from the early days. This is represented by the "A_s model" first published in 1971. Then, numerous technical breakthroughs including transplantation, stem cell culture, organ culture, intravital live-imaging, and lineage tracing have made SSCs an invaluable model for tissue stem cell research. Further progress is likely to come from emerging technologies, such as in vitro gametogenesis and single cell omics. An ensemble of these experimental systems has lead to the identification of the dynamic and heterogeneous nature of SSCs, underpinning their context-dependent and flexible behavior. In addition, active migration of SSCs over the open (or facultative) niche microenvironment of the seminiferous tubules is in stark contrast to stem cell regulations within anatomically defined niches such as in the *Drosophila* ovary and testis (see chapter "Germline stem cell homeostasis" by Nelson et al.). By revealing novel mechanisms of stem cell regulation, research on mouse spermatogenesis will continually make significant contributions to the understanding of general concepts in the tissue stem-cell field.

1. Introduction

1.1 Key questions about SSCs

Germ cells ensure successful reproduction and transmission of genetic information to the next generation. In particular, mammalian spermatogenesis, which occurs in the testis, represents a highly robust and high-throughput system that produces a huge number of sperm, in some species for as long as decades. The entire process of spermatogenesis involves self-renewing, amplifying, and differentiating compartments (Meistrich & van Beek, 1993). In the self-renewing compartment, stem cells (spermatogenic stem cells or spermatogonial stem cells, referred to as SSCs hereafter) maintain their number while producing differentiation-committed cells. Committed cells dramatically increase their number through a series of mitotic divisions. Then, mature sperm are produced after completion of meiosis and spermiogenesis (transformation), during which chromosomal recombination occurs and the epigenetic status of the genomes is prepared for fertilization and development of the next generation. As a result, as many as 100 million sperm are produced per day, for decades, in humans.

This massive production is supported by a robust stem-cell system. The study of SSCs has drawn the interest of many researchers as it is one of the central targets in mammalian reproductive and developmental biology. Led by the studies on mice and other rodents, this field is rapidly expanding to other species, including primates and humans. This review concentrates on mouse spermatogenesis, discussing the following key questions:

- How can we define SSCs and how do they behave in the testis?
- How are SSCs regulated in their tissue microenvironment?

The questions are of general interest as they are shared with many other tissue stem cell systems, making the study of mammalian spermatogenesis relevant beyond its important role in reproduction.

1.2 The stem-cell system in *Drosophila* spermatogenesis

To emphasize the characteristic features of mouse spermatogenesis, I will first provide a brief overview of *Drosophila* spermatogenesis (Fig. 1; refer also to chapter "Germline stem cell homeostasis" by Nelson et al., in this issue) (Fuller & Spradling, 2007; Spradling et al., 2011). At the distal tip of the *Drosophila* testis, germline stem cells (GSCs) are present as mononucleated germ cells tethered to specialized somatic niche cells, called the hub. Division of an GSC is a highly-controlled process, giving rise to one self-renewing GSC and one differentiating cell (a gonialblast), which maintain and lose their contact with the hub, respectively. Gonialblasts are destined for differentiation and after four rounds of mitotic division with incomplete cytokinesis, cysts of 16 interconnected cells (now called spermatogonia) enter meiosis synchronously (Fig. 1B). Importantly, these cell types are spatially arranged

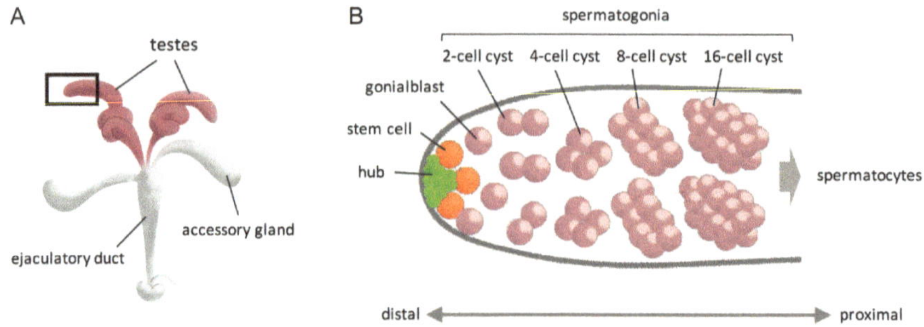

Fig. 1 Simplified views of *Drosophila* testes. (A) An overview of internal genitalia of *Drosophila*, showing a pair of blind-ended testes colored in magenta. The tip region (rectangle) is magnified in (B). (B) Diagram of a *Drosophila* testicular distal tip. Hub cells, located in the distal end, tether the stem cells (GSCs), while their immediate mononucleated differentiating progeny (gonialblasts) and their subsequent product of syncytial cysts (spermatogonia) are arranged in the order of differentiation progression along the distal-proximal axis of the organ. Note that many important elements are omitted, for a more detailed description see Fuller and Spradling (2007), Losick, Morris, Fox, and Spradling (2011), Spradling, Fuller, Braun, and Yoshida (2011).

along the distal-proximal axis of the testis in the order of differentiation progression; differentiating germ cells gradually move proximal from the distal tip where their ancestors, the GSCs, are located.

Thus, the *Drosophila* testis represents a stem cell-supported tissue from which many stem-cell concepts have been established (reviews including Fuller & Spradling, 2007; Losick et al., 2011; Spradling et al., 2011). These include an invariant asymmetric stem-cell division, a definitive niche microenvironment to which stem cells are tethered, and a polarized arrangement of differentiating cells (Fig. 1). *Drosophila* oogenesis shares similar features, where within a fixed stem cell niche a determined pattern of mostly asymmetric GSC divisions produces a new GSC and a differentiating daughter. Mouse and *Drosophila* spermatogenesis appear to represent opposite extremes of tissue stem cell systems. Hereafter, I will focus on the mouse SSCs that support persisting spermatogenesis in adults.

2. Overview of mouse spermatogenesis in adult testes

2.1 Spermatogenic cell types

Germ cells in mitotic, meiotic, and haploid stages (designated as spermatogonia, spermatocytes and spermatids, respectively) are found in mature mouse testis (Fig. 2) (Russell, Ettlin, Sinha Hikim, & Clegg, 1990). Spermatogonia include mononucleated cells and syncytial cysts. The latter

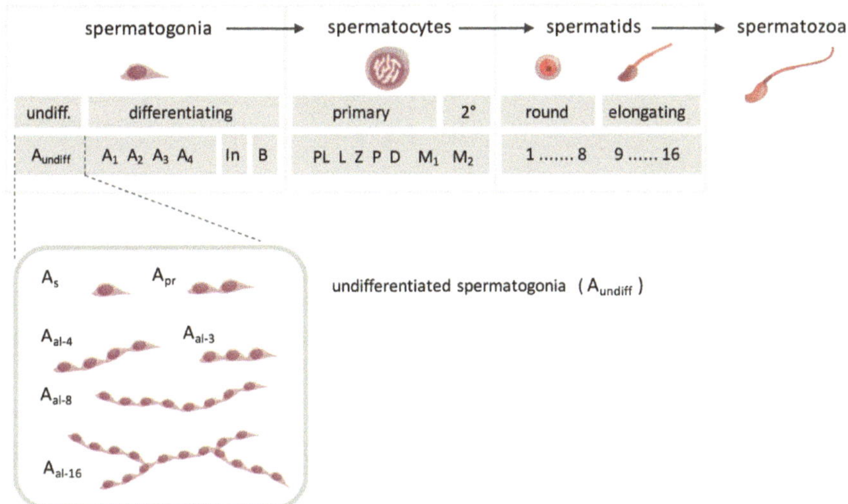

Fig. 2 Spermatogenic cell types found in the mature mouse testes. Spermatogonia (mitotic), spermatocyte (meiotic), and spermatids (haploid) are all observed in mature mouse testes, each containing multiple stages of differentiation. The nomenclature of the spermatogonial compartment is described in detail in the text. Undifferentiated spermatogonia (A_{undiff}) comprise the most primitive population including A_{single} (A_s, mononucleated cells), A_{paired} (A_{pr}, two-cell syncytia), and $A_{aligned}$ (A_{al}, syncytia of more than two cells). PL, L, Z, P, D, M1, and M2 in spermatocytes indicate, preleptotene, leptotene, zygotene, pachytene, diplotene, meiosis 1, and meiosis 2, respectively. Numbers 1 through 16 in spermatids indicate the steps of spermatid differentiation, of which 1–8 and 9–16 correspond to round and elongating spermatids, respectively.

are formed through incomplete cytokinesis, leaving the sister cells interconnected with cytoplasmic connection called intercellular bridges (ICB), a counterpart to the ring canals of *Drosophila*. There is an unfortunate inconsistency in the terminology: "spermatogonia" includes all mitotic cells in mice, while this term defines only interconnected syncytial cysts of mitotic cells in *Drosophila*. The number of cells included in individual syncytia, determined using whole-mount seminiferous tubule preparations, are 2^n (i.e., 2, 4, 8, 16, etc.) in most cases. However, a minority of syncytia are composed of other cell numbers (e.g., 3, 5, 6, 7, etc.), reflecting the fragmentation of syncytia through ICB breakdown, as described below. Such incomplete divisions, and resultant extension of syncytia, are repeated during the stages of mitotic amplification of spermatogonia and meiotic divisions of spermatocytes. Therefore, in rodents, as many as 1000 (or even more) spermatogonia are thought to be interconnected and synchronously undergo meiotic divisions, resulting in ~4000 connected spermatids that undergo spermiogenesis (Russell et al., 1990). The magnitude of mitotic amplification varies greatly between species; for example, 128-, 16-, and 8-cell long syncytia appear to enter meiosis in case of monkeys, humans, and quail, respectively (Di Persio et al., 2017; Fayomi & Orwig, 2018; Lin & Jones, 1992).

In mice, a minority population of *undifferentiated spermatogonia*, showing characteristic nuclear morphology with only a trace of heterochromatin (Chiarini-Garcia & Russell, 2001, 2002), is responsible for SSC function. Undifferentiated spermatogonia (also designated as A_{undiff}) compose a group of singly-isolated mononucleated cells called A_{single} (A_s), interconnected pairs of cells called A_{paired} (A_{pr}), and chains of three or more cells up to roughly 16—called $A_{aligned}$ (A_{al}) (Fig. 2). Undifferentiated spermatogonia play both self-renewing and transit-amplifying roles, while more advanced *differentiating spermatogonia* are essentially committed. The latter proceed with the differentiation steps of A_1, A_2, A_3, A_4, Intermediate (I_n), and B spermatogonia through synchronous mitotic divisions, then enter meiosis in the spermatocyte stages and spermiogenesis in the spermatid stages (Fig. 2). In this review special emphasis will be put on the dynamics of undifferentiated spermatogonia (A_{undiff}).

At first, the somewhat confusing nomenclature of mouse spermatogonia needs to be explained (Fig. 2). Historically, mouse spermatogonia have been classified into types A, I_n, and B, based on the relatively discrete appearances of their nuclei: Type A spermatogonia have been characterized by little heterochromatin, before accumulating massive amounts of heterochromatin in I_n and B prior to meiosis (Russell et al., 1990). However, while B and I_n are homogeneous populations, type A spermatogonia turned out to be highly heterogeneous. In particular, the majority of type A spermatogonia are committed and divide synchronously like I_n and B, these are termed A_1, A_2, A_3, and A_4 denoting specific differentiation steps. The remaining few A spermatogonia comprise a distinctive entity termed "A_{undiff}" (undifferentiated type A spermatogonia), including A_s, A_{pr} and A_{al}. These divide asynchronously, and show slightly different nuclear morphology from A_1 to A_4 by high-resolution light microscopy analysis of plastic-embedded tissues (Chiarini-Garcia & Russell, 2001). The nomenclature of "A_0" was also introduced to define a small population of spermatogonia similar to A_{undiff}, alongside the proposal that these are resting or reserve populations (see Meistrich & van Beek, 1993); however, given our current knowledge, this hypothesis is considered to be unlikely. Note that this nomenclature is valid for mice and other rodents, while considerable degrees of species differences are known, particularly in the extent of amplification and the composition of the undifferentiated fraction of spermatogonia (Di Persio et al., 2017; Fayomi & Orwig, 2018).

2.2 Seminiferous tubules

Murine spermatogenesis proceeds in *seminiferous tubules*, a shared structure across amniotes (Hilton & Turner, 1993; Lombardi, 1998; Yoshida, 2016, 2018a). This present article focuses on *convoluted* seminiferous tubules, which compose nearly the entire tubule segment harboring spermatogenesis (hereafter referred to as "seminiferous tubules" for simplicity) (Russell et al., 1990). The hinge region between convoluted seminiferous tubules and the rete testes (the common outlet of sperm from the testis) have characteristic tissue organizations, called *straight* seminiferous tubules and intermediate region (Aiyama et al., 2015; Perey, Clermont, & Leblond, 1961), but which will not be discussed in this review due to the interest of word limits.

A simple structure is observed uniformly throughout the entire length of the ~2 m (!) long seminiferous tubules, which is tightly packed within a single mouse testis (Fig. 3) (Hilton & Turner, 1993; Russell et al., 1990; Yoshida, 2018a). Sertoli cells and two layers of peritubular myoid and lymphatic endothelial cells along with a basement membrane compose the basic structural framework of seminiferous tubules. The prominent networking of tight junctions between Sertoli cells creates a barrier that separates the tubules into *basal* and *adlumnal* compartments. All stages of spermatogonia are located in the basal compartment, while spermatocytes and spermatids are located in the adluminal compartment in an orchestrated manner, illustrating a clear periphery-to-center polarity (Fig. 3). However, within the basal compartment, A_s, A_{pr} and A_{al} spermatogonia, and differentiating spermatogonia as well, are all attached to the basement membrane but are not at all arranged sequentially, in contrast with what is observed in *Drosophila* testis. This interesting distribution of mouse spermatogonia is tightly related to the SSC dynamics, as presented below.

2.3 The seminiferous epithelial cycle and spermatogenic wave

In seminiferous tubules, spermatogenesis proceeds periodically with a regular interval of 8.6 days (Russell et al., 1990), designated as the "*seminiferous epithelial cycle*" (Fig. 4) (Leblond & Clermont, 1952). Since the entire differentiation takes over a month to complete, combinations of four different stages of

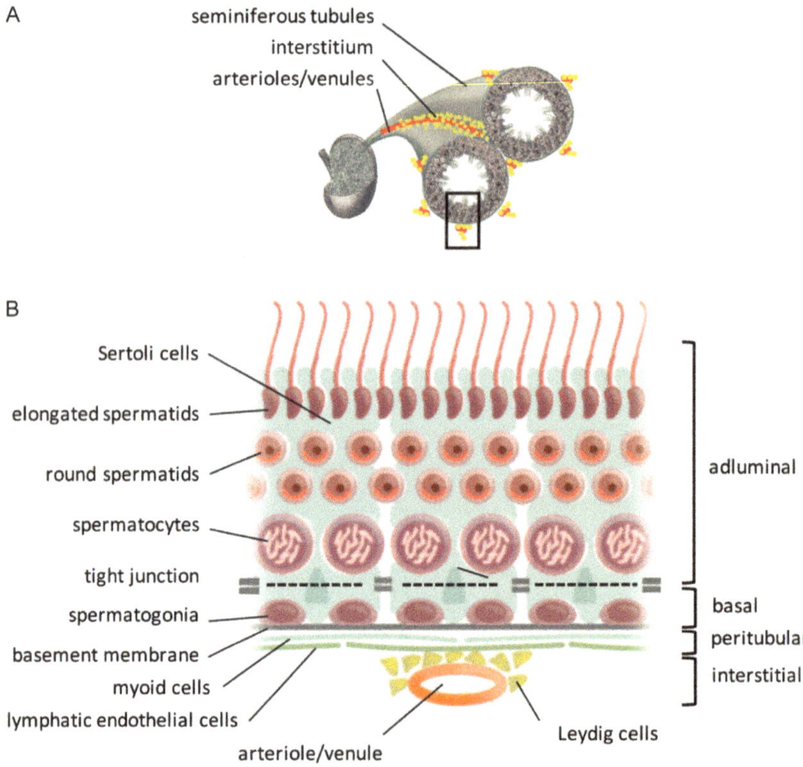

Fig. 3 Gross anatomy and tissue organization of mouse testes. (A) A simplified macroscopic view. In the testicular capsule, long convoluted seminiferous tubules (with a total length of up to 2 m per testes) are tightly packed. A network of blood vessels (red) and interstitial cells (yellow) comprise the interstitial compartment between tubules. Only arterioles and venules are shown (see text for details). The rectangle indicates the region magnified in (B). (B) Organization of the seminiferous epithelium showing the striated organization of spermatogenic cells within the anatomical framework composed of somatic cell types, separating into adluminal, basal, peritubular, and interstitial compartments.

differentiation are always observed (Fig. 4A). In mice, based on the resultant cellular associations, the entire cycle is separated into stages I–XII, with the originally-described stages II and III later combined into a single stage of II–III, leaving 11 stages (Fig. 4B) (Oakberg, 1956; Russell et al., 1990). Researchers new to the field might feel it to be impossible to come to like the seminiferous epithelial cycle. However, once gaining familiarity, most are fascinated by its beautiful organization. What is more charming is the fact that this temporal cycle is somewhat recapitulated by the spatial order

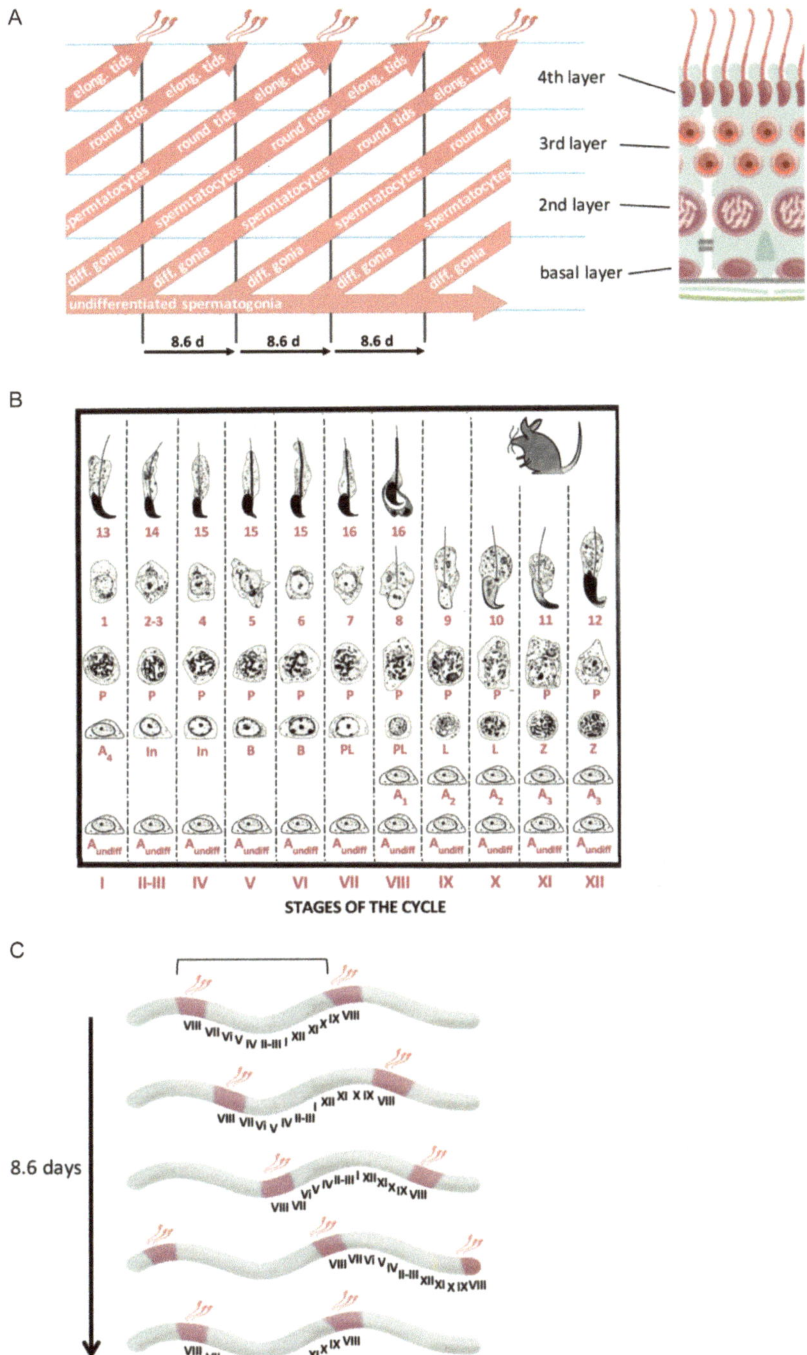

Fig. 4 See legend on next page.

along the length of the tubule, with the wave-like progression of spermatogenesis resembling a barber sign and known as the "*spermatogenic wave*" (Perey et al., 1961; Russell et al., 1990) (Fig. 4C).

Cycle and *wave* arrangements of seminiferous tubules have provided invaluable clues for the study of SSC dynamics over time, if limited to fixed samples, particularly in the early days of the field, when direct temporal tracing of cell behaviors (using live imaging or pulse labeling) was not yet available.

2.4 Interstitial compartment

Among seminiferous tubules lie interstitial tissues composed of blood vessels and other cells including Leydig cells and macrophages (Fig. 3) (Hilton & Turner, 1993; Russell et al., 1990; Yoshida, 2018a). This interstitium is not only responsible for the endocrine function of the testis through androgen production by Leydig cells, rather evidence has suggested its contribution to SSC regulation that occurs inside the seminiferous tubules.

Fig. 4 Seminiferous epithelial cycle and spermatogenic wave. (A) A schematic of the spermatogenic differentiation process with reference to particular position within seminiferous tubules. Note the wavy progression with a regular interval of 8.6 days, compared with 35 days for the total differentiation process from undifferentiated spermatogonia to the release of mature sperm (spermiation). As a result, four layers of different germ cell types are always observed in the cross section of seminiferous tubules: The basal layer includes all the spermatogonia (both A_{undiff} and differentiating spermatogonia) in the basal compartment, the second, third and the force or inner most layers are typically comprised of spermatocytes, round and elongated spermatids, respectively. Subsequently, sperm are released regularly once every 8.6 days from each particular region. Note that, round spermatids may be missing while both early and late spermatocytes are observed together in some stages of the seminiferous tubules (see (B)). (B) Cell associations of the seminiferous epithelial cycle in mice. Stages are historically separated into I–XII based on these associations (original stages II and III are often combined, designated as stage II–III). Refer to Fig. 1 for the nomenclature used. (C) Schematic of a spermatogenic wave. Seminiferous tubule segments showing different stages of seminiferous epithelium (B) are spatially arranged along the length of the tubule in chronological order (indicated by Roman numerals). As the cycle progresses, the entire spermatogenic process shows a wavy movement, as symbolically indicated by the site of spermiation in magenta. The same configuration will reappear after one cycle of 8.6 days. *Panel (B): Modified from Russell, L., Ettlin, R., Sinha Hikim, A., & Clegg, E. (1990). Histological and histopathological evaluation of the testis. Clearwater, Fl: Cache River Press, with permission.*

3. Advantages and available techniques in the study of SSCs

In addition to the unique tissue architecture of seminiferous tubules, the advantages of mouse spermatogenesis from a stem-cell biology point of view is emphasized by the fact that virtually all techniques available to date can be applied to the study of SSCs (Yoshida, 2012). Key advances of experimental techniques and/or analytical strategies have been continually strengthening our understanding of SSCs. Such techniques also have huge potential for application in medical and livestock sciences (Brinster, 2002, 2007; Kanatsu-Shinohara & Shinohara, 2007, 2013; Komeya & Ogawa, 2015; Nagamatsu & Hayashi, 2017).

3.1 Advantageous anatomy of seminiferous tubules

Seminiferous tubules have a highly simple and uniform structure over their total lengths of up to 2 m coiled within the mouse testis (Hilton & Turner, 1993; Russell et al., 1990; Yoshida, 2018a). SSCs are scattered over the basal compartment with strikingly constant average densities (Kitadate et al., 2019), making it simple to obtain large samples with high reproducibility and consistency. Furthermore, seminiferous tubules are easy to untangle and use for whole-mount preparations applicable for classical morphological investigations (Huckins & Kopriwa, 1969; Huckins & Oakberg, 1978) and newer techniques such as immunofluorescence (IF) or live observation in ex vivo preparations (Lee et al., 2015; Sakamoto et al., 2018; Shibuya, Morimoto, & Watanabe, 2014). This property enables the researcher to capture three-dimensional information comprehensively, without losing the larger-scale tissue arrangements (*cycle* and *wave*) and topology of spermatogonia, such as the number of interconnected cells in a single syncytium (Huckins, 1978).

3.2 Transplantation

Beside the hematopoietic stem cells, transplantation-based quantitative assays are also available for SSC research. This technique was established in 1994; when single-cell suspensions containing SSCs were shown to form repopulating colonies in host seminiferous tubules, several months after transplantation into germ cell-depleted seminiferous tubules (Brinster, 2002; Brinster & Avarbock, 1994; Brinster & Zimmermann, 1994). Since a separate post-transplantation repopulating colony is derived from a single transplanted cell (Kanatsu-Shinohara et al., 2006), SSC activity of the assayed fractions can be quantified, at least relatively, based on the number of the repopulating colonies upon transplantation (Nagano, 2003). This technology has great potential for medical and industrial purposes; it has been applied to other species, including monkeys and chickens (Hermann et al., 2012; Trefil et al., 2006).

3.3 In vitro SSC culture in 2D

In vitro cultivation has been developed that supports essentially infinite proliferation of spermatogonia in 2D on culture plates, retaining SSC activity as verified by transplantation assays (Kanatsu-Shinohara et al., 2003; Kubota, Avarbock, & Brinster, 2004). These cultures depend on the presence of signaling molecules, including GDNF (glial cell line-derived neurotropic factor), which is crucial for stem cell maintenance in vivo (Meng et al., 2000); and FGF (fibroblast growth factor), whose in vivo roles will be discussed later in this review. Using this experimental platform, detailed characterization of SSCs, including analyzing their responses to extracellular signals, have been conducted (literature including Kanatsu-Shinohara & Shinohara, 2013; Oatley, Avarbock, Telaranta, Fearon, & Brinster, 2006).

3.4 Ex vivo organ culture in 3D

In 2011, prolonged ex vivo culture systems for testis explants were developed, using air-liquid interface conditions (Sato et al., 2011), and more sophisticated microfluidic chambers (Komeya et al., 2016). Impressively, using these systems, continuous production of functional sperm can be maintained for several months (!). In addition to medical applications including ex vivo treatment of infertility (Sato et al., 2012), these systems have great potential to analyze the cellular properties of SSCs and mechanisms of spermatogenesis.

Organoid—in vitro-reconstructed 3D "mini-organs"—culture of testicular tissues (e.g., the seminiferous tubules) is another focus of ex vivo culture technology (Alves-Lopes & Stukenborg, 2018). What is of particular note is that air-liquid interface cultures of reaggregation of dispersed neonatal testis cells, with or without cultured spermatogonia, are shown to support the progression of meiosis inside the self-organized seminiferous tubule-like structures (Yokonishi et al., 2013).

3.5 Intravital live imaging

Long-term intravital live imaging of tissue stem cells was first achieved for mouse SSCs by taking advantage of the accessibility of SSCs from the surface of the testis and of the repetitive anatomy (Yoshida, Sukeno, & Nabeshima, 2007). At this moment, GFP-labeled spermatogonia have been continuously filmed for up to 3 days at single-cell resolution.

3.6 In vivo pulse-labeling fate analysis

Lineage tracing of pulse-labeled spermatogonia using tamoxifen-inducible cre (CreER)-loxP systems has been used in SSC research (Hayashi & McMahon, 2002; Yoshida, 2012). Here, the spatially-restricted distribution of single cell-derived cohorts in seminiferous tubules enables quantitative analysis of clonal fate at single-cell resolution in whole-mount preparations of seminiferous tubules.

3.7 Single-cell "omics" study

Recently, "omics" analyses, in particular single-cell omics (Wang & Bodovitz, 2010), provide powerful tools for analysis of the composition of spermatogenesis and SSCs. At this time, pioneering studies of single-cell transcriptomics in mouse and human spermatogenesis are emerging (Chen et al., 2018; Green et al., 2018; Hammoud et al., 2015; Hermann et al., 2015; La et al., 2018).

3.8 In vitro gametogenesis from pluripotent cells

In vitro gametogenesis is an important topic for biomedical research (Vogel, 2016) (chapter "Germ cell reprogramming" by Kurimoto and Saitou, in this issue). Primordial germ cell-like cells (PGC-LCs) can be induced in vitro from pluripotent, ES or iPS cells by mimicking the in vivo signaling environment, followed by differentiation into functional gametes, either to oocytes through cocultivation with embryonic ovary somatic cells, or to functional sperm through transplantation into seminiferous tubules (Hayashi, Ohta, Kurimoto, Aramaki, & Saitou, 2011; Hikabe et al., 2016; Saitou & Miyauchi, 2016). Cells showing SSC-like states have also been induced (Ishikura et al., 2016). In addition to high-impact medical applications, in vitro gametogenesis should provide invaluable platforms to study the otherwise unapproachable process of gamete formation.

3.9 Statistics, mathematics, and theory

Statistical, mathematical, and theoretical (biophysical) analyses have made a significant contribution to understand the rules and regulatory mechanisms of SSC dynamics, based on quantitative data sets obtained from the abovementioned strategies 3–5, 6, and 7. In accordance with the technical progress in acquiring multidimensional and quantitative data, these approaches should become more and more important for the study of SSC dynamics.

4. Identity and dynamics of SSCs

As new techniques have been developed and new data obtained, our knowledge and conception about SSCs has expanded (de Rooij & Russell, 2000; Yoshida, 2012). Of course, we are currently on a continual, maybe endless, road toward the ultimate understanding of SSCs. I will overview the position of the field as of early 2019.

4.1 The "A_s model"

Self-renewal and differentiation are processes dependent on the *progression of time*. However, it has been a big challenge to directly follow the fate of individual cells' behaviors over time inside multicellular animal bodies, seriously restricting our access to the in vivo dynamics of tissue stem cells. Nevertheless, in the 1950s to 1970s, the evolution of A_{undiff} population (A_s, A_{pr} and A_{al} spermatogonia) over time was investigated by static information obtained from fixed specimens. The bulk behavior of A_{undiff} population was captured by serial observations along the tubule length, given the unique organization of the "*cycle*" and "*wave*" (Sections 2 and 3) (Huckins, 1971b; Meistrich & van Beek, 1993). Among several hypotheses proposed at the time, a straightforward and simple one that became prevailing is called the "A_s (A_{single}) model." This model was first proposed in rats (Huckins, 1971a, 1971b) and then in mice (de Rooij, 1973; Oakberg, 1971).

Fig. 5A summarizes the original scheme of the "A_s model" model drawn by Huckins (1971b), based on the observed frequencies of A_s, A_{pr}, and A_{al} cells at various stages of the seminiferous epithelial cycle. In particular, A_s cells are observed at a low frequency throughout the seminiferous epithelial cycle. This cartoon also emphasizes the consistent frequencies of A_{undiff} in M phase (mitotic index), a result based on ^{3}H-thymidine incorporation assays. This "A_s model" is often presented by a simplified, and well-known,

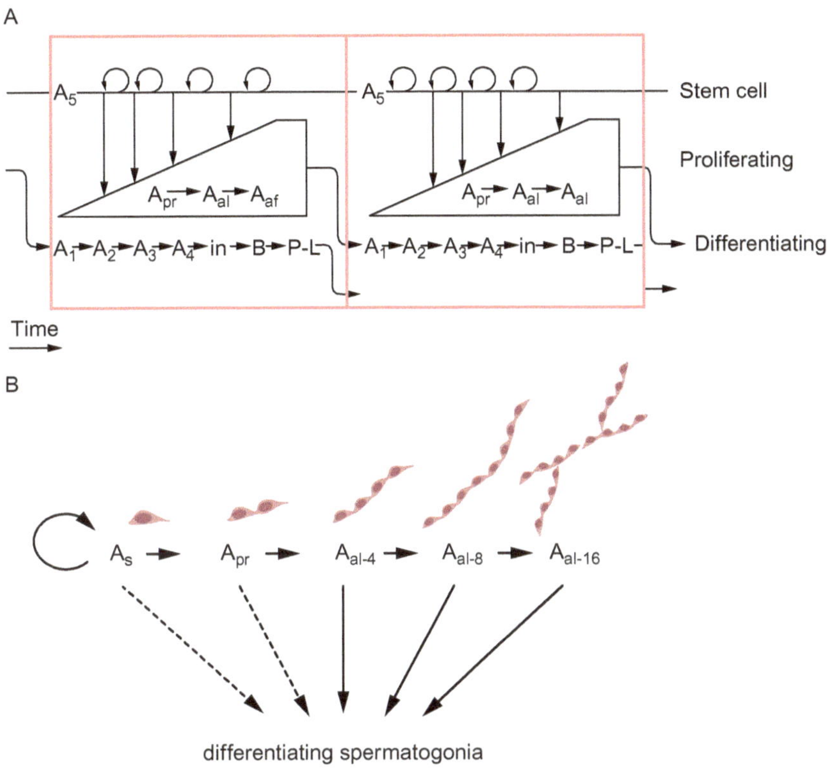

Fig. 5 The As model. (A) The original schematic proposal of Huckins (1971b), modified with permission. This model is based essentially on the number of A_s, A_{pr}, A_{al}, and more advanced cell types found in different stages of the seminiferous epithelial cycle (two cycles are shown as indicated by magenta rectangles). (B) A simplified diagram of the A_s model, indicating that the A_s is the only cell type that self-renews (indicated by a round arrow), while A_{pr} and A_{al} are committed. A transition to differentiating spermatogonia occurs in A_{al}, while a few A_{pr} and A_s are also proposed to become differentiating spermatogonia directly.

illustration as that shown in Fig. 5B (de Rooij & Russell, 2000; Meistrich & van Beek, 1993; Oatley & Brinster, 2008). Here, it is emphasized that A_s cells comprise the self-renewing compartment, while syncytial A_{pr} and A_{al} are irreversibly committed to differentiation. As a natural consequence, A_s cells are thought to either self-renew through complete division to generate two A_s cells, or commit to differentiate through incomplete division to form an A_{pr}. Subsequent mitotic divisions of A_{pr} are also thought to be incomplete and irreversible, each doubling increases the syncytial length to 4, 8, 16 and so on. This is consistent with the observation that most of the syncytia include 2^n cells, albeit with a few exceptions. Noteworthy, this model echoes the dynamics of *Drosophila* germline stem cells in the testis and ovary (Fig. 1) (Spradling et al., 2011), although the spatial arrangements are different between these organisms (Figs. 1 and 9).

The "A_s model" was established as a paradigm of SSC dynamics quite consistent with the frequencies of A_{undiff} with variable topologies during the seminiferous epithelial cycle. Yet, and as Huckins describes herself, this model is by nature not based on the direct observations of over-time behavior of spermatogonia (Huckins, 1971a, 1971b). Therefore, the aforementioned critical premises that A_{pr} and A_{al} are irreversibly committed and that their divisions are always incomplete to double the syncytial length warranted re-evaluations, once long-term in vivo observation of spermatogonial fates became technically available. Further, in this model, the heterogeneity of A_{undiff} was considered only with regard to their topology, which warrants detailed investigation of A_{undiff} heterogeneity, in particular with regard to their transcriptional states.

4.2 Heterogeneous and hierarchical transcriptional states within A_{undiff} population

Since around the beginning of this century, A_{undiff} populations have turned out to show heterogeneous gene expression (Song & Wilkinson, 2014; Yoshida, 2018b). Heterogeneity in their transcriptional states (a strong indicator of differentiation status) is related to their topological states, namely, the length of spermatogonial syncytia on which the "A_s model" is founded (Fig. 6). However, correlation between these two heterogeneities is not perfect at all. For example, the population of A_s cells includes multiple subpopulations showing different gene expression, while the same transcriptional states are found across A_s, A_{pr} and A_{al} (Grisanti et al., 2009; Nakagawa et al., 2010; Yoshida, 2018b; Zheng et al., 2009). Clearly, the heterogeneity of A_{undiff} is much more complex than had been originally thought, a notion that is being further confirmed by single-cell transcriptome analyses (Green et al., 2018; Hermann et al., 2015; La et al., 2018).

Fig. 6 is a rough summary of the expression of some, well-investigated genes in terms of their heterogeneous expression in the population of A_{undiff}. Within the population of A_{undiff}, genes expressed in relatively less mature subfractions include *Gfrα1, Ret, Nanos2, Id4, Shisa6, T, Pdx1, Eomes, Lhx1*, as well as *Pax7, Bmi1*, and *Erbb3* (Abid et al., 2014; Aloisio et al., 2014; Chan et al., 2014; Hofmann et al., 2005; Kitadate et al., 2019; Komai et al., 2014; La et al., 2018; Meng et al., 2000; Nakagawa et al., 2010; Sada et al., 2009; Suzuki et al., 2009; Tokue et al., 2017); those

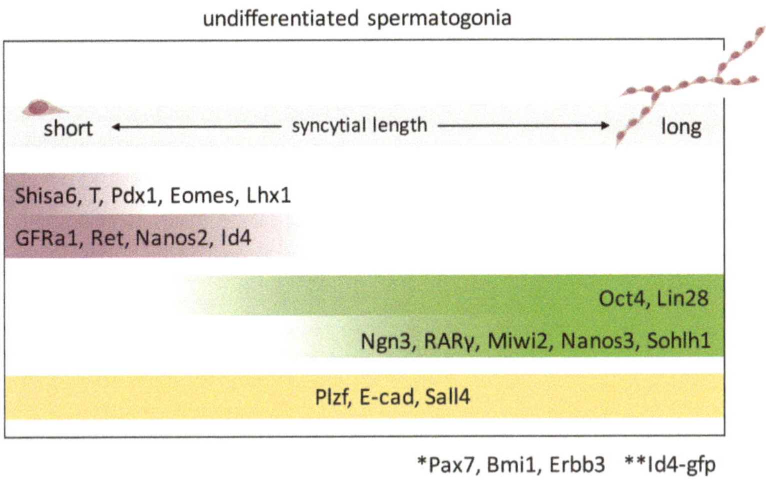

Fig. 6 Heterogeneity of undifferentiated spermatogonia (A_{undiff}) in terms of transcriptional states. Illustration of heterogeneous gene expression among the A_{undiff} compartment, based on expression of endogenous protein and/or mRNA, some of which have been verified with consistent expression of reporters from transgenic and knock-in alleles. This diagram is consistent with a recent single-cell transcriptome analysis focused on A_{undiff} population (La et al., 2018). Summarized from literature including (Abid et al., 2014; Aloisio et al., 2014; Buaas et al., 2004; Carrieri et al., 2017; Chakraborty et al., 2014; Chan et al., 2014; Costoya et al., 2004; Gely-Pernot et al., 2012; Hobbs et al., 2012; Hofmann, Braydich-Stolle, & Dym, 2005; Ikami et al., 2015; Kitadate et al., 2019; Komai et al., 2014; La et al., 2018; Meng et al., 2000; Nakagawa, Sharma, Nabeshima, Braun, & Yoshida, 2010; Sada, Suzuki, Suzuki, & Saga, 2009; Suzuki et al., 2012; Suzuki, Sada, Yoshida, & Saga, 2009; Tokuda, Kadokawa, Kurahashi, & Marunouchi, 2007; Tokue et al., 2017; Yoshida et al., 2004; Zheng, Wu, Kaestner, & Wang, 2009). *Pax7 and Bmi1 are suggested to show restricted expression in some A_s cells largely on the basis of the result of CreER mediated recombination (Aloisio et al., 2014; Komai et al., 2014), while their endogenous transcripts are widely detected over A_{undiff} at low levels (La et al., 2018; Tokue et al., 2017). Errb3 expression has been detected by immunofluorescence in subset A_s cells in a seminiferous stage-specific manner in rat (Abid et al., 2014), while its expression in mice (and relation with other genes) remain elusive. **Id4-gfp transgene is observed to show highly restricted expression in a subset of GFRα1$^+$ A_s cells (Chan et al., 2014), while the endogenous Id4 transcript and protein are found in a wider manner similar to GFRα1 (including A_s, A_{pr}, and A_{al} cells) (Garbuzov et al., 2018; Kitadate et al., 2019; La et al., 2018).

enriched in more mature fractions of A_{undiff} include *Oct4*, *Lin28*, *Ngn3*, *RARγ*, *Miwi2*, *Nanos3*, and *Sohlh1* (Carrieri et al., 2017; Chakraborty et al., 2014; Gely-Pernot et al., 2012; Ikami et al., 2015; La et al., 2018; Nakagawa et al., 2010; Suzuki et al., 2009; Suzuki et al., 2012; Yoshida et al., 2004; Zheng et al., 2009). Some of the latter also are expressed in differentiating spermatogonia. The entire A_{undiff} population is marked by the expression of *E-cadherin*, *Plzf*, and *Sall4* (Buaas et al., 2004; Costoya et al., 2004; Hobbs et al., 2012; Tokuda et al., 2007).

4.3 Extension of the A_s model

Based on the observation that the compartment of A_s spermatogonia shows heterogeneous and hierarchical gene expression, it was proposed that particular subsets of A_{undiff} define the stem cells, extending the classical A_s model. In 2014, *Id4*, *Bmi1*, and *Pax7* were reported to delineate the subset of A_s cells, leading to proposals that SSCs could be defined as *Id4*$^+$, *Bmi1*$^+$, or *Pax7*$^+$ A_s spermatogonia (Aloisio et al., 2014; Chan et al., 2014; Komai et al., 2014).

However, confusingly, it appears that populations delineated by these genes are not identical with each other. Further, many different subfractions of A_{undiff}, delineated by either undifferentiated and differentiating marker genes (e.g., *GFRα1*, *Id4*, *Pdx1*, *Ngn3*, and *Miwi2*) are found to show variable but considerable levels of SSC functions, in transplantation or pulse-labeling assays (Carrieri et al., 2017; Grisanti et al., 2009; Hara et al., 2014; Helsel et al., 2017; La et al., 2018; Nakagawa, Nabeshima, & Yoshida, 2007; Sun, Xu, Zhao, & Degui Chen, 2015). Therefore, the potential to act as SSCs are shared by wide ranges of cells across the A_{undiff} population. These observations warranted a reevaluation of the premise that SSCs are a strictly defined entity that is marked by the expression of specific genes. Such reevaluation was further motivated by the outcome of functional assessment of SSCs over time described below.

4.4 Clonal fate variability: Population asymmetry and neutral competition

Pulse labeling and intravital live-imaging studies enabled, for the first time, the precise tracing of fate behavior of individual cells in undisturbed tissue. This provided a significant improvement over averaged, population-level behaviors extrapolated from a sequence of static data obtained from fixed specimens.

First of all, clonal fate behavior of sparsely labeled SSCs (e.g., GFRα1$^+$ cells) revealed that individual SSCs indeed had highly variable fates, in terms of the number of self-renewing and differentiating cells contained in a single SSC-derived cohort (Fig. 7) (Hara et al., 2014): Some SSCs produced more than one stem cells to expand their proportion within the stem cell pool,

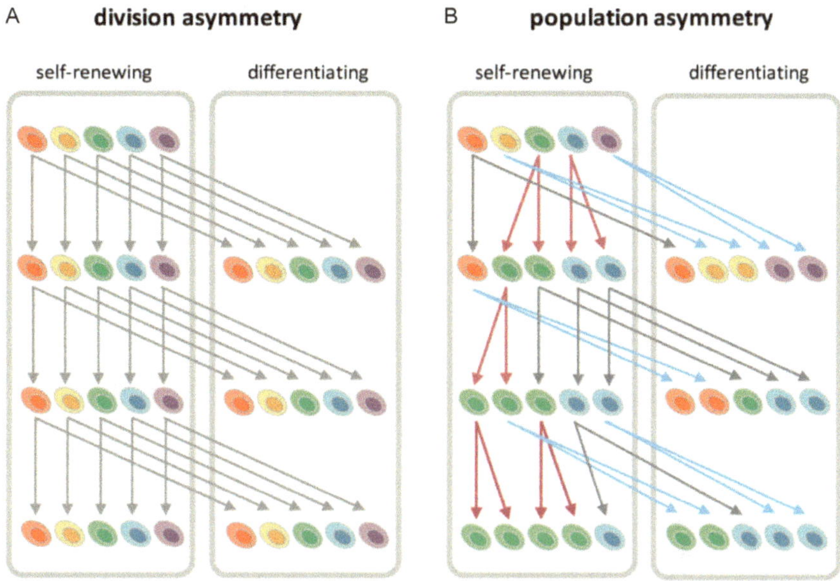

Fig. 7 Division asymmetry versus population asymmetry. (A) Diagrams of steady-state stem-cell dynamics of "*division asymmetry*." Here, all the divisions of stem cells (cells in self-renewing compartment) invariantly give rise to the asymmetric outcome of one self-renewing and one differentiating daughter cell. As a result, the clonal repertoire of the self-renewing pool (indicated by different colors), and differentiating cells as well, remains unchanged. (B) In "*population asymmetry*," the balance between self-renewal and differentiation is achieved at the population level, while each stem-cell division gives rise to variable outcomes, i.e., two self-renewing daughters (magenta arrows), two differentiating daughters (blue arrows), or one self-renewing and one differentiating daughter (gray arrows). In this scheme, the stem cells' clonal repertoire drifts over time, becoming dominated by fewer clones, and eventually dominated by a single clone. For simplicity, the syncytial formation of spermatogonia is not included in these drawings. However, the cell division in this cartoon corresponds to complete division or syncytial fragmentation in case of spermatogenesis, given that all spermatogonia in a syncytium share an identical state, refer to Hara et al. (2014) for a deeper analysis of the involvement of syncytial states of spermatogonia.

while others gave rise to only differentiating cells and went extinct. Importantly, however, their averaged, population-level behavior demonstrated a perfect balance between self-renewal and differentiation, a hallmark behavior of any self-renewing compartment in homeostasis (Hara et al., 2014; Klein, Nakagawa, Ichikawa, Yoshida, & Simons, 2010). Furthermore, such fate variance among clones is precisely predicted by a simple mathematical model in which every SSC has a stochastic fate behavior (Hara et al., 2014). Consistent clonal drifts are also found in long term observation after pulse-labelling of other fractions of A_{undiff} (*Ngn3*$^+$, *Nanos2*$^+$, *Bmi1*$^+$ or *Pax7*) (Aloisio et al., 2014; Klein et al., 2010; Komai et al., 2014; Sada et al., 2009).

These observations indicate that homeostasis of SSCs is achieved only at a population level, as a collective behavior of SSCs, while each individual SSC behaves stochastically. This paradigm of stem cell dynamics is designated as "*population asymmetry*," under which SSCs show continuous *neutral competition* with each other (Fig. 7B) (Klein & Simons, 2011; Simons & Clevers, 2011). This is contrary to the paradigm of "*division asymmetry*," in which all stem cells would, in principle, persist eternally through invariant asymmetric divisions, each division giving rise to one self-renewing and one differentiating daughter cell (Fig. 7A). Mouse spermatogenesis is one of the stem-cell systems in which population asymmetry was first discovered (Klein et al., 2010; Nakagawa et al., 2007), together with interfollicular epithelia and intestinal crypts (Clayton et al., 2007; Doupe, Klein, Simons, & Jones, 2010; Lopez-Garcia, Klein, Simons, & Winton, 2010; Snippert et al., 2010). Subsequently, this class of stem cell dynamics has been found in many other tissues, and it is now recognized as a universal rule (Klein & Simons, 2011; Krieger & Simons, 2015).

4.5 Syncytial fragmentation: Interchange between topologically different states

Using intravital live-imaging studies of GFP-labeled A_{undiff} (GFRα1$^+$ and Ngn3$^+$), fragmentation of syncytial spermatogonia (A_{pr} and A_{al}) was observed to occur through intercellular bridge breakdown (Fig. 8A) (Hara et al., 2014; Nakagawa et al., 2010), a process that had not been thought to occur according to the "A_s model." However, fragmentation of syncytia had been proposed to occur based on the existence of syncytia with cell numbers other than 2^n, particularly after insult by irradiation (Erickson, 1981). Of particular note, based on a series of intravital live-imaging data, most (i.e., 33 out of 35) of the GFRα1$^+$ A_s cell divisions were observed to be incomplete, leaving the daughter pairs interconnected for at least 37 h after division. Depleted A_s cells were replenished by the fragmentation of A_{pr}, and more frequently, $A_{al\text{-}4}$, indicating a continual exchange between the topologically different states A_s, A_{pr} and A_{al} (Hara et al., 2014).

Fragmentation of spermatogonial syncytia and their role in the maintenance of the A_{undiff} pool is apparent from the frequent fragmentation of GFRα1$^+$ A_{al}, (Hara et al., 2014) (Fig. 8A), as well as that of Ngn3$^+$ A_{al} albeit less frequently (Nakagawa et al., 2010). It has been proposed that "fragmentation of an A_{pr} 'could also be due to' delayed cytokinesis of a false

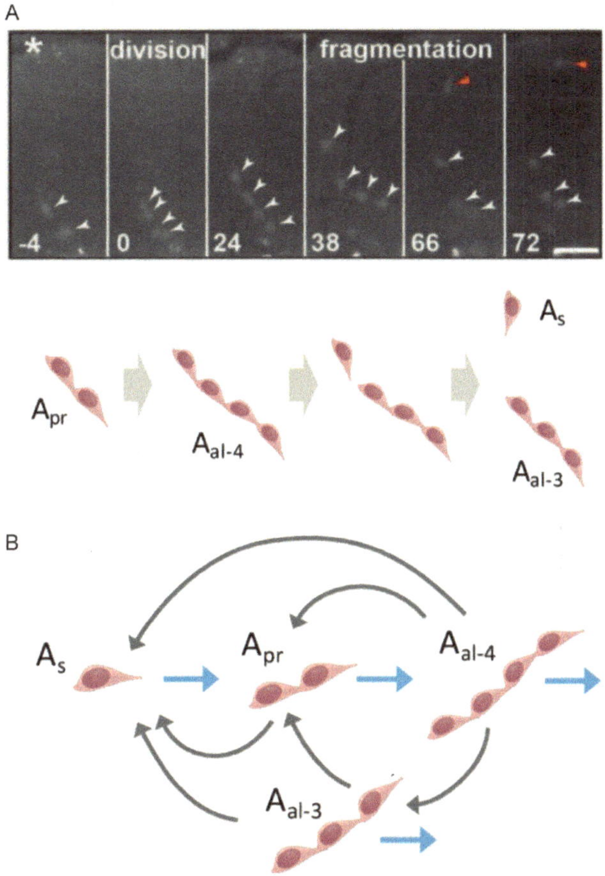

Fig. 8 An example of syncytial fragmentation and proposed dynamics of the A_{undiff} pool. (A) Selected video frames of fragmentation of a GFRα1-GFP$^+$ A_{al-4} into A_s and an A_{al-3} spermatogonia. Arabic numerals indicate the hours relative to the formation of an A_{al-4} through incomplete division of an A_{pr}. Bar, 30 μm. (B) Diagram of the suggested dynamics of the GFRα1$^+$ A_{undiff} population, maintained through incomplete divisions (rightward blue arrows) and fragmentation (leftward gray arrows). See text and Hara et al. (2014) for detail. *Panel (A): Reproduced with permission from Hara, K., Nakagawa, T., Enomoto, H., Suzuki, M., Yamamoto, M., Simons, B.D., Yoshida, S. (2014). Mouse spermatogenic stem cells continually interconvert between equipotent singly isolated and syncytial states. Cell Stem Cell 14, 658–672.*

A_{pr} intermediate during complete division of an A_s" (de Rooij, 2010). However, this interpretation may represent semantic differences describing the essentially same biological process, which may also apply to the fragmentation of A_{al}: Accumulated molecular evidence suggests that intercellular bridges are stabilized connection between sister cells that appears in mitotic telophase, caused by the prevention of cytoplasmic abscission at the midbody (Greenbaum, Ma, & Matzuk, 2007). It is worth mentioning here that similar fragmentation occurs in mouse female germline development (Lei & Spradling, 2013; Pepling & Spradling, 2001). Further, delayed cytokinesis during GSC division and fragmentation of 4- or 8-germ cell-cysts during de-differentiation are also observed in *Drosophila* (Brawley & Matunis, 2004; Kai & Spradling, 2004; Lenhart & DiNardo, 2015). It may be a general rule of germ cell development across species that germ cells may interchange between different topological states by reversibly controlling the stability of telophase structures during division.

4.6 Context dependent interchange between different transcriptional states

SSCs, like other tissue stem cells, play active roles in processes including tissue maintenance, regeneration, and growth. Therefore, SSCs have been defined practically depending on experimental context (Yoshida, 2012, 2018b; Yoshida, Nabeshima, & Nakagawa, 2007). First, the most fundamental function of SSCs is likely to support persisting homeostatic turnover of spermatogenesis; this process can be assessed by techniques including pulse-labeling and live-imaging in undisturbed testes. Second, SSCs contribute to regeneration after insult by radiation or chemicals, which have been assessed by lineage tracing. Further, an extreme regenerative condition can be induced by transplantation. In addition, using in vitro culture, important factors such as signaling molecules in the media can be manipulated freely.

The context dependence of tissue stem-cell behavior, which became a generic feature of tissue stem cells, was first illustrated by the study of SSCs (Nakagawa et al., 2007). Here, fate behaviors of the same pulse-labeled subpopulation of A_{undiff} were compared under different in vivo contexts: While pulse-labeled $Ngn3^+$ cells largely differentiated with rare contributions to the long-term stem cell pool in homeostasis, the same labeled population was recruited to the stem-cell pool with significantly higher frequencies during regeneration following insult or transplantation. $Miwi2^+$ cells, which show a significant overlap with $Ngn3^+$ cells, also showed consistent transplantable SSC activity and, notably, their depletion caused delayed regeneration (Carrieri et al., 2017). In these situations, conversion from $Ngn3^+$ or $Miwi2^+$ to $GFR\alpha1^+$ states were observed (Carrieri et al., 2017; Nakagawa et al., 2007).

Similarly, the Id4-based SSC hypothesis has been revised. Initially it was proposed that Id4 expression is restricted to a definitive fraction of A_s cells, which irreversibly gives rise to differentiating cells that are Id4-negative (Chan et al., 2014). A new study proposes that self-renewing, $Id4^{high}$ ($Id4\text{-}gfp^{bright}$) A_s cells and differentiation-primed, $Id4^{low}$ ($Id4\text{-}gfp^{dim}$) A_s and A_{pr} cells represent interconvertable states (called ultimate and transitory SSCs, respectively). These new findings are based on differences in the post-transplantation repopulation potential between $Id4\text{-}gfp^{bright}$ and $Id4\text{-}gfp^{dim}$ fractions (Helsel et al., 2017; Lord & Oatley, 2017). Curiously, the *Id4-gfp* transgene appears to not faithfully reflect the endogenous *Id4* expression (mRNA and protein, as well as a GFP reporter knocked into the *Id4* locus), which is detected widely in GFRα1$^+$ A_s, A_{pr} and A_{al} cells (Garbuzov et al., 2018; Kitadate et al., 2019; La et al., 2018) (Fig. 6). Thus, the validity of *Id4-gfp* transgenic reporters as a marker for definitive SSCs with the topology of A_s spermatogonia is uncertain.

Finally, in cultured spermatogonia, the differentiation-primed Oct4$^+$ state and the self-renewal-primed Oct4$^-$/Pdx1$^+$ state are shown to reversibly converted under different conditions (La et al., 2018).

Collectively, these findings support the idea that SSCs can interconvert between multiple states, so that different fractions of A_{undiff} contribute to the stem cell functions in different tissue—or extracellular—contexts. Similar *reversions* of differentiation-primed cells to the self-renewing pool have been observed prominently under regenerative conditions in other tissues as well (i.e., mouse intestinal epithelium or *Drosophila* GSCs) (Brawley & Matunis, 2004; Kai & Spradling, 2004; van Es et al., 2012). Context-dependent state change is likely to be a common feature of tissue stem cells.

4.7 Dynamic heterogeneity: A paradigm of stem-cell homeostasis?

A model for SSC dynamics in homeostasis was put forward in 2014, proposing that the entire GFRα1$^+$ population composes *a single equipotent self-renewing pool*, in which cells continually interconvert between the A_s, A_{pr} and A_{al} topological states (Fig. 8B) (Hara et al., 2014). This model was consolidated by the high predictive capacity of a minimal biophysical model, synthesized from the intravital live-imaging data, to capture a wide range of acquired data (e.g., clonal fates over short to long term) in undisturbed testes, assuming that incomplete division and syncytial fragmentation of GFRα1$^+$ cells occur stochastically (Hara et al., 2014; Klein et al., 2010). These findings are consistent with a model whereby stem cells support homeostasis through continual interconversion between multiple states showing different topologies. This hypothesis challenges the conventional view of tissue stem cells in homeostasis; namely, that stem cells comprise a specifically defined, homogeneous population. It will be interesting to see whether the idea of "*dynamic heterogeneity*" will be extended to stem cells from other tissues (Krieger & Simons, 2015).

4.8 Future questions about the SSC identity

As described above, emerging evidence has revealed the heterogenic, dynamic, and flexible nature of A_{undiff} spermatogonia. With this come an increasing challenge to capture the dynamics of this population in its entirety. Given the context-dependency of SSCs, it will be necessary to investigate the A_{undiff} subpopulations under consistent experimental conditions over extended periods in homeostasis (i.e., in undisturbed adult tissue) and compared with other non-steady state contexts. The captured dynamics will for certain inform our concept of SSCs.

5. Microenvironmental regulation of SSCs

There is a second key question: how are SSCs regulated in their tissue microenvironment? This question addresses the understanding of the stem cell niche in a broader context. As discussed below, mouse seminiferous tubules represent an open (or facultative) stem cell niche, in contrast to the closed (or definitive) niche, represented by *Drosophila* GSCs (Morrison & Spradling, 2008; Stine & Matunis, 2013; Yoshida, 2018a).

5.1 Scattered distribution of A_{undiff} in the basal compartment

As described above, all stages of spermatogonia reside in the basal compartment of seminiferous tubules (Fig. 3). Therefore, all the aforementioned dynamic behaviors of spermatogonia, i.e., the maintenance of the self-renewing pool, the production of differentiation-destined progenitors, and the state changes between A_{undiff} subfractions happen in one microenvironment (Yoshida, 2018a). It should be emphasized that A_{undiff} and its subsets show a dispersed localization and intermingle between differentiating progeny (differentiating spermatogonia) in an apparently unorganized fashion (Fig. 9). This is in contrast to the highly organized arrangement in the *Drosophila* testis (Fig. 1).

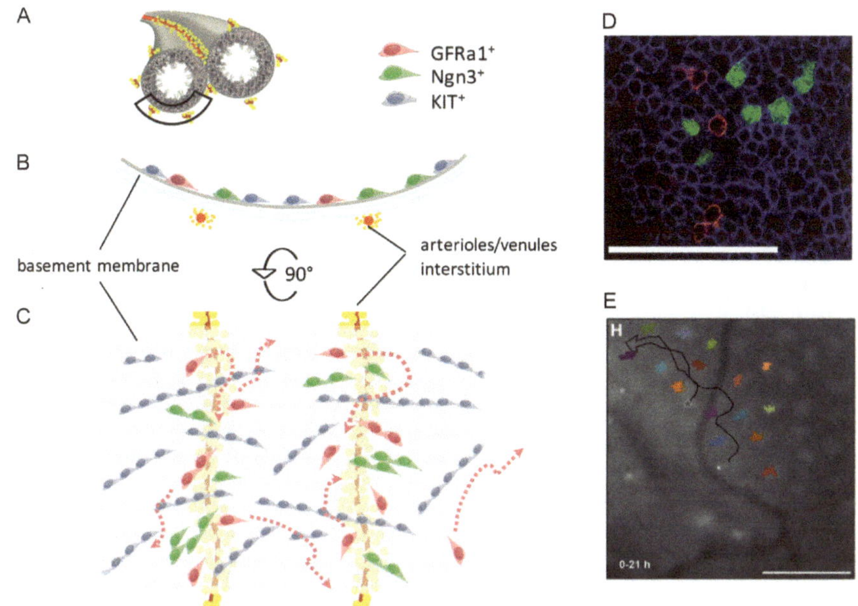

Fig. 9 Open niche microenvironment of the basal compartment for A_{undiff}. (A) Overall view of the seminiferous tubules and intervening arterioles and venules (red) and interstitium (yellow). Other classes of blood vessels (capillaries and thicker vessels) are not shown. (B) A simplified view of the area indicated in (A), Only GFRα1$^+$, Ngn3$^+$ and Kit$^+$ spermatogonia (magenta, green and blue, respectively) are shown, while spermatocytes, spermatids and Sertoli cells are not shown. (C) Diagram of the basal compartment, viewed from the inside of the tubule, rotated by 90 degrees relative to (B). Note that on the basement membrane, GFRα1$^+$ and Ngn3$^+$ cells are biased to the area near the interstitium (yellow) accompanied by arterioles and venules (red) beneath the basement membrane. These spermatogonia, in particular GFRα1$^+$ cells, are continually moving (indicated by dotted magenta arrows), although the correlation between their paths and neighboring spermatogonia remain unclear, due to current unavailability of multi-colored imaging of differentially labeled spermatogonial subsets. Other cell types are omitted as in (B). (D) Immunofluorescent detection of GFRα1$^+$ (magenta), Ngn3-GFP$^+$ (green) and Kit$^+$ (blue) spermatogonia in the basal compartment, showing mixing at a single-cell level. Bar, 100 μm. (E) Trajectory of a single GFRα1$^+$ spermatogonium over 21 h (black), overlaid on the actual image from time 0. Note that the spermatogonium weaves its way between immotile Sertoli cells, whose trajectories are shown in colored lines. *Panel (D): Modified from Ikami, K., Tokue, M., Sugimoto, R., Noda, C., Kobayashi, S., Hara, K., et al. (2015). Hierarchical differentiation competence in response to retinoic acid ensures stem cell maintenance during mouse spermatogenesis.* Development (Cambridge, England), 142, *1582–1592 with permission. Panel (E): Reproduced from Hara, K., Nakagawa, T., Enomoto, H., Suzuki, M., Yamamoto, M., Simons, B.D., et al. (2014). Mouse spermatogenic stem cells continually interconvert between equipotent singly isolated and syncytial states.* Cell Stem Cell, 14, *658–672.*

5.2 Biased distribution toward interstitium and blood vessels

Macroscopically, localization of the A_{undiff} population is biased to regions adjacent to the interstitium accompanying the vascular network of arterioles and venules. This bias was first reported for the entire A_{undiff} population identified based on nuclear morphology (Chiarini-Garcia, Hornick, Griswold, & Russell, 2001; Chiarini-Garcia, Raymer, & Russell, 2003). Then, a similar bias was found in the localization of A_{undiff} subsets based on marker expression (Ngn3$^+$, GFRα1$^+$, and Pdx1$^+$) (Hara et al., 2014; La et al., 2018; Yoshida, Sukeno, & Nabeshima, 2007). While Id4-gfp$^+$ cells are shown not to prefer blood vessel-proximal region (Chan et al., 2014), this inconsistency might have been due to different topological criteria (i.e., counting thicker vessels without surrounding interstitium (Yoshida, 2018a)). Indeed, based on immunofluorescent detection of endogenous protein, Id4$^+$ cells have been reported to be preferentially located near the interstitium (Kitadate et al., 2019).

In contrast to A_{undiff}, differentiating spermatogonia (Kit$^+$) show a relatively even distribution over the basal compartment regardless of the interstitium (Chiarini-Garcia et al., 2001, 2003). This predicted relocation during the transition from A_{undiff} to differentiating spermatogonia (Chiarini-Garcia et al., 2003) was also demonstrated by direct intravital live-imaging (Yoshida, Sukeno, & Nabeshima, 2007).

5.3 Motion of spermatogonia

One of the most characteristic features of A_{undiff} in seminiferous tubules is their active motion, which has been directly observed by intravital live-imaging (Fig. 9C) (Hara et al., 2014; Yoshida, Sukeno, & Nabeshima, 2007). *Gfrα1*-GFP$^+$ cells were recorded weaving their way between Sertoli cells (and likely between differentiating spermatogonia, as well) (Fig. 9E). In these videos, a majority of GFRα1$^+$ cells crawl over the basement membrane near the interstitium, while others migrate further away (Hara et al., 2014). After transition to Ngn3$^+$, they remain located near the interstitium, followed by their rather even scattering over the basal compartment (Nakagawa et al., 2010; Yoshida, Sukeno, & Nabeshima, 2007).

5.4 Spatiotemporal distribution of extracellular signaling factors in an open (facultative) niche microenvironment

Exemplified by the seminiferous tubules, microenvironments harboring scattered and migrating stem cells mixed with their differentiating progeny are a hallmark of "*open niches*" (Stine & Matunis, 2013; Yoshida, 2018a). In the open niche of seminiferous tubules, A_{undiff} maintain their own pool and their composition for the long term, while producing differentiating progeny to generate sperm. The mechanism of stem-cell regulation in such open niche environment is a long-lasting question.

The fates of A_{undiff} (whether to self-renew or differentiate) are regulated by an ensemble of extracellular signals (Kanatsu-Shinohara & Shinohara, 2013). In this present review, I will focus on a subset of these factors that has been best characterized. GDNF (glial cell line-derived neurotropic factor) is the best characterized factor promoting self-renewal (Hofmann, 2008; Meng et al., 2000). Since GDNF is a ligand acting through the GFRα1-Ret receptor complex, GFRα1$^+$ cells are the primary targets of this ligand; Ret is expressed in a similar but broader population to GFRα1$^+$. GDNF is the key factor for successful in vitro maintenance of SSCs (Kanatsu-Shinohara, Ogonuki, et al., 2003; Kubota, Avarbock, & Brinster, 2003), while heterozygous loss causes significant spermatogenesis defects (Meng et al., 2000). While Sertoli cells have been identified as a prominent source of GDNF (Garcia, Parekh, Gandhi, Sinha, & Hofmann, 2017; Hofmann, 2008), cell type-specific disruption experiments also suggest the importance of GDNF expression in peritubular myoid cells (Chen, Willis, & Eddy, 2016).

The best characterized differentiation-promoting factor is retinoic acid (RA). RA promotes the irreversible progression of the transition of A_{undiff} to differentiating spermatogonia. RA also induces the spermatogonia-to-spermatocyte transition, i.e., entry into meiosis (Hogarth & Griswold, 2010). It appears that Sertoli cells and different stages of germ cells may cooperatively control RA levels in seminiferous tubules, based on the expression of enzymes catalyzing RA biosynthesis from its dietary precursor, vitamin A (Endo, Freinkman, de Rooij, & Page, 2017; Sugimoto, Nabeshima, & Yoshida, 2012; Vernet et al., 2006).

It has been shown that the Wnt/β-Catenin pathway also regulates A_{undiff} by promoting GFRα1$^+$ to Ngn3$^+$ differentiation in conjunction with proliferation (Chassot et al., 2017; Takase & Nusse, 2016; Tokue et al., 2017). Among the Wnt family members, Wnt6 is considered the most likely natural ligand affecting A_{undiff} in the testis based on its prominent expression in Sertoli cells, although this remains to be verified experimentally (Takase & Nusse, 2016; Tokue et al., 2017).

Interestingly, GDNF, RA, and Wnt6 are all predominantly produced by Sertoli cells but show periodical fluctuation along the seminiferous epithelial cycle with peaks at different stages (Fig. 10A). The expression patterns of the

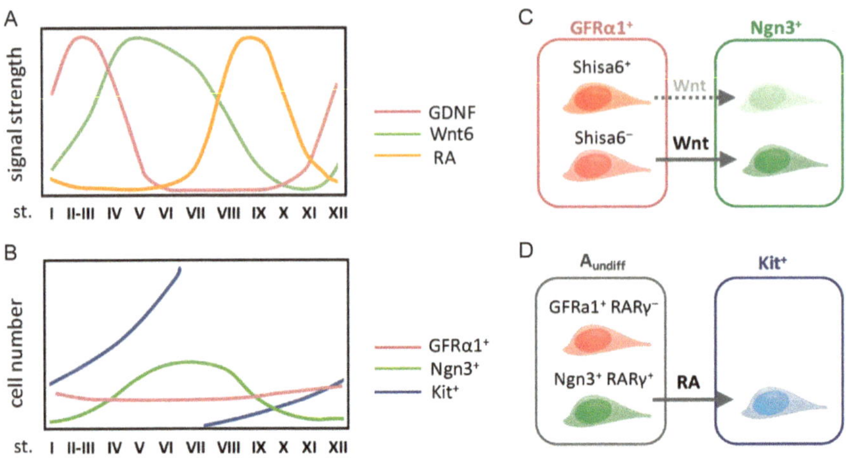

Fig. 10 Temporal fluctuation of the microenvironments and mechanisms of asymmetric fate outcome therein. (A–B) Diagram of periodic fluctuations in the levels of GDNF, Wnt6, and RA (A) and the numbers of GFRα1$^+$, Ngn3$^+$ and Kit$^+$ spermatogonia (B) during the seminiferous epithelial cycle (stages shown by Roman numerals), based on the reports by Endo et al. (2017), Sato et al. (2011), Sharma and Braun (2018), Sugimoto et al. (2012), Tokue et al. (2017). (C–D) Model for the mechanisms underlying the heterogeneous fate outcome of A_{undiff} in the spatially homogeneous microenvironment, in which cells are evenly bathed with Wnt or RA. The heterogeneous expression of Shisa6, a Wnt pathway inhibitor (C) or RARγ, a RA receptor (D), provide heterogeneous susceptibilities to these extracellular signals.

endogenous protein and knock-in reporter (for GDNF) and mRNAs (for Wnt6 and RA-synthesizing enzymes) suggest that, at peak expression, these factors are produced in a spatially ubiquitous manner around the circumference of the tubule (Johnston, Olivas, DiCandeloro, & Wright, 2011; Sato, Aiyama, et al., 2011; Sharma & Braun, 2018; Sugimoto et al., 2012; Tokue et al., 2017; Vernet et al., 2006). Of note, the temporal changes in the expression of these factors appear to be well-correlated with the frequency of GFRα1$^+$, Ngn3$^+$ and Kit$^+$ cells in tune with the seminiferous epithelial cycle (Fig. 10A and B).

FGF (fibroblast growth factor) is another class of ligands promoting self-renewal. FGFs comprise a huge family of 22 (or 23) members that activate receptor complexes including FGFR1, FGFR2, FGFR3, and FGFR4. The effects of FGFs on SSCs (or A_{undiff}) have been demonstrated by administering FGF2 in vitro and in vivo (Hasegawa & Saga, 2014; Kanatsu-Shinohara, Ogonuki, et al., 2003; Masaki et al., 2018; Takashima et al., 2015). On the other hand, the physiological FGF ligands and their expression were only

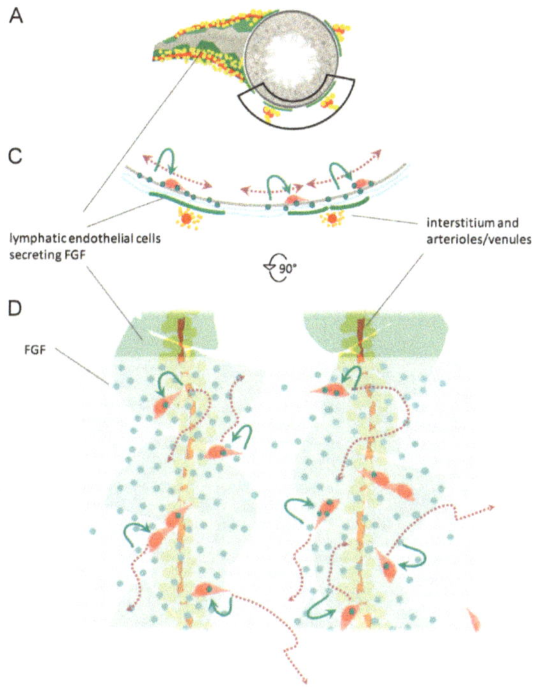

Fig. 11 Diagram of the mitogen competition model for regulation of stem-cell homeostasis in the basal compartment. (A) FGF expression is observed in a subset of lymphatic endothelial cells located near the interstitium and arterioles/venules (green). (B) Characteristics of GFRα1$^+$ spermatogonia crucial for this model: active migration (dotted magenta arrow) and consumption of FGF (green dots) after internalization (green arrow). (C–D) Simplified views of the area indicated in (A), showing GFRα1$^+$ spermatogonia, basement membrane (gray), and underlying peritubular myoid and lymphatic endothelial cell layers (light blue), as well as interstitium (yellow) and arterioles/venules (red). Lymphatic endothelial cells secreting FGF are highlighted in green. Myoid cells and lymphatic endothelial cells that do not express FGF are omitted in (D) for simplicity. Note that GFRα1$^+$ cells migrate over the basement membrane and consume FGF from the tissue, as shown in (B). (E) Fates of GFRα1$^+$ cells dependent on their FGF consumption. Those that consume larger and smaller amounts of FGF are primed for self-renewal and differentiation, respectively, reflecting the expression of FGF target genes. See Kitadate et al. (2019) for details.

characterized recently. It has been reported that FGF5, FGF4, and FGF8 are expressed in large and flattened lymphatic endothelial cells covering the outer surface of the tubules near the interstitium, as schematically shown in Fig. 11A, C and D (Kitadate et al., 2019). The expression of FGF5 persists during the seminiferous epithelial cycle. Therefore, FGF appears to be produced in a spatially heterogeneous (biased to the proximity of interstitium) and temporally stable (throughout the seminiferous epithelial cycle) manner, unlike GDNF, RA or Wnt. Of note, the A_{undiff} distribution shows spatial correlation with FGF5 expression (Kitadate et al., 2019). However, A_{undiff} subsets (e.g., GFRα1$^+$ and Ngn3$^+$) are still interspersed among differentiating spermatogonia in these regions. Therefore, it remains unlikely that the asymmetric fate outcome of SSCs is caused by local heterogeneity of FGF.

The aforementioned characters of seminiferous tubules and SSCs support the idea of an "open niche." Yet, it remains a deep-rooted thought that there may be unidentified, cryptic microenvironments that act as definitive (closed) niches. However, recent studies, summarized below, have suggested novel mechanisms that may achieve SSC homeostasis in an open niche, without assuming cryptic definitive sub-environments. These mechanisms could, in principle, answer how SSCs follow different fates—self-renewal and differentiation—in a uniform microenvironment and how the SSC pool size is kept constant, despite that SSCs are migratory and scattered between differentiating progeny.

5.5 Heterogeneous susceptibility to extracellular signals

The kinetics of extracellular signal strengths and the number of A_{undiff} subpopulations during the seminiferous epithelial cycle (Fig. 10A and B) support the idea that the decrease of GDNF and increase of Wnt6 promote the GFRα1$^+$ to Ngn3$^+$ transition (Tokue et al., 2017). However, that does not explain why a significant proportion of the cells remain in the GFRα1$^+$ state, despite all cells apparently being bathed in Wnt6. Tokue et al. (2017) identified Shisa6 as a cell-intrinsic Wnt inhibitor with restricted expression in a subset of GFRα1$^+$ cells (see also Fig. 6). Therefore, the fate asymmetry of GFRα1$^+$ cells in this homogeneous and differentiation-inducing microenvironment may be a result of heterogeneous susceptibility to Wnt signaling, a hypothesis also supported by the phenotypes observed in the *Shisa6* mutants (Tokue et al., 2017) (Fig. 10C). Given that GFRα1$^+$ cells are not totally depleted in the absence of Shisa6, redundant and/or independent mechanisms likely exist in the maintenance of GFRα1$^+$ spermatogonia.

A similar mechanism of heterogeneous susceptibility has been proposed for the Ngn3$^+$ to Kit$^+$ transition induced by RA (Ikami et al., 2015). In this study, using vitamin A-deficient models in which RA signal is artificially decreased followed by its restoration (van Pelt & de Rooij, 1990), combined with cre-mediated lineage tracing, it was shown that RA induced the differentiation of Ngn3$^+$ (also expressing the RA receptor RARγ; Fig. 6), but not GFRα1$^+$ (mostly RARγ^-). Furthermore, ectopic expression of RARγ in GFRα1$^+$ cells caused their differentiation into Kit$^+$ cells, consistent with a lack of RARγ expression protecting the GFRα1$^+$ cells from differentiation induction by RA. Interestingly, GFRα1$^+$ cells express another RA receptor, RARα (Tokue et al., 2017) and are capable of responding to RA in vitro (Lord, Oatley, & Oatley, 2018), suggesting that in the in vivo context the role of RARα must be restricted. This is consistent with the RARα mutant phenotypes (Gely-Pernot et al., 2012). In addition to the apparently lower level of RARα expression (Vernet et al., 2006), there is also a large discrepancy (~100-fold) in sensitivity to RA between these receptors: the half-maximal effective concentrations (EC_{50}) of RARγ and RARα are 0.5–2 nM and 20–200 nM, respectively (Astrom, Pettersson, Krust, Chambon, & Voorhees, 1990; Brand et al., 1988; Giguere et al., 1990). Since the measured in vivo RA concentrations in the testis lies in between these sensitivities (~10–20 nM; Endo et al., 2017; Kane, Chen, Sparks, & Napoli, 2005; Obrochta, Kane, & Napoli, 2014), the major part of RA signaling is likely mediated by RARγ.

5.6 Density regulation through the mechanism of mitogen competition

How is the pool size of the self-renewing stem cells kept constant in the open niche microenvironment? This question moves to the forefront when we consider that A_{undiff} are scattered and actively migrating (Fig. 9). Indeed, the number of A_{undiff} and its subsets remain highly constant over the long term, albeit fluctuating along with the seminiferous epithelial cycle (Fig. 10B). Typically, besides local heterogeneities reflecting the adjacent interstitium and blood vessels, GFRα1$^+$ cells have strikingly even average densities over >1 cm of tubules (Kitadate et al., 2019). Furthermore, partial killing of spermatogonia including GFRα1$^+$ cells elicits a regeneration process that eventually reinstates the homeostasis with the original density of GFRα1$^+$ cells in 2–3 months (Kanatsu-Shinohara et al., 2003; Kitadate et al., 2019), indicating that SSC homeostasis is regulated in a manner that achieves constant average densities.

Recently, a mechanism has been proposed that explains this density homeostasis, called a *"mitogen competition model"* (Fig. 11) (Kitadate et al., 2019). In this model, A_{undiff} (in particular GFRα1$^+$ cells) uptake and consume FGF secreted by lymphatic endothelial cells, while migrating over the basement membrane (Fig. 11B–D). GFRα1$^+$ cells that have taken up and consumed larger amounts of FGF will self-renew with high probabilities, while those that have consumed less FGFs will be more inclined to differentiate, through the action of genes whose expression is up- or down-regulated downstream of FGF signaling (Fig. 11E). As a result, GFRα1$^+$ cells could effectively compete with each other for a limited supply of FGF, and the density homeostasis would be self-organized. The observed high density of GFRα1$^+$ cells near the FGF source could also be self-organized in this model, without assuming a chemoattractive mechanism (Kitadate et al., 2019).

This simple model was proposed based on the following findings: First, in mutants of *Fgf5*, *Fgf8*, *Fgf4*, and their compounds, as well as in transgenic mice with an extra copy of *Fgf5*, GFRα1$^+$ cells developed homeostasis, with their density linearly correlated with the total dose of *Fgf* genes. Second, FGF5 protein was detected in the digestive organelles (i.e., multivesicular bodies and lysosomes) of GFRα1$^+$ cells. Importantly, a minimal theoretical model based on this scenario faithfully predicts a wide range of experimental data: In addition to the linear correlation between FGF supply and the steady-state density of GFRα1$^+$ cells, this mathematical model quantitatively predicts the characteristic decaying oscillation of GFRα1$^+$ cell density during post-insult regeneration before restoration of the original cell numbers (Kitadate et al., 2019). In addition, theoretical assessments suggest that this simple model may capture stem-cell dynamics in different biological contexts (Jorg, Kitadate, Yoshida, & Simons, 2019).

To the best of my knowledge, this mitogen competition model may be the first quantitative proposal to explain the underlying mechanism of stem cell-pool homeostasis in an open-niche microenvironment. This model warrants critical evaluation in the future.

5.7 Next questions

As described above, recent models and data should ignite further consideration and investigation of the mechanisms of SSC regulation in an open-niche microenvironment. I consider the following to be must-ask questions. First, understanding the mechanisms underlying the stable and reproducible generation of SSC subpopulations that show different susceptibilities to extracellular factors are of primary importance. This issue is related to the dynamic and heterogeneous nature of the A_{undiff} compartment, composed of multiple interchangeable states, discussed in this review. In this regard, it is interesting to note that the mitogen competition model has the potential to stabilize the proportions of multiple states, in the case that cells in different states compete for the same mitogen(s) (Kitadate et al., 2019). Second, does the mechanism of mitogen competition work with other factors regulating SSC fate besides FGF? In particular, it would be an interesting to know whether SSC regulation by GDNF, which shows temporal fluctuation, is also achieved by the same paradigm of mitogen competition. Third, can competition for limiting factors also be mediated by mechanisms other than direct consumption? Here, the possible mechanisms may include secretion of mitogen inhibitors by SSCs, which is reminiscent of the mechanisms of bacterial quorum sensing (Miller & Bassler, 2001). Fourth, do these proposed mechanisms of heterogeneous susceptibility to signals and competition for mitogens serve as a model for other tissue stem-cell systems?

6. Conclusion

I have overviewed the current endeavors toward answering key questions about mouse SSCs. Given the advances in research, our understanding of SSCs is continually expanded and, of course, currently proposed models (including those from our group) need to be continually re-evaluated as well.

In this regard, the rapidly developing technology of single cell omics will be of particular importance in the next stages and should provide detailed information about SSC heterogeneity. The current major drawback of this technology may be the lack of information about the position and topology of cells within the tissue. However, these could be overcome by technical breakthroughs (such as Frieda et al., 2017; Lubeck, Coskun, Zhiyentayev, Ahmad, & Cai, 2014) or by combination with other methods. The next round of technical evolutions may bring about a totally new understanding of SSCs. The potential of the technologies of organ culture, organoid culture, and in vitro gametogenesis is likely to go beyond our current imagination. These experimental platforms will become most powerful when they are combined with analytical tools for SSC behavior over time, such as live imaging and pulse labeling. Finally, the importance of statistical and/or mathematical analyses should significantly increase, to allow the best exploitation of the massive data sets provided by omics and imaging studies etc.

Based on the unique nature of seminiferous tubules, the study of mouse SSCs will continue to make significant contributions to the field of tissue stem cell research and to that of developmental biology in general.

Acknowledgments

The author expresses deep appreciation to the exciting collaborations and discussions conducted with all our laboratory members and colleagues. I also acknowledge the constructive comments by Takehiko Ogawa (Yokohama City University) to this review article. Studies from our lab described in this document have been financially supported, in part, by the Grant-in-Aid for Scientific Research (KAKENHI) program from the MEXT and JSPS, the PRESTO program from Japan Science and Technology Agency, and the AMED-CREST program from AMED. There is no conflict of interests to declare.

References

Abid, S. N., Richardson, T. E., Powell, H. M., Jaichander, P., Chaudhary, J., Chapman, K. M., et al. (2014). A-single spermatogonia heterogeneity and cell cycles synchronize with rat seminiferous epithelium stages VIII-IX. *Biology of Reproduction, 90*, 32.

Aiyama, Y., Tsunekawa, N., Kishi, K., Kawasumi, M., Suzuki, H., Kanai-Azuma, M., et al. (2015). A niche for GFRα1-positive spermatogonia in the terminal segments of the seminiferous tubules in hamster testes. *Stem cells (Dayton, Ohio), 33*, 2811–2824.

Aloisio, G. M., Nakada, Y., Saatcioglu, H. D., Pena, C. G., Baker, M. D., Tarnawa, E. D., et al. (2014). PAX7 expression defines germline stem cells in the adult testis. *The Journal of Clinical Investigation, 124*, 3929–3944.

Alves-Lopes, J. P., & Stukenborg, J. B. (2018). Testicular organoids: A new model to study the testicular microenvironment in vitro? *Human Reproduction Update, 24*, 176–191.

Astrom, A., Pettersson, U., Krust, A., Chambon, P., & Voorhees, J. J. (1990). Retinoic acid and synthetic analogs differentially activate retinoic acid receptor dependent transcription. *Biochemical and Biophysical Research Communications, 173*, 339–345.

Brand, N., Petkovich, M., Krust, A., Chambon, P., de The, H., Marchio, A., et al. (1988). Identification of a second human retinoic acid receptor. *Nature, 332*, 850–853.

Brawley, C., & Matunis, E. (2004). Regeneration of male germline stem cells by spermatogonial dedifferentiation in vivo. *Science, 304*, 1331–1334.

Brinster, R. L. (2002). Germline stem cell transplantation and transgenesis. *Science, 296*, 2174–2176.

Brinster, R. L. (2007). Male germline stem cells: From mice to men. *Science, 316*, 404–405.

Brinster, R. L., & Avarbock, M. R. (1994). Germline transmission of donor haplotype following spermatogonial transplantation. *Proceedings of the National Academy of Sciences of the United States of America, 91*, 11303–11307.

Brinster, R. L., & Zimmermann, J. W. (1994). Spermatogenesis following male germ-cell transplantation. *Proceedings of the National Academy of Sciences of the United States of America, 91*, 11298–11302.

Buaas, F. W., Kirsh, A. L., Sharma, M., McLean, D. J., Morris, J. L., Griswold, M. D., et al. (2004). Plzf is required in adult male germ cells for stem cell self-renewal. *Nature Genetics, 36*, 647–652.

Carrieri, C., Comazzetto, S., Grover, A., Morgan, M., Buness, A., Nerlov, C., et al. (2017). A transit-amplifying population underpins the efficient regenerative capacity of the testis. *The Journal of Experimental Medicine, 214*, 1631–1641.

Chakraborty, P., Buaas, F. W., Sharma, M., Snyder, E., de Rooij, D. G., & Braun, R. E. (2014). LIN28A marks the spermatogonial progenitor population and regulates its cyclic expansion. *Stem cells (Dayton, Ohio), 32*, 860–873.

Chan, F., Oatley, M. J., Kaucher, A. V., Yang, Q. E., Bieberich, C. J., Shashikant, C. S., et al. (2014). Functional and molecular features of the Id4+ germline stem cell population in mouse testes. *Genes & Development, 28*, 1351–1362.

Chassot, A. A., Le Rolle, M., Jourden, M., Taketo, M. M., Ghyselinck, N. B., & Chaboissier, M. C. (2017). Constitutive WNT/CTNNB1 activation triggers spermatogonial stem cell proliferation and germ cell depletion. *Developmental Biology*, *426*, 17–27.
Chen, L. Y., Willis, W. D., & Eddy, E. M. (2016). Targeting the GDNF gene in peritubular myoid cells disrupts undifferentiated spermatogonial cell development. *Proceedings of the National Academy of Sciences of the United States of America*, *113*, 1829–1834.
Chen, Y., Zheng, Y., Gao, Y., Lin, Z., Yang, S., Wang, T., et al. (2018). Single-cell RNA-seq uncovers dynamic processes and critical regulators in mouse spermatogenesis. *Cell Research*, *28*, 879–896.
Chiarini-Garcia, H., Hornick, J. R., Griswold, M. D., & Russell, L. D. (2001). Distribution of type A spermatogonia in the mouse is not random. *Biology of Reproduction*, *65*, 1179–1185.
Chiarini-Garcia, H., Raymer, A. M., & Russell, L. D. (2003). Non-random distribution of spermatogonia in rats: Evidence of niches in the seminiferous tubules. *Reproduction*, *126*, 669–680.
Chiarini-Garcia, H., & Russell, L. D. (2001). High-resolution light microscopic characterization of mouse spermatogonia. *Biology of Reproduction*, *65*, 1170–1178.
Chiarini-Garcia, H., & Russell, L. D. (2002). Characterization of mouse spermatogonia by transmission electron microscopy. *Reproduction*, *123*, 567–577.
Clayton, E., Doupe, D. P., Klein, A. M., Winton, D. J., Simons, B. D., & Jones, P. H. (2007). A single type of progenitor cell maintains normal epidermis. *Nature*, *446*, 185–189.
Costoya, J. A., Hobbs, R. M., Barna, M., Cattoretti, G., Manova, K., Sukhwani, M., et al. (2004). Essential role of Plzf in maintenance of spermatogonial stem cells. *Nature Genetics*, *36*, 653–659.
de Rooij, D. G. (1973). Spermatogonial stem cell renewal in the mouse. I. Normal situation. *Cell and Tissue Kinetics*, *6*, 281–287.
de Rooij, D. G. (2010). Morphometric description of spermatogonial stem cells and expansion of their clonal derivatives. In K. E. Orwig & B. P. Hermann (Eds.), *Male germline stem cells: Developmental and regenerative potential*. Springer.
de Rooij, D. G., & Russell, L. D. (2000). All you wanted to know about spermatogonia but were afraid to ask. *Journal of Andrology*, *21*, 776–798.
Di Persio, S., Saracino, R., Fera, S., Muciaccia, B., Esposito, V., Boitani, C., et al. (2017). Spermatogonial kinetics in humans. *Development (Cambridge, England)*, *144*, 3430–3439.
Doupe, D. P., Klein, A. M., Simons, B. D., & Jones, P. H. (2010). The ordered architecture of murine ear epidermis is maintained by progenitor cells with random fate. *Developmental Cell*, *18*, 317–323.
Endo, T., Freinkman, E., de Rooij, D. G., & Page, D. C. (2017). Periodic production of retinoic acid by meiotic and somatic cells coordinates four transitions in mouse spermatogenesis. *Proceedings of the National Academy of Sciences of the United States of America*, *114*, E10132–E10141.
Erickson, B. H. (1981). Survival and renewal of murine stem spermatogonia following 60Co gamma radiation. *Radiation Research*, *86*, 34–51.
Fayomi, A. P., & Orwig, K. E. (2018). Spermatogonial stem cells and spermatogenesis in mice, monkeys and men. *Stem Cell Research*, *29*, 207–214.
Frieda, K. L., Linton, J. M., Hormoz, S., Choi, J., Chow, K. K., Singer, Z. S., et al. (2017). Synthetic recording and in situ readout of lineage information in single cells. *Nature*, *541*, 107–111.
Fuller, M. T., & Spradling, A. C. (2007). Male and female Drosophila germline stem cells: Two versions of immortality. *Science*, *316*, 402–404.
Garbuzov, A., Pech, M. F., Hasegawa, K., Sukhwani, M., Zhang, R. J., Orwig, K. E., et al. (2018). Purification of GFRα1+ and GFRα1− spermatogonial stem cells reveals a niche-dependent mechanism for fate determination. *Stem Cell Reports*, *10*, 553–567.

Garcia, T. X., Parekh, P., Gandhi, P., Sinha, K., & Hofmann, M. C. (2017). The NOTCH ligand JAG1 regulates GDNF expression in Sertoli cells. *Stem Cells and Development, 26*, 585–598.

Gely-Pernot, A., Raverdeau, M., Celebi, C., Dennefeld, C., Feret, B., Klopfenstein, M., et al. (2012). Spermatogonia differentiation requires retinoic acid receptor gamma. *Endocrinology, 153*, 438–449.

Giguere, V., Shago, M., Zirngibl, R., Tate, P., Rossant, J., & Varmuza, S. (1990). Identification of a novel isoform of the retinoic acid receptor gamma expressed in the mouse embryo. *Molecular and Cellular Biology, 10*, 2335–2340.

Green, C. D., Ma, Q., Manske, G. L., Shami, A. N., Zheng, X., Marini, S., et al. (2018). A comprehensive roadmap of murine spermatogenesis defined by single-cell RNA-Seq. *Developmental Cell, 46*. 651–667.e610.

Greenbaum, M. P., Ma, L., & Matzuk, M. M. (2007). Conversion of midbodies into germ cell intercellular bridges. *Developmental Biology, 305*, 389–396.

Grisanti, L., Falciatori, I., Grasso, M., Dovere, L., Fera, S., Muciaccia, B., et al. (2009). Identification of spermatogonial stem cell subsets by morphological analysis and prospective isolation. *Stem Cells (Dayton, Ohio), 27*, 3043–3052.

Hammoud, S. S., Low, D. H., Yi, C., Lee, C. L., Oatley, J. M., Payne, C. J., et al. (2015). Transcription and imprinting dynamics in developing postnatal male germline stem cells. *Genes & Development, 29*, 2312–2324.

Hara, K., Nakagawa, T., Enomoto, H., Suzuki, M., Yamamoto, M., Simons, B. D., et al. (2014). Mouse spermatogenic stem cells continually interconvert between equipotent singly isolated and syncytial states. *Cell Stem Cell, 14*, 658–672.

Hasegawa, K., & Saga, Y. (2014). FGF8-FGFR1 signaling acts as a niche factor for maintaining undifferentiated spermatogonia in the mouse. *Biology of Reproduction, 91*, 145.

Hayashi, S., & McMahon, A. P. (2002). Efficient recombination in diverse tissues by a tamoxifen-inducible form of Cre: A tool for temporally regulated gene activation/inactivation in the mouse. *Developmental Biology, 244*, 305–318.

Hayashi, K., Ohta, H., Kurimoto, K., Aramaki, S., & Saitou, M. (2011). Reconstitution of the mouse germ cell specification pathway in culture by pluripotent stem cells. *Cell, 146*, 519–532.

Helsel, A. R., Yang, Q. E., Oatley, M. J., Lord, T., Sablitzky, F., & Oatley, J. M. (2017). ID4 levels dictate the stem cell state in mouse spermatogonia. *Development (Cambridge, England), 144*, 624–634.

Hermann, B. P., Mutoji, K. N., Velte, E. K., Ko, D., Oatley, J. M., Geyer, C. B., et al. (2015). Transcriptional and translational heterogeneity among neonatal mouse spermatogonia. *Biology of Reproduction, 92*, 54.

Hermann, B. P., Sukhwani, M., Winkler, F., Pascarella, J. N., Peters, K. A., Sheng, Y., et al. (2012). Spermatogonial stem cell transplantation into rhesus testes regenerates spermatogenesis producing functional sperm. *Cell Stem Cell, 11*, 715–726.

Hikabe, O., Hamazaki, N., Nagamatsu, G., Obata, Y., Hirao, Y., Hamada, N., et al. (2016). Reconstitution in vitro of the entire cycle of the mouse female germ line. *Nature, 539*, 299–303.

Hilton, B. T., & Turner, T. T. (1993). The seminiferous tubular microenvironment. In C. Desjardins & L. L. Ewing (Eds.), *Cell and molecular biology of the testis* (pp. 238–265). New York: Oxford University Press.

Hobbs, R. M., Fagoonee, S., Papa, A., Webster, K., Altruda, F., Nishinakamura, R., et al. (2012). Functional antagonism between Sall4 and Plzf defines germline progenitors. *Cell Stem Cell, 10*, 284–298.

Hofmann, M. C. (2008). GDNF signaling pathways within the mammalian spermatogonial stem cell niche. *Molecular and Cellular Endocrinology, 288*, 95–103.

Hofmann, M. C., Braydich-Stolle, L., & Dym, M. (2005). Isolation of male germ-line stem cells; influence of GDNF. *Developmental Biology*, *279*, 114–124.
Hogarth, C. A., & Griswold, M. D. (2010). The key role of vitamin A in spermatogenesis. *The Journal of Clinical Investigation*, *120*, 956–962.
Huckins, C. (1971a). The spermatogonial stem cell population in adult rats. I. Their morphology, proliferation and maturation. *The Anatomical Record*, *169*, 533–557.
Huckins, C. (1971b). The spermatogonial stem cell population in adult rats. II. A radioautographic analysis of their cell cycle properties. *Cell and Tissue Kinetics*, *4*, 313–334.
Huckins, C. (1978). Spermatogonial intercellular bridges in whole-mounted seminiferous tubules from normal and irradiated rodent testes. *The American Journal of Anatomy*, *153*, 97–121.
Huckins, C., & Kopriwa, B. M. (1969). A technique for the radioautography of germ cells in whole mounts of seminiferous tubules. *The Journal of Histochemistry and Cytochemistry*, *17*, 848–851.
Huckins, C., & Oakberg, E. F. (1978). Morphological and quantitative analysis of spermatogonia in mouse testes using whole mounted seminiferous tubules. II. The irradiated testes. *The Anatomical Record*, *192*, 529–542.
Ikami, K., Tokue, M., Sugimoto, R., Noda, C., Kobayashi, S., Hara, K., et al. (2015). Hierarchical differentiation competence in response to retinoic acid ensures stem cell maintenance during mouse spermatogenesis. *Development (Cambridge, England)*, *142*, 1582–1592.
Ishikura, Y., Yabuta, Y., Ohta, H., Hayashi, K., Nakamura, T., Okamoto, I., et al. (2016). In vitro derivation and propagation of spermatogonial stem cell activity from mouse pluripotent stem cells. *Cell Reports*, *17*, 2789–2804.
Johnston, D. S., Olivas, E., DiCandeloro, P., & Wright, W. W. (2011). Stage-specific changes in GDNF expression by rat Sertoli cells: A possible regulator of the replication and differentiation of stem spermatogonia. *Biology of Reproduction*, *85*, 763–769.
Jorg, D. J., Kitadate, Y., Yoshida, S., & Simons, B. D. (2019). Competition for stem cell fate determinants as a mechanism for tissue homeostasis. *arXiv*. 1901.03903.
Kai, T., & Spradling, A. (2004). Differentiating germ cells can revert into functional stem cells in *Drosophila* melanogaster ovaries. *Nature*, *428*, 564–569.
Kanatsu-Shinohara, M., Inoue, K., Miki, H., Ogonuki, N., Takehashi, M., Morimoto, T., et al. (2006). Clonal origin of germ cell colonies after spermatogonial transplantation in mice. *Biology of Reproduction*, *75*, 68–74.
Kanatsu-Shinohara, M., Ogonuki, N., Inoue, K., Miki, H., Ogura, A., Toyokuni, S., et al. (2003). Long-term proliferation in culture and germline transmission of mouse male germline stem cells. *Biology of Reproduction*, *69*, 612–616.
Kanatsu-Shinohara, M., & Shinohara, T. (2007). Culture and genetic modification of mouse germline stem cells. *Annals of the New York Academy of Sciences*, *1120*, 59–71.
Kanatsu-Shinohara, M., & Shinohara, T. (2013). Spermatogonial stem cell self-renewal and development. *Annual Review of Cell and Developmental Biology*, *29*, 163–187.
Kanatsu-Shinohara, M., Toyokuni, S., Morimoto, T., Matsui, S., Honjo, T., & Shinohara, T. (2003). Functional assessment of self-renewal activity of male germline stem cells following cytotoxic damage and serial transplantation. *Biology of Reproduction*, *68*, 1801–1807.
Kane, M. A., Chen, N., Sparks, S., & Napoli, J. L. (2005). Quantification of endogenous retinoic acid in limited biological samples by LC/MS/MS. *The Biochemical Journal*, *388*, 363–369.
Kitadate, Y., Jorg, D. J., Tokue, M., Maruyama, A., Ichikawa, R., Tsuchiya, S., et al. (2019). Competition for mitogens regulates spermatogenic stem cell homeostasis in an open niche. *Cell Stem Cell*, *24*. 79–92.e76.

Klein, A. M., Nakagawa, T., Ichikawa, R., Yoshida, S., & Simons, B. D. (2010). Mouse germ line stem cells undergo rapid and stochastic turnover. *Cell Stem Cell*, *7*, 214–224.

Klein, A. M., & Simons, B. D. (2011). Universal patterns of stem cell fate in cycling adult tissues. *Development (Cambridge, England)*, *138*, 3103–3111.

Komai, Y., Tanaka, T., Tokuyama, Y., Yanai, H., Ohe, S., Omachi, T., et al. (2014). Bmi1 expression in long-term germ stem cells. *Scientific Reports*, *4*, 6175.

Komeya, M., Kimura, H., Nakamura, H., Yokonishi, T., Sato, T., Kojima, K., et al. (2016). Long-term ex vivo maintenance of testis tissues producing fertile sperm in a microfluidic device. *Scientific Reports*, *6*, 21472.

Komeya, M., & Ogawa, T. (2015). Spermatogonial stem cells: Progress and prospects. *Asian Journal of Andrology*, *17*, 771–775.

Krieger, T., & Simons, B. D. (2015). Dynamic stem cell heterogeneity. *Development (Cambridge, England)*, *142*, 1396–1406.

Kubota, H., Avarbock, M. R., & Brinster, R. L. (2003). Spermatogonial stem cells share some, but not all, phenotypic and functional characteristics with other stem cells. *Proceedings of the National Academy of Sciences of the United States of America*, *100*, 6487–6492.

Kubota, H., Avarbock, M. R., & Brinster, R. L. (2004). Growth factors essential for self-renewal and expansion of mouse spermatogonial stem cells. *Proceedings of the National Academy of Sciences of the United States of America*, *101*, 16489–16494.

La, H. M., Makela, J. A., Chan, A. L., Rossello, F. J., Nefzger, C. M., Legrand, J. M. D., et al. (2018). Identification of dynamic undifferentiated cell states within the male germline. *Nature Communications*, *9*, 2819.

Leblond, C. P., & Clermont, Y. (1952). Definition of the stages of the cycle of the seminiferous epithelium in the rat. *Annals of the New York Academy of Sciences*, *55*, 548–573.

Lee, C. Y., Horn, H. F., Stewart, C. L., Burke, B., Bolcun-Filas, E., Schimenti, J. C., et al. (2015). Mechanism and regulation of rapid telomere prophase movements in mouse meiotic chromosomes. *Cell Reports*, *11*, 551–563.

Lei, L., & Spradling, A. C. (2013). Mouse primordial germ cells produce cysts that partially fragment prior to meiosis. *Development (Cambridge, England)*, *140*, 2075–2081.

Lenhart, K. F., & DiNardo, S. (2015). Somatic cell encystment promotes abscission in germline stem cells following a regulated block in cytokinesis. *Developmental Cell*, *34*, 192–205.

Lin, M., & Jones, R. C. (1992). Renewal and proliferation of spermatogonia during spermatogenesis in the Japanese quail, Coturnix coturnix japonica. *Cell and Tissue Research*, *267*, 591–601.

Lombardi, J. (1998). *Gametes and their production, comparative vertebrate reproduction* (pp. 109–153). New York: Springer Science+Business Media New York.

Lopez-Garcia, C., Klein, A. M., Simons, B. D., & Winton, D. J. (2010). Intestinal stem cell replacement follows a pattern of neutral drift. *Science*, *330*, 822–825.

Lord, T., & Oatley, J. M. (2017). A revised A_{single} model to explain stem cell dynamics in the mouse male germline. *Reproduction*, *154*, R55–R64.

Lord, T., Oatley, M. J., & Oatley, J. M. (2018). Testicular architecture is critical for mediation of retinoic acid responsiveness by undifferentiated spermatogonial subtypes in the mouse. *Stem Cell Reports*, *10*, 538–552.

Losick, V. P., Morris, L. X., Fox, D. T., & Spradling, A. (2011). Drosophila stem cell niches: A decade of discovery suggests a unified view of stem cell regulation. *Developmental Cell*, *21*, 159–171.

Lubeck, E., Coskun, A. F., Zhiyentayev, T., Ahmad, M., & Cai, L. (2014). Single-cell in situ RNA profiling by sequential hybridization. *Nature Methods*, *11*, 360–361.

Masaki, K., Sakai, M., Kuroki, S., Jo, J. I., Hoshina, K., Fujimori, Y., et al. (2018). FGF2 has distinct molecular functions from GDNF in the mouse Germline niche. *Stem Cell Reports*, *10*, 1782–1792.

Meistrich, M. L., & van Beek, M. E. (1993). Spermatogonial stem cells. In C. Desjardins & L. L. Ewing (Eds.), *Cell and molecular biology of the testis* (pp. 266–295). New York, NY: Oxford University Press.

Meng, X., Lindahl, M., Hyvonen, M. E., Parvinen, M., de Rooij, D. G., Hess, M. W., et al. (2000). Regulation of cell fate decision of undifferentiated spermatogonia by GDNF. *Science*, *287*, 1489–1493.

Miller, M. B., & Bassler, B. L. (2001). Quorum sensing in bacteria. *Annual Review of Microbiology*, *55*, 165–199.

Morrison, S. J., & Spradling, A. C. (2008). Stem cells and niches: Mechanisms that promote stem cell maintenance throughout life. *Cell*, *132*, 598–611.

Nagamatsu, G., & Hayashi, K. (2017). Stem cells, in vitro gametogenesis and male fertility. *Reproduction*, *154*, F79–F91.

Nagano, M. C. (2003). Homing efficiency and proliferation kinetics of male germ line stem cells following transplantation in mice. *Biology of Reproduction*, *69*, 701–707.

Nakagawa, T., Nabeshima, Y., & Yoshida, S. (2007). Functional identification of the actual and potential stem cell compartments in mouse spermatogenesis. *Developmental Cell*, *12*, 195–206.

Nakagawa, T., Sharma, M., Nabeshima, Y., Braun, R. E., & Yoshida, S. (2010). Functional hierarchy and reversibility within the murine spermatogenic stem cell compartment. *Science*, *328*, 62–67.

Oakberg, E. F. (1956). Duration of spermatogenesis in the mouse and timing of stages of the cycle of the seminiferous epithelium. *The American Journal of Anatomy*, *99*, 507–516.

Oakberg, E. F. (1971). Spermatogonial stem-cell renewal in the mouse. *The Anatomical Record*, *169*, 515–531.

Oatley, J. M., Avarbock, M. R., Telaranta, A. I., Fearon, D. T., & Brinster, R. L. (2006). Identifying genes important for spermatogonial stem cell self-renewal and survival. *Proceedings of the National Academy of Sciences of the United States of America*, *103*, 9524–9529.

Oatley, J. M., & Brinster, R. L. (2008). Regulation of spermatogonial stem cell self-renewal in mammals. *Annual Review of Cell and Developmental Biology*, *24*, 263–286.

Obrochta, K. M., Kane, M. A., & Napoli, J. L. (2014). Effects of diet and strain on mouse serum and tissue retinoid concentrations. *PLoS One*, *9*, e99435.

Pepling, M. E., & Spradling, A. C. (2001). Mouse ovarian germ cell cysts undergo programmed breakdown to form primordial follicles. *Developmental Biology*, *234*, 339–351.

Perey, B., Clermont, Y., & Leblond, C. P. (1961). The wave of the seminiferous epithelium in the rat. *Developmental Dynamics*, *108*, 47–77.

Russell, L., Ettlin, R., Sinha Hikim, A., & Clegg, E. (1990). *Histological and histopathological evaluation of the testis*. Clearwater, Fl: Cache River Press.

Sada, A., Suzuki, A., Suzuki, H., & Saga, Y. (2009). The RNA-binding protein NANOS2 is required to maintain murine spermatogonial stem cells. *Science*, *325*, 1394–1398.

Saitou, M., & Miyauchi, H. (2016). Gametogenesis from pluripotent stem cells. *Cell Stem Cell*, *18*, 721–735.

Sakamoto, S., Thumkeo, D., Ohta, H., Zhang, Z., Huang, S., Kanchanawong, P., et al. (2018). mDia1/3 generate cortical F-actin meshwork in Sertoli cells that is continuous with contractile F-actin bundles and indispensable for spermatogenesis and male fertility. *PLoS Biology*, *16*, e2004874.

Sato, T., Aiyama, Y., Ishii-Inagaki, M., Hara, K., Tsunekawa, N., Harikae, K., et al. (2011). Cyclical and patch-like GDNF distribution along the basal surface of Sertoli cells in mouse and hamster testes. *PLoS One*, *6*, e28367.

Sato, T., Katagiri, K., Gohbara, A., Inoue, K., Ogonuki, N., Ogura, A., et al. (2011). In vitro production of functional sperm in cultured neonatal mouse testes. *Nature*, *471*, 504–507.

Sato, T., Yokonishi, T., Komeya, M., Katagiri, K., Kubota, Y., Matoba, S., et al. (2012). Testis tissue explantation cures spermatogenic failure in c-Kit ligand mutant mice. *Proceedings of the National Academy of Sciences of the United States of America*, *109*, 16934–16938.

Sharma, M., & Braun, R. E. (2018). Cyclical expression of GDNF is required for spermatogonial stem cell homeostasis. *Development (Cambridge, England)*, *145*, dev151555.

Shibuya, H., Morimoto, A., & Watanabe, Y. (2014). The dissection of meiotic chromosome movement in mice using an in vivo electroporation technique. *PLoS Genetics*, *10*, e1004821.

Simons, B. D., & Clevers, H. (2011). Strategies for homeostatic stem cell self-renewal in adult tissues. *Cell*, *145*, 851–862.

Snippert, H. J., van der Flier, L. G., Sato, T., van Es, J. H., van den Born, M., Kroon-Veenboer, C., et al. (2010). Intestinal crypt homeostasis results from neutral competition between symmetrically dividing Lgr5 stem cells. *Cell*, *143*, 134–144.

Song, H. W., & Wilkinson, M. F. (2014). Transcriptional control of spermatogonial maintenance and differentiation. *Seminars in Cell & Developmental Biology*, *30*, 14–26.

Spradling, A., Fuller, M. T., Braun, R. E., & Yoshida, S. (2011). Germline stem cells. *Cold Spring Harbor Perspectives in Biology*, *3*, a002642.

Stine, R. R., & Matunis, E. L. (2013). Stem cell competition: Finding balance in the niche. *Trends in Cell Biology*, *23*, 357–364.

Sugimoto, R., Nabeshima, Y., & Yoshida, S. (2012). Retinoic acid metabolism links the periodical differentiation of germ cells with the cycle of Sertoli cells in mouse seminiferous epithelium. *Mechanisms of Development*, *128*, 610–624.

Sun, F., Xu, Q., Zhao, D., & Degui Chen, C. (2015). Id4 marks spermatogonial stem cells in the mouse testis. *Scientific Reports*, *5*, 17594.

Suzuki, H., Ahn, H. W., Chu, T., Bowden, W., Gassei, K., Orwig, K., et al. (2012). SOHLH1 and SOHLH2 coordinate spermatogonial differentiation. *Developmental Biology*, *361*, 301–312.

Suzuki, H., Sada, A., Yoshida, S., & Saga, Y. (2009). The heterogeneity of spermatogonia is revealed by their topology and expression of marker proteins including the germ cell-specific proteins Nanos2 and Nanos3. *Developmental Biology*, *336*, 222–231.

Takase, H. M., & Nusse, R. (2016). Paracrine Wnt/beta-catenin signaling mediates proliferation of undifferentiated spermatogonia in the adult mouse testis. *Proceedings of the National Academy of Sciences of the United States of America*, *113*, E1489–E1497.

Takashima, S., Kanatsu-Shinohara, M., Tanaka, T., Morimoto, H., Inoue, K., Ogonuki, N., et al. (2015). Functional differences between GDNF-dependent and FGF2-dependent mouse spermatogonial stem cell self-renewal. *Stem Cell Reports*, *4*, 489–502.

Tokuda, M., Kadokawa, Y., Kurahashi, H., & Marunouchi, T. (2007). CDH1 is a specific marker for undifferentiated spermatogonia in mouse testes. *Biology of Reproduction*, *76*, 130–141.

Tokue, M., Ikami, K., Mizuno, S., Takagi, C., Miyagi, A., Takada, R., et al. (2017). SHISA6 confers resistance to differentiation-promoting Wnt/beta-catenin signaling in mouse spermatogenic stem cells. *Stem Cell Reports*, *8*, 561–575.

Trefil, P., Micakova, A., Mucksova, J., Hejnar, J., Poplstein, M., Bakst, M. R., et al. (2006). Restoration of spermatogenesis and male fertility by transplantation of dispersed testicular cells in the chicken. *Biology of Reproduction*, *75*, 575–581.

van Es, J. H., Sato, T., van de Wetering, M., Lyubimova, A., Nee, A. N., Gregorieff, A., et al. (2012). Dll1 + secretory progenitor cells revert to stem cells upon crypt damage. *Nature Cell Biology*, *14*, 1099–1104.

van Pelt, A. M., & de Rooij, D. G. (1990). Synchronization of the seminiferous epithelium after vitamin A replacement in vitamin A-deficient mice. *Biology of Reproduction*, *43*, 363–367.

Vernet, N., Dennefeld, C., Rochette-Egly, C., Oulad-Abdelghani, M., Chambon, P., Ghyselinck, N. B., et al. (2006). Retinoic acid metabolism and signaling pathways in the adult and developing mouse testis. *Endocrinology*, *147*, 96–110.
Vogel, G. (2016). Mouse eggs made in the lab. *Science*, *354*, 1521.
Wang, D., & Bodovitz, S. (2010). Single cell analysis: The new frontier in 'omics'. *Trends in Biotechnology*, *28*, 281–290.
Yokonishi, T., Sato, T., Katagiri, K., Komeya, M., Kubota, Y., & Ogawa, T. (2013). In vitro reconstruction of mouse seminiferous tubules supporting germ cell differentiation. *Biology of Reproduction*, *89*, 15.
Yoshida, S. (2012). Elucidating the identity and behavior of spermatogenic stem cells in the mouse testis. *Reproduction*, *144*, 293–302.
Yoshida, S. (2016). From cyst to tubule: Innovations in vertebrate spermatogenesis. *Wiley Interdisciplinary Reviews: Developmental Biology*, *5*, 119–131.
Yoshida, S. (2018a). Open niche regulation of mouse spermatogenic stem cells. *Development, Growth & Differentiation*, *60*, 542–552.
Yoshida, S. (2018b). Regulatory mechanism of spermatogenic stem cells in mice: Their dynamic and context-dependent behavior. In K. Kobayashi, T. Kitano, Y. Iwao, & M. Kondo (Eds.), *Reproductive & developmental strategies* (pp. 47–67). Tokyo: Springer.
Yoshida, S., Nabeshima, Y., & Nakagawa, T. (2007). Stem cell heterogeneity: Actual and potential stem cell compartments in mouse spermatogenesis. *Annals of the New York Academy of Sciences*, *1120*, 47–58.
Yoshida, S., Sukeno, M., & Nabeshima, Y. (2007). A vasculature-associated niche for undifferentiated spermatogonia in the mouse testis. *Science*, *317*, 1722–1726.
Yoshida, S., Takakura, A., Ohbo, K., Abe, K., Wakabayashi, J., Yamamoto, M., et al. (2004). Neurogenin3 delineates the earliest stages of spermatogenesis in the mouse testis. *Developmental Biology*, *269*, 447–458.
Zheng, K., Wu, X., Kaestner, K. H., & Wang, P. J. (2009). The pluripotency factor LIN28 marks undifferentiated spermatogonia in mouse. *BMC Developmental Biology*, *9*, 38.

CHAPTER EIGHT

Sperm success and immunity

Stuart Wigby[a,*], Susan S. Suarez[b], Brian P. Lazzaro[b], Tommaso Pizzari[a], Mariana F. Wolfner[b]
[a]Edward Grey Institute, Department of Zoology, University of Oxford, Oxford, United Kingdom
[b]Departments of Biomedical Sciences (SSS), Entomology (BPL), and Molecular Biology and Genetics (MFW), Cornell University, Ithaca, NY, United States
*Corresponding author: e-mail address: stuart.wigby@zoo.ox.ac.uk

Contents

1. Introduction 288
2. Sperm success and the male immune system 288
   2.1 Expensive ejaculates and trade-offs with immunity 288
   2.2 Antagonistic pleiotropy between sperm success and immune capacity 292
   2.3 Vertebrate male sperm auto-immunity 292
   2.4 Effects of infections in males, including STIs, on sperm success 293
3. Sperm success and the female immune system 294
   3.1 Female reproductive tract immunity effects on sperm success 295
   3.2 Immunity and sperm storage in females 299
   3.3 The role of female immunity in post-copulatory sexual selection: Immune-mediated cryptic female choice 301
4. Conclusions 304
Acknowledgments 306
References 306

Abstract

The moment of the fertilization of an egg by a spermatozoon—the point of "sperm success"—is a key milestone in the biology of sexually reproducing species and is a fundamental requirement for offspring production. Fertilization also represents the culmination of a suite of sexually selected processes in both sexes and is commonly used as a landmark to measure reproductive success. Sperm success is heavily dependent upon interactions with other key aspects of male and female biology, with the immune system among the most important. The immune system is vital to maintaining health in both sexes; however, immune reactions can also have antagonistic effects on sperm success. The effects of immunity on sperm success are diverse, and may include trade-offs in the male between investment in the production or protection of sperm, as well as more direct, hostile, immune responses to sperm within the female, and potentially the male, reproductive tract. Here, we review current understanding of where the biology of immunity and sperm meet, and identify the gaps in our knowledge.

Current Topics in Developmental Biology, Volume 135
ISSN 0070-2153
https://doi.org/10.1016/bs.ctdb.2019.04.002

1. Introduction

The production of sperm via spermatogenesis is well-understood from many perspectives (e.g., reviewed in Nishimura & L'Hernault, 2017). However, within the male, there is a potential trade-off between the energy demands required for making high-quality sperm and the demands of protecting the soma from infectious disease. Moreover, the female reproductive tract is a potentially hostile environment that contains molecules and cells that protect the female from infection but that may also be harmful to sperm. These defenses include a low pH and reactive oxygen species, an abundance of antimicrobial peptides, scavenging macrophages, and, in vertebrates, antibodies generated by the acquired immune system. Although they may be immunologically protective in intent, these chemical and cellular defenses have the potential to impair sperm viability or function (i.e., sperm success). Here, we discuss the interactions between sperm and the immune system, delineating areas in which function of one can impair function of the other (summarized in Fig. 1). The main themes we consider are (1) energetic trade-offs between investment by males in making sperm or by males and females in maintaining viable sperm versus investment in the immune system; (2) shared genetics, such as where specific genes have pleiotropic sperm- and immune-related functions; (3) the impact of infections in males and females, both somatic and sexually transmitted, on sperm; (4) immune responses directly against the sperm in males and females; and (5) the role of immunity in sperm selection by female choice.

2. Sperm success and the male immune system

In the next sections, we describe how relationships and interactions between sperm and the male immune system can be both harmful (Sections 2.1–2.4) and helpful (Section 2.4) to sperm success (Fig. 1).

2.1 Expensive ejaculates and trade-offs with immunity

Given the relatively small size of sperm cells in most species (among the smallest cells in the body) in comparison to ova (among the largest), the production of sperm was once thought to be energetically cheap (Gee, 1999). However, biologists have since begun to realize that (1) males are typically limited in their ejaculate production rate; (2) females often mate with multiple males, which leads to sperm competition, in which rival

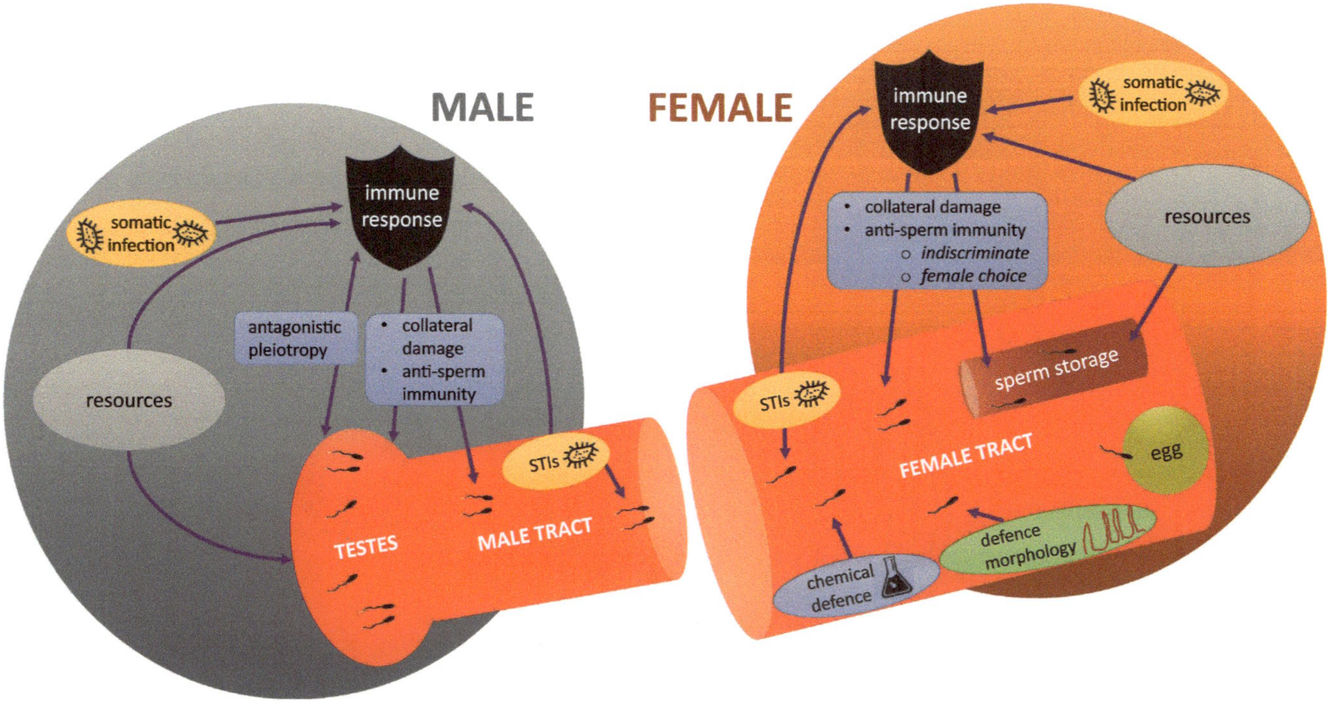

Fig. 1 Schematic of the interplay between immunity and sperm success in males and females. Resources may trade-off between the immune system and the production of sperm in males, and the protection of sperm in storage in females, leading to sub-maximal trait values. Trade-offs could also come about genetically: genes of shared male immune and sperm-related function could generate antagonistic pleiotropy, preventing simultaneous maximization of both traits. STIs can harm sperm directly, and thus immune defense against STIs could protect sperm. However, immune responses to STIs or somatic infections could also harm sperm as a side-effect of their defensive actions, causing collateral damage to sperm. Immune responses may negatively impact sperm directly, particularly in vertebrates via anti-sperm antibodies, both in females, and in males as an auto-immune response. Female immune responses may indiscriminately harm sperm, which could contribute to infertility, or they may form part of adaptive female choice, selecting out the sperm of disfavored males after mating, but before fertilization. Within the female reproductive tract sperm interact with the chemical and morphological environment—which is likely be designed to inhibit pathogen proliferation, and may help or hinder the sperm—on the way to sperm storage sites and fertilization.

ejaculates compete for access to eggs; and (3) sperm competition favours ejaculates with larger numbers of sperm (and potentially more seminal fluid) to increase their probability of winning fertilizations (Birkhead & Pizzari, 2002; Cameron, Day, & Rowe, 2007; Dewsbury, 1982; Parker, 1970; Perry, Sirot, & Wigby, 2013; Wigby & Chapman, 2004). These findings have led to the theory that males should strategically allocate sperm and seminal fluid across different matings in order to maximize their reproductive returns (Cameron et al., 2007; Parker, 1990; Parker et al., 1996). This newer perspective has now gained extensive empirical support from studies across a range of vertebrate and invertebrate species (Garbaczewska, Billeter, & Levine, 2013; Hopkins, Sepil, & Wigby, 2017; Kelly & Jennions, 2011; Sirot, Wolfner, & Wigby, 2011; Wigby et al., 2009). To increase their reproductive success, males (particularly in polyandrous or polygynandrous species) must devote considerable resources into developing large testes and accessory glands, and into using those organs to produce large volumes of sperm and seminal fluid. Together, these factors point toward non-trivial costs of ejaculate production which, in extreme cases, can lead to situations where females become sperm-limited because they fail to receive sufficient sperm from a single mating to fertilize all their eggs (Wedell, Gage, & Parker, 2002).

The fundamental assumption of life-history theory is that resources allocated to one trait will trade-off against those needed for other traits (Zera & Harshman, 2001). Ejaculates are no exception: the cost of producing them is expected to decrease resources available for other important life-history traits. The idea of such trade-offs has received empirical support from studies that suggest that investment in ejaculates compromises traits involved in attracting mates, such as ornaments or weapons (Parker & Pizzari, 2010; Simmons & Emlen, 2006; Simmons, Lüpold, & Fitzpatrick, 2017). Nevertheless, some of the most compelling evidence for trade-offs involving ejaculates comes from negative correlations between ejaculate quality and somatic immunity. The idea of an antagonistic relationship between male sexual and immunological physiology has been widely recognized in evolutionary biology, since Folstad and Karter (1992) suggested the "immunocompetence handicap hypothesis." In vertebrates, testosterone is vital for the development of male reproductive traits, but it also suppresses immunity. The immunocompetence handicap hypothesis posits that these antagonistic effects could make male sexual characteristics reliable indicators of general male genetic quality or condition—and thus targets for sexually selected female preferences—because high levels of sexual characteristics provide evidence that those males can cope with testosterone-mediated immune suppression. A similar logic can apply to primary male sexual characteristics, whereby high testosterone levels should be associated with large testes and high-quality ejaculates, yet result in reduced immune function. In support of these ideas, negative relationships have been found between the size of the testes and the size of the spleen—which has important immune functions—in bats, Cape ground squirrels *Xerus inauris,* and Arctic char *Salvelinus alpinus* (Hosken & O'Shea, 2001; Liljedal, Folstad, & Skarstein, 1999; Manjerovic & Waterman, 2012; Roberts, Buchanan, & Evans, 2004). These data are broadly consistent with a testosterone-mediated ejaculate-immunity trade-off (assuming that testes and spleen size are reliable indicators of general testosterone and immunity levels, respectively). Nevertheless, the data do not uniformly support the concept of costly ejaculates. While a positive relationship between testes size and lice burden was found in the Cape ground squirrel (Manjerovic & Waterman, 2012), nematode infection (also an indication of depressed immunity) was negatively associated with ejaculate quality in arctic char (Liljedal et al., 1999). The complete picture is therefore one in which the consequences of investment in ejaculates for realized resistance to parasites may vary according to the taxa of the host and/or parasite.

The potential for trade-offs in males between immunity and the ejaculate is not limited to vertebrates. Studies on insects, which lack testosterone, also provide evidence in support of the immunocompetence handicap hypothesis. For example, removal of sexual selection in dung fly *Scatophaga stercoraria* populations, via experimental evolution under enforced monogamy, results in the evolution of decreased testis size, but increased phenoloxidase activity (Hosken, 2001). Since phenoloxidase is an important component of the insect innate immune system, these results are consistent with a testes-immunity trade-off. In field crickets *Teleogryllus oceanicus*, negative genetic and phenotypic correlations between sperm viability and lysozyme activity (Simmons & Roberts, 2005), and a diet-dependent reduction in viability under immune challenge (Simmons, 2011), lend support to the existence of within-individual trade-offs in resource allocation. Again supporting such trade-offs, in decorated crickets, *Gryllodes sigillatus*, immune challenged males make smaller spermatophores (which are discrete, sperm-containing ejaculate packages), while inducing elevated spermatophore production experimentally reduces immune capacity (Gershman et al., 2010; Kerr, Gershman, & Sakaluk, 2010). The commonality of negative relationships between measures of immunity and ejaculate quality suggests a widespread trade-off between these traits, with compromised immunity being a cost of producing a high quality, competitive ejaculate. Nevertheless, the molecular-genetic mechanisms that underpin these potential trade-offs remain largely opaque.

2.2 Antagonistic pleiotropy between sperm success and immune capacity

Another reason why sperm success and male immune capacity may show a negative association is because of constraints on fitness maximization imposed by genetic architecture. Antagonistic pleiotropy occurs when a single gene affects multiple traits related to fitness, and some allelic variants promote fitness through one trait while compromising fitness through others. Antagonistic pleiotropy could prevent males from simultaneously reaching individual optima for sperm function and immunity if genes have shared immunity and sperm-related functions. An example of this may be Thioester-containing Protein 1 (TEP1), which contributes to immune defense against malaria parasites in the mosquito *Anopheles gambiae* (Blandin et al., 2004). The same protein is also involved in the removal of aberrant sperm in the testes, which is important for ensuring male fertility. Remarkably, the allelic variant that promotes male fertility also renders mosquitoes more susceptible to malaria (Pompon & Levashina, 2015). If such phenomena are widespread, sperm-immunity trade-offs could arise through shared genetic function and antagonistic pleiotropy, as opposed to through resource allocation. This particular example may also provide evidence of genetic antagonism between the sexes, because only female *A. gambiae* are exposed to malaria parasites via blood-feeding, since males do not feed on blood.

2.3 Vertebrate male sperm auto-immunity

Roughly 1000 genes in the mouse genome have been detected as being active primarily in germ cells in the testis (Miyata et al., 2016). The protein products of many of these genes have been detected on the surface of sperm, serving as potential antigens (Tokuhiro et al., 2012). The existence of sperm-specific molecules creates the possibility that, in organisms with acquired immunity (i.e., vertebrates), males could generate antibodies against their own sperm, leading to infertility by targeting their own developing or mature sperm cells. Male vertebrates possess a range of traits to prevent the development of anti-sperm auto-antibodies. These traits include the blood-testis barrier, which prevents the immune system from encountering germ cell-specific antigens that arise during spermatogenesis (Cheng & Mruk, 2012). If this system breaks down, anti-sperm antibodies can be produced, causing immune infertility (Bohring & Krause, 2003). The "sperm protection hypothesis" suggests that immunosuppression by testosterone directly benefits sperm by protecting them from the male's immune system (Hillgarth, Ramenofsky, & Winfield, 1997). A meta-analysis of human studies lends empirical support to this theory, as immunosuppressive corticosteroid treatments reduce anti-sperm antibodies and improve sperm performance (Skau & Folstad, 2005). More generally, selection may act to balance immunocompetence versus the risk of autoimmunity to sperm and potentially other cells. Such compromises are expected to be particularly acute in vertebrates given their acquired immune systems. Invertebrates lack any highly-specific self/non-self-recognition mechanisms, making it seem unlikely that they would recognize sperm as "foreign." Nevertheless, invertebrates are subject to the general problem of autoimmunity (Schmid-Hempel, 2005), although it is unclear whether sperm would be a particular autoimmune target.

2.4 Effects of infections in males, including STIs, on sperm success

The previous sections have presented a number of ways in which a male's immune system could negatively impact sperm success. The male immune system can also promote sperm success by fighting against sperm-damaging sexually transmitted infections (STIs). STIs are relevant because—by their nature—the causative pathogens are often present in the reproductive organs and thus in close proximity to sperm. STIs have been most studied in humans, and in other animals of economic importance, but evidence suggests that they are taxonomically widespread (Knell & Webberley, 2004; Lockhart, Thrall, & Antonovics, 1996). Some STIs seem to be particularly harmful to sperm. These sperm-damaging STIs include viruses present in human semen, including papillomavirus and the human immunodeficiency virus (HIV) (Dejucq & Jégou, 2001; Lai et al., 1997; Le Tortorec & Dejucq-Rainsford, 2010). Urogenital infections in men can decrease male fertility through various mechanisms. Infections may disrupt local immune regulation and thus be associated with the development of auto-antibodies to sperm, such as those described in Section 2.3. Infections may also cause an obstruction preventing the normal passage of the ejaculate, or may disrupt the function of male accessory glands and thus reduce the quality or quantity of seminal fluid. Additionally, pathogens may damage sperm directly, or via inflammation induced by the infection (Schuppe et al., 2017; Wald, 2005). Some researchers have hypothesized that STI-induced infertility in either sex is an adaptive strategy of STIs (Apari, de Sousa, & Müller, 2014). This argument is based on the idea that sperm success, resulting in fertilization (and in mammals, pregnancy), often contributes to the inhibition of female sexual activity for a period of time. Given that sexual activity is required for STIs present in females to reach new hosts, preventing fertilization can elevate sexual activity and thus STI transmission opportunities. Alternatively, if STIs reduce female polyandry—perhaps by reducing overall health—then males could potentially gain fitness from infecting their mates, if the reduction in sperm competition outweighed the various costs of the STI (Knell & Webberley, 2004; Wardlow & Agrawal, 2018). Similarly, if infection with an STI elevates short-term female reproductive output, benefitting the male that delivers the infection, then males can theoretically evolve lower resistance to STIs (Johns, Henshaw, Jennions, & Head, 2019). However, considerably more work is required to establish whether STI-mediated effects on fertility or behavior represent adaptive strategies for pathogens or their hosts, or are simply side-effects of other actions that are under selection.

Deleterious effects of infections on sperm success need not be local to the reproductive organs. Systemic infections are likely to reduce the overall condition of a male and thus diminish the pool of resources available for producing high quality sperm. As mentioned in Section 2.1, immune-challenged decorated cricket males make smaller spermatophores (Kerr et al., 2010), consistent with an immune-mediated reduction in condition that, in turn, impacts sperm. Negative effects of immune system activation on sperm function have also been seen in great tits *Parus major* (Losdat et al., 2011) and guppies *Poecilia reticulata* (Devigili et al., 2017), lending further support to the idea. However, a similar test in the house sparrow *Passer domesticus*, while demonstrating a negative impact of infection on testosterone, failed to detect harm to sperm quality (Needham et al., 2017). Thus, whether somatic infections negatively impact sperm is likely specific to the host or parasite species, or dependent on the strength of infection.

3. Sperm success and the female immune system

In the next sections, we describe how interactions between sperm and the female immune system can potentially be harmful to sperm (Sections 3.1 and 3.2) but also can be important in determining which particular sperm are successful (Section 3.3) (Fig. 1).

3.1 Female reproductive tract immunity effects on sperm success

The organs of the female reproductive tract are subject to being colonized by pathogens and can also provide a pathway into the body cavity, where pathogen proliferation can wreak havoc, causing major health and fitness consequences (Soper, 2010). The female therefore has multiple adaptations to impede the invasion and proliferation of pathogens in the reproductive tract, or their ascent within it. Physical impediments in the female tract include (1) production of a cleansing outward flow of fluid (in mammals, at least; Mullins & Saacke, 1989), (2) secretion of a thick (viscoelastic) mucus (also in mammals, at least; Taherali, Varum, & Basit, 2018, (3) construction of narrow labyrinths of passages (seen in Muro et al., 2016; Suarez, Brockman, & Lefebvre, 1997; Yaniz et al., 2014). Chemical impediments include (4) acidification of the vaginal fluid in mammals to a level that kills bacterial pathogens (O'Hanlon, Moench, & Cone, 2013) and may also damage sperm. Interestingly, this acidity is known to be produced by bacteria resident in the vagina, particularly *Lactobacillus* spp. (O'Hanlon et al., 2013). (5) Last, and our primary focus, are immunological impediments to pathogen survival. These can include components of the innate immune system (including inflammatory responses, reactive oxygen species, and antimicrobial peptides) that could potentially damage sperm. In addition, because sperm, like pathogens, contain proteins (i.e., antigens) that are foreign to the female, vertebrate females have the potential to develop antibodies to sperm proteins. Potentially, these various types of impediments, presumably evolved for use against pathogens, could block the survival of sperm or their ability to reach the site of fertilization.

Sperm are more adept at circumventing many of these impediments than are most pathogens. First, due to the shape of the sperm and their mechanism of swimming, hydrodynamic interactions of mammalian sperm with the fluid flow in the female tract enable the sperm to orient and swim "upstream" in the tract, toward the site of fertilization (Miki & Clapham, 2013; Tung et al., 2015a, 2015b). Muscle and organ movements in the tract can also facilitate the movement of sperm toward storage sites (in *Drosophila;* Adams & Wolfner, 2007; Avila & Wolfner, 2009; Mattei et al., 2015; and reviewed in Suarez & Pacey, 2006). The evolution of shapes, energetics, or swimming behaviors that allow individual sperm to move faster also include mechanisms by which sperm can cooperate (reviewed in Pizzari & Foster, 2008; Immler, 2008), and in some cases bundle together and swim faster than single sperm (e.g., hooks on deer mouse *Peromyscus maniculatus* sperm head; Moore, Dvorakova, Jenkins, & Breed, 2002). Second, mammalian sperm with normal motility and morphology readily swim through the mucus that fills the cervix of estrous females, as demonstrated in cattle and humans (Anilkumar et al., 2001; Katz, Slade, & Nakajima, 1997). This ability is based, at least in part, on physical hydrodynamic interactions between sperm and the viscoelasticity of the cervical mucus (Katz et al., 1997). Further along the tract, mucous secretions have been detected in the lumen of the oviduct (Jansen, 1980; Jansen & Bajpai, 1982; Suarez et al., 1997). Transillumination of mouse oviducts reveals that sperm swim quickly within the lumen, despite the presence of the secretions (Suarez, 1987). Third, again in mammals, there is evidence that some labyrinth-like structures that pose physical impediments to most pathogens actually serve as guides to sperm. For example, the microgrooves that line the cervical canal in cattle can provide privileged pathways through the cervix for sperm. These pathways cannot be accessed by some pathogens that cause STIs, such as trichomoniasis (Tung et al., 2015a). Fourth, while sperm are sensitive to destruction by the acidity of the fluid of the vagina, they are protected by pH buffers in the seminal plasma (Fox, Meldrum, & Watson, 1973; O'Hanlon et al., 2013; Owen & Katz, 2005). Additionally, male-derived seminal proteins can help sperm reach and enter (*Drosophila*) or bind to (bovine) sperm storage sites (Avila & Wolfner, 2009; Gwathmey, Ignotz, & Suarez, 2003; Gwathmey et al., 2006; Neubaum & Wolfner, 1999; Bloch Qazi & Wolfner, 2003). In mammals, either the sperm leave the vagina rapidly by swimming into the more neutral mucus in the cervical canal (e.g., humans, cattle; Sobrero & Macleod, 1962) or the male bypasses the vagina altogether and deposits semen directly into the cervix or uterus (e.g., rodents, pigs; Bedford & Yanagimachi, 1992; Hunter, 1981). Finally, in some taxa, males provide their own physical protection for sperm: either placing them within the female inside a spermatophore sac, or covering the sperm with a "coat" that is removed later, during storage (Friedländer & Gitay, 1972; Wedell, 1993).

In producing immune responses, the female is faced with a balance between protecting herself against pathogen invasion versus potentially compromising fertilization. The responses of the innate immune system, including inflammatory responses, rapidly generate protection against pathogens and serve as a first line of defense. This system is activated by small molecular motifs conserved in pathogens, such as bacterial lipopolysaccharides and peptidoglycans, and depends on receptors present on immune cells, particularly Toll-like receptors in mammals and peptidoglycan recognition proteins in insects (e.g., reviewed in Dziarski & Gupta, 2006; Kurata, 2014; Sheldon, Owens, & Turner, 2017). Such molecular motifs are not present on the surface of sperm, but bacteria are commonly introduced to the female tract by the male during mating (Cottell et al., 2000; Schulze et al., 2018; Sidaway, 2016) and could trigger an innate immune response. In addition to inflammatory responses triggered by incidentally introduced microbes, cells of the *Drosophila* female tract, as well as mucosal epithelial cells and neutrophils of the mammalian female tract, produce anti-microbial peptides that can kill some bacteria and viruses (Amjadi et al., 2014). In *Drosophila* at least, further production of these anti-microbial peptides is induced by mating (Kapelnikov et al., 2008; McGraw, Clark, & Wolfner, 2008), which may serve as a first line of defense for the sperm against microbes that may be present in the female reproductive tract, or introduced during mating. However, recent data have indicated that the vagina, uterus, and oviduct of mammals all contain their own resident, and presumably largely commensal, bacterial biomes. Furthermore, sequencing of 16S rDNA from these microbiomes indicates that the composition of bacterial species in the vagina, uterus, and oviduct differ, despite sharing several species (Chen et al., 2017; Moreno & Franasiak, 2017). The effect of these resident bacteria on immunological activity and fertility is not well understood.

In vertebrates, the female tract can also activate acquired immune responses, which enlist specific antibodies in the defense against pathogens. As in males, females can develop anti-sperm antibodies, but in females the sperm surface proteins in the reproductive tract are non-self and thus clear potential antigens (Tokuhiro et al., 2012). Although sperm-surface antigens could theoretically always provoke an immune response, only 2–3% of women are estimated to produce anti-sperm antibodies (Clark & Schust, 2013). However, in this subset of women anti-sperm antibodies have long been identified as a likely cause of infertility (Dondero et al., 1993; Meaker, 1922). A higher frequency of immunoglobulins IgG and IgA is found on cervicovaginal secretions and sera of women with fertility problems in at least some populations, providing further evidence of their potential role in infertility (Mahdi et al., 2011; Lu, Huang, & Lu, 2008; but see Vujisić et al., 2005). In estrous cervical mucus, anti-sperm antibodies are usually at levels far below those in the bloodstream (Stern et al., 1992). However, mucus samples containing anti-sperm antibodies are known to inhibit movement of sperm (Chiu & Chamley, 2004), suggesting a potential infertility mechanism.

Involvement of acquired immune responses in differential fertilization success is also well established in birds. In domestic fowl, *Gallus domesticus*, female exposure to sperm leads to an immune response, which can reduce the fertilizing efficiency of inseminations. For example, experimental injection of testicular material into hens depresses the fertility of females for the following 12–67 days (McCartney, 1923). This response to sperm exposure is associated with rising levels of sperm antibodies in blood plasma and specifically in serum levels of sperm agglutinins, which modulate sperm agglutination in vitro (Hosoda, Abe, & Otsuka, 1964; Wentworth & Mellen, 1963; see Haley & Abplanalp, 1970 and Burke, Rieser, & Shoffner, 1971 for similar patterns in turkeys, *Meleagris gallopavo*, and Japanese quails, *Coturnix japonica*, respectively). However, it remains unclear whether these same patterns of agglutination occur in vivo, or whether agglutination harms sperm success.

Avian IgY (the equivalent of mammalian IgG) appears to play a particularly important role in female anti-sperm responses in birds. For example, an excess of sperm recovered from the vagina of domestic fowl hens following insemination are bound by IgY compared to sperm recovered from the distal end of the female reproductive tract (the infundibulum), where fertilization occurs, suggesting that IgY may be implicated in targeting sperm (Steele & Wishart, 1992). Interestingly, antibodies bound to sperm are also found in experimentally inseminated hens who had never before mated, suggesting that immunity against other antigens (e.g., pathogens) may have cross-reactivity, causing sperm targeting (Bakst, 2011). Similarly, turkey hens with low fertility expressed IgY on the epithelium of their sperm storage tubules, a site that is typically immunosuppressed (Bakst, 2011; Das, Isobe, & Yoshimura, 2008) and where IgY is not expressed in hens with high fertility (Kirk et al., 1989). In some lines of domestic fowl, female infertility is associated with infiltration of lymphocytes in the sperm storage tubules as well as high numbers of antigen-presenting cells (Ia + cells), and T cells (CD4 +, CD8 +) (Das, Nagasaka, & Yoshimura, 2005), while in Japanese quails, insemination is followed by a rapid increase of leukocytes (neutrophils and lymphocytes) in the utero-vaginal junction, some of which are in close contact with the inseminated sperm, suggesting that female immune reactions could negatively affect sperm following mating (Higaki et al., 1995).

To avoid the problem of anti-sperm antibodies causing infertility, it seems likely that the innate immune system is especially active as a defensive system in the female tract of mammals, as it is in insects such as *Drosophila.*

The male-derived strategy of moving sperm rapidly out of the vagina, or bypassing the vagina altogether in species such as bedbugs, where males inseminate directly into the body cavity (but see Reinhardt, Naylor, & Siva-Jothy, 2003), is also thought to help evade local female immune responses (Suarez, 2015). Combined with the deposition of large numbers of sperm this strategy could also increase the chance that some sperm will survive those attacks. In some insect species, such as *Drosophila pseudoobscura* and some lepidopterans, the male introduces two types of sperm into the female: "eusperm," which are functional for fertilization, and the more numerous "parasperm," which are thought to act as decoys against harmful molecules in the reproductive tract and perhaps titrate out such molecules (Holman & Snook, 2006, 2008; Swallow & Wilkinson, 2002). Mammals also appear to use a "strength-in-numbers" strategy: large numbers of inseminated sperm reach the mammalian uterus, and from there thousands or tens of thousands pass into the lower oviduct (Overstreet & Cooper, 1978; Suarez et al., 1997). In response to the large numbers of sperm that enter the uterus, an innate immune response eventually builds up in the uterine cavity. Large numbers of neutrophils enter the cavity to trap and kill sperm by extruding DNA extracellular traps (Alghamdi & Foster, 2005; Zambrano et al., 2016) and phagocytosing sperm (Hong et al., 2017). In this way, the neutrophils eventually clear sperm from the uterine cavity, helping to prepare the uterus for implantation (Troedsson et al., 2001).

3.2 Immunity and sperm storage in females

Although sperm-use often begins relatively rapidly after copulation, in many species sperm can be stored for some period before use—typically days or weeks—but in some species up to years (reviewed in Orr & Brennan, 2015; Orr & Zuk, 2012; Schnakenberg, Siegal, & Bloch Qazi, 2012). In particular, females of invertebrate taxa, such as insects, commonly store sperm in specialized organs, such as seminal receptacles and spermathecae, that bud off the main conduit in the female tract, and may provide protected sites for the sperm. However, many female mammals also contain sperm reservoirs, where sperm can be maintained alive prior to ovulation, for up to a few days in most species (up to 5 days in humans; Holt & Fazeli, 2016), or as long as throughout the winter during hibernation in the some bats (Racey et al., 1987; Suarez & Pacey, 2006). In storage reservoirs, sperm speed and numbers would not likely be sufficient to protect sperm against immune responses. Given that the female immune system has the potential to destroy sperm, storage organs must therefore protect sperm from the female's own immunity, presumably by being areas of low immunological activity (reviewed in Orr & Brennan, 2015). Although mammalian sperm are held in the storage reservoir of many (if not all) species by binding to the mucosal epithelium, and thus are exposed to fluid contents of the female tract, here they seem protected against attack by the innate immune system. In cattle, binding of sperm to oviductal mucosal epithelium in vitro upregulates expression of anti-inflammatory cytokines (*TGFB1* and *IL10*) in the epithelial cells (Yousef et al., 2016) which may help to protect sperm from harmful immune action. Still, much remains to be learned about how the immune system may be repressed in the oviduct during sperm storage and how epithelium binding helps maintain vitality of the sperm (Suarez, 2016).

Several studies on insects provide evidence of interplay between female immunity and sperm in storage. Experiments using the field cricket *T. oceanicus* provide evidence that experimental bacterial infections in females reduce the viability of sperm in storage (McNamara, van Lieshout, & Simmons, 2014), while in *D. melanogaster*, stimulation of the immune system using a pathogen mimic (peptidoglycan) results in reduced viability of stored sperm (Radhakrishnan & Fedorka, 2012). Effects of infections on stored sperm viability could arise through one of several mechanisms. First, the impact of infection on stored sperm could represent a resource trade-off in female investment between immune responses versus investment in maintaining the viability of sperm in storage. However, in *D. melanogaster*, sperm survive well in spermathecae dissected out of the female body. Given that removed organs are cut-off from somatic resources this finding argues against a trade-off (Radhakrishnan & Fedorka, 2012). Second, it may be that the bacteria interact with—and harm—the sperm directly. In vitro studies using bed bug *Cimex lectularius* sperm show that incubation with environmental microbes damages sperm (Otti, McTighe, & Reinhardt, 2013), raising the potential they may also be able to do so in vivo. Finally, sperm may be harmed as a side-effect of the female's immune response, for example, if the by-products are especially toxic to sperm (McNamara et al., 2014; Radhakrishnan & Fedorka, 2012). Males may thus act to suppress the female immune system. Evidence in the cricket, *Allonemobius socius*, is consistent with the idea that the male ejaculate interferes with the female immune system, potentially to help prevent sperm from being harmed (Fedorka & Zuk, 2005). Studies of *D. melanogaster* have also revealed an increase in the frequency of empty sperm storage organs in females that had mated to males whose genitalia had been dipped in peptidoglycan, which thus might approximate a male infected with an STI (Radhakrishnan & Fedorka, 2012). These data suggest that females eject the sperm of these "infected" males from their storage organs, perhaps to remove dead sperm or avoid further infection, raising the possibility that STI-infected males suffer an additional loss of sperm success via female rejection of their sperm from storage.

Comparable patterns have been seen in other insect species. A negative relationship between sperm storage and immunity occurs in a species with extreme long-term sperm storage: the ant, *Atta colombica*. *Atta queens* can live for decades, and it is thought that their reproductive success is at least partly limited by the ability to store sufficient numbers of sperm, which are required to fertilize eggs to produce workers for the colony. In this species, females found with more sperm in storage, and who store the sperm of more males, display a lower encapsulation response (the surrounding of pathogens and parasites with layers of dead melanized hemocytes), which is consistent with costs arising from storing sperm in general, and from encountering multiple male genotypes (Baer, Armitage, & Boomsma, 2006). However, being a correlational result, it is also possible that females who display lower encapsulation responses tend to mate with more males and store more sperm. Either way, the work highlights that sperm storage and immunity are often intimately—and negatively—associated.

3.3 The role of female immunity in post-copulatory sexual selection: Immune-mediated cryptic female choice

The high potential for female immune responses to sperm, as outlined above, presents the opportunity to generate post-copulatory sexual selection. As described earlier, where females mate with more than one male (which is very common; reviewed in Taylor, Price, & Wedell, 2014), the sperm of different males can compete for fertilization opportunities in sperm competition (Parker & Pizzari, 2010; Wigby & Chapman, 2004). However, this situation also creates the conditions necessary for cryptic female choice (CFC) whereby females generate a systematic bias in the paternity share of competing males (Eberhard, 1996; Firman et al., 2017), and there are good reasons to believe that CFC can occur via the female immune response. Here, we briefly discuss: (1) the scope for immune-mediated CFC across taxa, (2) the adaptive significance and functional causes of immune-mediated CFC, (3) underpinning immunological mechanisms, and (4) evolutionary consequences.

The scope for female immunity to drive CFC is strongly dependent on taxon and reproductive system. In external fertilizers, where partner interactions are limited or altogether absent, female immunity is expected to have a limited role in post-copulatory sexual selection, although immune factors on the egg could potentially influence fertilization patterns via differential sperm chemotaxis (Eisenbach & Giojalas, 2006; Evans et al., 2012). Female immunity is expected to be more important in internal fertilizers, particularly those with prolonged female sperm storage and acquired immune responses where responses have the potential to be dynamic and change over time.

Three non-mutually exclusive scenarios could drive immune-mediated CFC. Immune-mediated CFC may represent an adaptive female strategy for (1) her own viability, by neutralizing ejaculates that contain pathogens, or in response to ejaculate traits with harmful effects on females (Morrow & Innocenti, 2012) or (2) increasing the fitness of her offspring, by selecting sires that maximize offspring viability or reproductive success. Alternatively (3) immune-mediated CFC could be a non-adaptive by-product of some other naturally-selected function, e.g., resistance against certain pathogens/parasites may generate an immune predisposition against the sperm of certain male phenotypes. A bias in paternity share is selected for only in the second hypothesis, while it is a non-adaptive consequence in the other two.

One class of genes that has been hypothesized to play a particularly important role in immune-mediated CFC is the Major Histocompatibility Complex (MHC), also known as Human Leukocyte Antigen (HLA) complex in humans. The MHC plays a fundamental role in vertebrate acquired immunity by enabling antigen presentation to T cells and self versus non-self recognition. The extreme polymorphism of the MHC has long attracted evolutionary explanations, including a role for sexual selection (Edwards & Hedrick, 1998). In principle, females may be able to optimize the MHC of their offspring by biasing fertilization in favor of sperm or ejaculates with certain MHC haplotypes (Firman et al., 2017; Milinski, 2006; Ziegler, Kentenich, & Uchanska-Ziegler, 2005). Consistent with this idea, Zheng et al. (2001) found that insemination upregulated MHC class II genes, in certain regions of the female reproductive tract of domestic fowl hens. MHC class II genes code for molecules involved in the recognition and presentation of extracellular antigens by antigen-presenting cells, and thus represent the class of MHC that one would expect to be involved in recognition of sperm antigens. Løvlie et al. (2013) found that female red junglefowl, *Gallus gallus*, retain more sperm following natural mating by males that are less MHC-similar, suggesting that female fowl select against the sperm of genetically related males to avoid inbreeding, as indicated by previous work (Pizzari, Løvlie, & Cornwallis, 2004). However, further research has failed to find strong evidence that MHC similarity plays an important role in female sperm selection above and beyond inbreeding avoidance (e.g., Collet, 2010). The idea of MHC-based sperm selection finds further empirical support in the externally-fertilizing three-spined stickleback, *Gasterosteus aculeatus,* where fertilization is biased in favor of combinations of gametic MHC haplotypes that maximize offspring viability (Lenz et al., 2018). However, in whitefish, *Coregonus* sp., there is no evidence for MHC effects (Wedekind et al., 2004). One possible explanation of the discrepancy between the Lenz et al. (2018) and Wedekind et al. (2004) studies is that the latter considered MHC genotypes of male and female partners, in contrast to variation in MHC haplotypes of individual gametes, which was used in the former study. An effect of individual MHC sperm haplotypes consistent with the results of Lenz et al. (2018) would require MHC haploid expression in mature sperm cells, although evidence for this remains limited. Thus, the importance of MHC in immune-mediated CFC requires considerably more investigation.

Regardless of underpinning mechanisms, patterns of sperm selection via female immunity may have far-reaching evolutionary implications. Males are expected to evolve strategies to escape female immune responses. For example, it is possible that the abundance of seminal fluid proteins with antimicrobial properties detected in the male seminal fluid of different species (Borziak et al., 2016; Dorus, Skerget, & Karr, 2012) may be partly explained by males attempting to lower the microbial load of an ejaculate in order to reduce the risk of triggering female immune reactions. Other strategies may include masking the antigenicity of the sperm surface through components of the glycocalyx, the matrix of hundreds of glycoproteins and glycolipids that encapsulates the entire sperm cell and plays a key role in sperm transport within the female reproductive tract. The sperm glycocalyx is rich in sialic acids, which tend to occupy the outermost part of the sperm surface. In mammals, highly sialylated glycoproteins such as beta defensins can play a key role for protecting sperm against female antibody binding (Yudin et al., 2005) and phagocytosis by macrophages (Ma et al., 2016). Similarly, in domestic fowl, experimental desialylation leaves sperm functionally intact in vitro but prevents them from moving through the female vagina following insemination (Steele & Wishart, 1992). These patterns indicate that sialic decoration of the sperm glycocalyx may have evolved at least partly as a male adaptation to evade anti-sperm female immune responses (Ma et al., 2016).

These tight coevolutionary immune interactions between sperm and female reproductive tract may in turn contribute post-copulatory pre-zygotic reproductive isolation and ultimately speciation. For example, Ghaderi et al. (2011) recently presented compelling evidence of prezygotic reproductive barriers mediated by female immune responses against the sperm glycocalyx in mammals. Humans lost the ability to produce a type of sialic acid common in other mammals, the *N*-glycolylneuraminic acid (Neu5Gc), approximately 3 million years ago, and are now capable of producing Neu5Gc-specific antibodies. Female mice immunized to express anti-Neu5Gc antibodies suffer from lower fertility when mated with transgenic Neu5Gc-positive males. Moreover, human anti-Neu5Gc antibodies target Neu5Gc antigens on the sperm of chimpanzees, *Pan troglodytes*, and kill these sperm. Thus, female immunity against Neu5Gc-carrying sperm may have played an important role in human evolution by creating reproductive isolation in ancestral polymorphic hominin populations (Ghaderi et al., 2011). Female immune responses to sperm have also been advocated to explain patterns of post-copulatory pre-zygotic reproductive isolation in bird populations, such as between Japanese quails and domestic fowl, mediated by female differential sperm agglutination (Haley & Abplanalp, 1970), and between collared flycatchers, *Ficedula albicollis*, and pied flycatchers, *F. hypoleuca*, mediated by acquired immunity (Cramer et al., 2016).

Collectively, these empirical results indicate that female anti-sperm antibody responses have the potential to systematically bias the outcome of sperm competition, with potentially far-reaching ramifications for inter-sexual coevolution and speciation. These responses, however, are likely to be complex and temporally dynamic, and to date, the extent of immune-mediated CFC remains to be defined.

4. Conclusions

It is clear that sperm success and immunity are linked in varied ways, from the onset of spermatogenesis in the male, through to the moment of fertilization in the female (in internal fertilizers). However, our review has identified many gaps that remain in our current knowledge. For example, although several lines of evidence suggest that males often face resource trade-offs between investing in immunity versus investing in sperm production the mechanisms underpinning resource trade-offs remain opaque. One potentially fruitful future approach may be to use isotope labeling of molecular components of food, for example, heavy lysine (de Godoy, 2014), and track where those molecules end up. By quantifying the amount of consumed material in testes versus immunoactivity organs, it may be possible to test the idea that resource use by sperm and the immune system is a zero-sum game. A similar approach may help us better understand the costs of sperm storage in females.

Although testosterone has long been invoked as a potential hormonal mediator of the immunity versus sperm trade-off in vertebrates, our understanding in invertebrates lags behind. Juvenile hormone is a broad scale mediator of immunity and reproduction in insects (Schwenke, Lazzaro, & Wolfner, 2016), yet its relevance for direct impacts on sperm remains little studied, and ripe for investigation. Similarly, whereas anti-sperm immunity in both females and males clearly plays a key role in fertility in vertebrates, we know very little about whether immune systems directly impact sperm in invertebrates. Again, this is an area ripe for investigation, with studies at cellular level likely to provide us the biggest advances in knowledge.

Antagonistic pleiotropy, or more broad negative epistatic effects between genes involved in immunity and sperm, remains almost entirely the subject of speculation, with very little data to go on. As our general functional genetic understanding of model organisms improves, we may stumble upon examples, but the field lacks the same investment as has been seen in aging, where 100s of candidates of antagonistic pleiotropy between senescence and early life-benefits have been identified (Austad & Hoffman, 2018). Even a much more modest effort to understand the shared genetics of immunity and male sperm success would provide a considerable new insight into the mechanisms linking immunity and sperm, with potentially large benefits to the fields of disease and reproductive biology.

STIs clearly impact fertility in many cases, often via harmful effects on sperm. A key current question is whether STIs harm sperm as an adaptive strategy, to promote their own transmission, or merely as a side effects of other aspects of their biology. Answering this question will likely require development of mathematical theory (e.g., Johns et al., 2019) as well as careful empirical experimentation, particularly in short-lived lab systems that allow observation of experimental evolution.

It is certain that the immune systems of both the female and male present a risk to sperm, particularly in species with acquired immunity. It is perhaps remarkable that males and females avoid complete destruction of the sperm and remain fertile. Although we have some understanding of how females and males protect sperm from their own immune system, much remains to be discovered about sperm protection and about what makes the mechanisms break down, leading to infertility. The intriguing possibility that immune targeting of sperm could be co-opted by females to discriminate among the sperm of different males has far-reaching ramifications for evolution. Progress in identifying the discrimination among specific sperm phenotypes and genotypes by the female immune system may require novel approaches such as labeling the sperm of individual males, or haploid subsets of sperm, in order to directly visualize their fate within the female immune environment.

Acknowledgments

We thank Ruth Lehmann for the invitation that allowed us to probe this interesting topic together. The literature in this field is large, and we apologize to those authors whose excellent work could not be cited here due to space limitations. We thank the following sources for support: NIH grants R01-HD038921 (M.F.W.) and R01-HD059060 (M.F.W. and AG Clark), a BBSRC fellowship to S.W. (BB/K014544/1), and an industrial LINK award from the BBSRC and Aviagen® to T.P. (BB/L009587/1).

References

Adams, E. M., & Wolfner, M. F. (2007). Seminal proteins but not sperm induce morphological changes in the *Drosophila melanogaster* female reproductive tract during sperm storage. *Journal of Insect Physiology*, *53*, 319–331.

Alghamdi, A. S., & Foster, D. N. (2005). Seminal DNase frees spermatozoa entangled in neutrophil extracellular traps. *Biology of Reproduction*, *73*(6), 1174–1181.

Amjadi, F., et al. (2014). Role of the innate immunity in female reproductive tract. *Advanced Biomedical Research*, *3*(1), 1.

Anilkumar, R., et al. (2001). Correlation between the spermatozoal characteristics and sperm penetration distance in polyacrylamide gel and bovine cervical mucus. *Theriogenology*, *55*(2), 685–691.

Apari, P., de Sousa, J. D., & Müller, V. (2014). Why sexually transmitted infections tend to cause infertility: An evolutionary hypothesis. *PLoS Pathogens*, *10*(8), 1–5.

Austad, S. N., & Hoffman, J. M. (2018). Is antagonistic pleiotropy ubiquitous in aging biology? *Evolution, Medicine, and Public Health*, *1*, 287–294.

Avila, F. W., & Wolfner, M. F. (2009). Acp36DE is required for uterine conformational changes in mated *Drosophila* females. *Proceedings of the National Academy of Sciences of the United States of America*, *106*(37), 15796–15800.

Baer, B., Armitage, S. A. O., & Boomsma, J. J. (2006). Sperm storage induces an immunity cost in ants. *Nature*, *441*(7095), 872–875.

Bakst, M. R. (2011). Physiology and endocrinology symposium: Role of the oviduct in maintaining sustained fertility in hens. *Journal of Animal Science*, *89*(5), 1323–1329.

Bedford, J. M., & Yanagimachi, R. (1992). Initiation of sperm motility after mating in the rat and hamster. *Journal of Andrology*, *13*(5), 444–449.

Birkhead, T. R., & Pizzari, T. (2002). Postcopulatory sexual selection. *Nature Reviews Genetics*, *3*(4), 262–273.

Blandin, S., et al. (2004). Complement-like protein TEP1 is a determinant of vectorial capacity in the malaria vector *Anopheles gambiae*. *Cell*, *116*(5), 661–670.

Bloch Qazi, M. C. B., & Wolfner, M. F. (2003). An early role for the *Drosophila melanogaster* male seminal protein Acp36DE in female sperm storage. *The Journal of Experimental Biology*, *206*, 3521–3528.

Bohring, C., & Krause, W. (2003). Immune infertility: Towards a better understanding of sperm (auto)-immunity: The value of proteomic analysis. *Human Reproduction*, *18*(5), 915–924.

Borziak, K., et al. (2016). The seminal fluid proteome of the polyandrous Red junglefowl offers insights into the molecular basis of fertility, reproductive ageing and domestication. *Scientific Reports*, *6*, 35864.

Burke, W. H., Rieser, J. W., & Shoffner, R. N. (1971). The effects of isoimmunization with semen, on fertility in the Turkey hen. *Poultry Science*, *50*(6), 1841–1847.

Cameron, E., Day, T., & Rowe, L. (2007). Sperm competition and the evolution of ejaculate composition. *The American Naturalist*, *169*(6), E158–E172.

Chen, C., et al. (2017). The microbiota continuum along the female reproductive tract and its relation to uterine-related diseases. *Nature Communications*, *8*(1), 875.

Cheng, C. Y., & Mruk, D. D. (2012). The blood-testis barrier and its implications for male contraception. *Pharmacological Reviews*, *64*(1), 16–64.

Chiu, W. W.-C., & Chamley, L. W. (2004). Clinical associations and mechanisms of action of antisperm antibodies. *Fertility and Sterility*, *82*(3), 529–535.

Clark, G. F., & Schust, D. J. (2013). Manifestations of immune tolerance in the human female reproductive tract. *Frontiers in Immunology*, *4*, 26.

Collet, J. M. M. (2010). *The operation of sexual selection in the red junglefowl*. Oxford University. D. Phil. Thesis.

Cottell, E., et al. (2000). Are seminal fluid microorganisms of significance or merely contaminants? *Fertility and Sterility*, *74*(3), 465–470.

Cramer, E. R. A., et al. (2016). Females discriminate against heterospecific sperm in a natural hybrid zone. *Evolution*, *70*(8), 1844–1855.

Das, S. C., Isobe, N., & Yoshimura, Y. (2008). Mechanism of prolonged sperm storage and sperm survivability in hen oviduct: A review. *American Journal of Reproductive Immunology*, *60*(6), 477–481.

Das, S. C., Nagasaka, N., & Yoshimura, Y. (2005). Changes in the localization of antigen presenting cells and T cells in the utero-vaginal junction after repeated artificial insemination in laying hens. *Journal of Reproduction and Development*, *51*(5), 683–687.

de Godoy, L. M. F. (2014). SILAC yeast: From labeling to comprehensive proteome quantification. In D. Martins-de-Souza (Ed.), *Shotgun proteomics: Methods and protocols* (pp. 81–109). New York, NY: Springer New York.

Dejucq, N., & Jégou, B. (2001). Viruses in the mammalian male genital tract and their effects on the reproductive system. *Microbiology and Molecular Biology Reviews*, *65*(2), 208–231.

Devigili, A., et al. (2017). Postcopulatory cost of immune system activation in *Poecilia reticulata*. *Ethology Ecology and Evolution*, *29*(3), 266–279.

Dewsbury, D. A. (1982). Ejaculate cost and male choice. *The American Naturalist*, *119*, 601–610.

Dondero, F., et al. (1993). Immunological infertility in humans. *Experimental and Clinical Immunogenetics*, *10*(2), 65–72.

Dorus, S., Skerget, S., & Karr, T. L. (2012). Proteomic discovery of diverse immunity molecules in mammalian spermatozoa. *Systems Biology in Reproductive Medicine*, *58*(4), 218–228.

Dziarski, R., & Gupta, D. (2006). The peptidoglycan recognition proteins (PGRPs). *Genome Biology*, *7*(8), 232–244.

Eberhard, W. G. (1996). *Female control: Sexual selection by cryptic female choice*. Princeton and Chichester: Princeton University Press.

Edwards, S. V., & Hedrick, P. W. (1998). Evolution and ecology of MHC molecules: From genomics to sexual selection. *Trends in Ecology & Evolution*, *13*(8), 305–311.

Eisenbach, M., & Giojalas, L. C. (2006). Sperm guidance in mammals—An unpaved road to the egg. *Nature Reviews Molecular Cell Biology*, 7(4), 276–285.

Evans, J. P., et al. (2012). Assessing the potential for egg chemoattractants to mediate sexual selection in a broadcast spawning marine invertebrate. *Proceedings of the Royal Society B: Biological Sciences*, *279*(1739), 2855–2861.

Fedorka, K. M., & Zuk, M. (2005). Sexual conflict and female immune suppression in the cricket. *Allonemobious socius, Journal of Evolutionary Biology*, *18*, 1515–1522.

Firman, R. C., et al. (2017). Postmating female control: 20 years of cryptic female choice. *Trends in Ecology & Evolution*, *32*(5), 368–382.

Folstad, I., & Karter, A. J. (1992). Parasites, bright males, and the immunocompetence handicap. *American Naturalist*, *139*, 603–622.

Fox, C. A., Meldrum, S. J., & Watson, B. W. (1973). Continuous measurement by radio-telemetry of vaginal pH during human coitus. *Journal of Reproduction and Fertility*, *33*(1), 69–75.

Friedländer, M., & Gitay, H. (1972). The fate of the normal-anucleated spermatozoa in inseminated females of the silkworm *Bombyx mori*. *Journal of Morphology*, *138*(1), 121–129.

Garbaczewska, M., Billeter, J. C., & Levine, J. D. (2013). *Drosophila melanogaster* males increase the number of sperm in their ejaculate when perceiving rival males. *Journal of Insect Physiology*, *59*(3), 306–310.

Gee, H. (1999). Size and the single sex cell. *Nature*. https://doi.org/10.1038/news991125-4.

Gershman, S. N., et al. (2010). Give 'til it hurts: Trade-offs between immunity and male reproductive effort in the decorated cricket, *Gryllodes sigillatus*. *Journal of Evolutionary Biology*, *23*(4), 829–839.

Ghaderi, D., et al. (2011). Sexual selection by female immunity against paternal antigens can fix loss of function alleles. *Proceedings of the National Academy of Sciences of the United States of America*, *108*(43), 17743–17748.

Gwathmey, T. M., Ignotz, G. G., & Suarez, S. S. (2003). PDC-109 (BSP-A1/A2) promotes bull sperm binding to oviductal epithelium in vitro and may be involved in forming the oviductal sperm reservoir. *Biology of Reproduction*, *69*, 809–815.

Gwathmey, T. M., et al. (2006). Bovine seminal plasma proteins PDC-109, BSP-A3, and BSP-30-kDa share functional roles in storing sperm in the oviduct. *Biology of Reproduction*, *75*, 501–507.

Haley, L. E., & Abplanalp, H. (1970). Possible immunological basis for a reduction of fertility in cross-mating fowl with Japanese quail. *Journal of Reproduction and Fertility*, *23*(3), 375–381.

Higaki, K., et al. (1995). Localization of spermatozoa and leukocytes in vagina and uterovaginal junction after copulation in Japanese quail (*Coturnix coturnix japonica*). *Japanese Poultry Science*, *32*(6), 387–393.

Hillgarth, N., Ramenofsky, M., & Winfield, J. (1997). Testosterone and sexual selection. *Behavioral Ecology*, *8*(1), 108–109.

Holman, L., & Snook, R. R. (2006). Spermicide, cryptic female choice and the evolution of sperm form and function. *Journal of Evolutionary Biology*, *19*(5), 1660–1670.

Holman, L., & Snook, R. R. (2008). A sterile sperm caste protects brother fertile sperm from female-mediated death in *Drosophila pseudoobscura*. *Current Biology*, *18*, 292–296.

Holt, W. V., & Fazeli, A. (2016). Sperm storage in the female reproductive tract. *Annual Review of Animal Biosciences*, *4*(1), 291–310.

Hong, J., et al. (2017). Strong inhibition of neutrophil-sperm interaction in cattle by selective phosphatidylinositol 3-kinase inhibitors. *Biology of Reproduction*, *97*(5), 671–687.

Hopkins, B. R., Sepil, I., & Wigby, S. (2017). Seminal fluid. *Current Biology*, *27*(11), 404–405.

Hosken, D. J. (2001). Sex and death: Microevolutionary trade-offs between reproductive and immune investment in dung flies. *Current Biology*, *11*(10), R379–R380.
Hosken, D. J., & O'Shea, J. E. (2001). Sperm production and immune function in two Australian bats, *Chalinolobus morio* and *Nyctophilus geoffroyi*. *Ethology Ecology and Evolution*, *13*(2), 173–180.
Hosoda, T., Abe, T., & Otsuka, S. (1964). Serological studies on the insemination of domestic fowl. *Japanese Poultry Science*, *1*(1), 33–38.
Hunter, R. H. (1981). Sperm transport and reservoirs in the pig oviduct in relation to the time of ovulation. *Journal of Reproduction and Fertility*, *63*(1), 109–117.
Immler, S. (2008). Sperm competition and sperm cooperation: The potential role of diploid and haploid expression. *Reproduction*, *135*(3), 275–283.
Jansen, R. P. (1980). Cyclic changes in the human fallopian tube isthmus and their functional importance. *American Journal of Obstetrics and Gynecology*, *136*(3), 292–308.
Jansen, R. P., & Bajpai, V. K. (1982). Oviduct acid mucus glycoproteins in the estrous rabbit: Ultrastructure and histochemistry. *Biology of Reproduction*, *26*(1), 155–168.
Johns, S., Henshaw, J. M., Jennions, M. D., & Head, M. L. (2019). Males can evolve lower resistance to sexually transmitted infections to infect their mates and thereby increase their own fitness. *Evolutionary Ecology*, *33*(2), 149–172.
Kapelnikov, A., et al. (2008). Mating induces an immune response and developmental switch in the *Drosophila* oviduct. *Proceedings of the National Academy of Sciences of the United States of America*, *105*(37), 13912–13917.
Katz, D. F., Slade, D. A., & Nakajima, S. T. (1997). Analysis of pre-ovulatory changes in cervical mucus hydration and sperm penetrability. *Advances in Contraception: The Official Journal of the Society for the Advancement of Contraception*, *13*(2–3), 143–151.
Kelly, C. D., & Jennions, M. D. (2011). Sexual selection and sperm quantity: Meta-analyses of strategic ejaculation. *Biological Reviews of the Cambridge Philosophical Society*, *86*(4), 863–884.
Kerr, A. M., Gershman, S. N., & Sakaluk, S. K. (2010). Experimentally induced spermatophore production and immune responses reveal a trade-off in crickets. *Behavioral Ecology*, *21*(3), 647–654.
Kirk, T. A., et al. (1989). The relationship of infertility to antibody production in the uterovaginal sperm storage tubules of Turkey breeder hens. *Theriogenology*, *31*(5), 955–961.
Knell, R. J., & Webberley, K. M. (2004). Sexually transmitted diseases of insects: Distribution, evolution, ecology and host behaviour. *Biological Reviews of the Cambridge Philosophical Society*, *79*, 557–581.
Kurata, S. (2014). Peptidoglycan recognition proteins in *Drosophila* immunity. *Developmental and Comparative Immunology*, *42*(1), 36–41.
Lai, Y. M., et al. (1997). The effect of human papillomavirus infection on sperm cell motility. *Fertility and Sterility*, *67*(6), 1152–1155.
Le Tortorec, A., & Dejucq-Rainsford, N. (2010). HIV infection of the male genital tract—Consequences for sexual transmission and reproduction. *International Journal of Andrology*, *33*(1), e98–e108.
Lenz, T. L., et al. (2018). Cryptic haplotype-specific gamete selection yields offspring with optimal MHC immune genes. *Evolution*, *72*(11), 2478–2490.
Liljedal, S., Folstad, I., & Skarstein, F. (1999). Secondary sex traits, parasites, immunity and ejaculate quality in the Arctic charr. *Proceedings of the Royal Society B: Biological Sciences*, *266*(1431), 1893–1898.
Lockhart, A. B., Thrall, P. H., & Antonovics, J. (1996). Sexually transmitted diseases in animals: Ecological and evolutionary implications. *Biological Reviews*, *71*(3), 415–471.
Losdat, S., et al. (2011). Immune activation reduces sperm quality in the great tit. *PLoS One*, *6*(7), e22221.

Løvlie, H., et al. (2013). Cryptic female choice favours sperm from major histocompatibility complex-dissimilar males. *Proceedings of the Royal Society of London B*, *280*, 20131296.

Lu, J.-C., Huang, Y.-F., & Lu, N.-Q. (2008). Antisperm immunity and infertility. *Expert Review of Clinical Immunology*, *4*(1), 113–126.

Ma, X., et al. (2016). Sialylation facilitates the maturation of mammalian sperm and affects its survival in female uterus. *Biology of Reproduction*, *94*(6), 123. 1–10.

Mahdi, B. M., et al. (2011). Frequency of antisperm antibodies in infertile women. *Journal of Reproduction & Infertility*, *12*(4), 261.

Manjerovic, M. B., & Waterman, J. M. (2012). Immunological sex differences in socially promiscuous African ground squirrels. *PLoS One*, 7(6), 1–5.

Mattei, A. L., et al. (2015). Integrated 3D view of postmating responses by the *Drosophila melanogaster* female reproductive tract, obtained by micro-computed tomography scanning. *Proceedings of the National Academy of Sciences of the United States of America*, *112*(27), 8475–8480.

McCartney, J. L. (1923). Studies on the mechanism of sterilization of the female by spermotoxin. *American Journal of Physiology-Legacy Content*, *63*(2), 207–217.

McGraw, L. A., Clark, A. G., & Wolfner, M. F. (2008). Post-mating gene expression profiles of female *Drosophila melanogaster* in response to time and to four male accessory gland proteins. *Genetics*, *179*, 1395–1408.

McNamara, K. B., van Lieshout, E., & Simmons, L. W. (2014). Females suffer a reduction in the viability of stored sperm following an immune challenge. *Journal of Evolutionary Biology*, *27*(1), 133–140.

Meaker, S. R. (1922). Some aspects of the problem of sterility. *The Boston Medical and Surgical Journal*, *187*(15), 535–539.

Miki, K., & Clapham, D. E. (2013). Rheotaxis guides mammalian sperm. *Current Biology*, *23*(6), 443–452.

Milinski, M. (2006). The major histocompatibility complex, sexual selection, and mate choice. *Annual Review of Ecology, Evolution, and Systematics*, *37*, 159–186.

Miyata, H., et al. (2016). Genome engineering uncovers 54 evolutionarily conserved and testis-enriched genes that are not required for male fertility in mice. *Proceedings of the National Academy of Sciences of the United States of America*, *113*(28), 7704–7710.

Moore, H., Dvorakova, K., Jenkins, N., & Breed, W. (2002). Exceptional sperm cooperation in the wood mouse. *Nature*, *418*(6894), 174–177.

Moreno, I., & Franasiak, J. M. (2017). Endometrial microbiota—New player in town. *Fertility and Sterility*, *108*(1), 32–39.

Morrow, E. H., & Innocenti, P. (2012). Female postmating immune responses, immune system evolution and immunogenic males. *Biological Reviews*, *87*(3), 631–638.

Mullins, K. J., & Saacke, R. G. (1989). Study of the functional anatomy of bovine cervical mucosa with special reference to mucus secretion and sperm transport. *The Anatomical Record*, *225*(2), 106–117.

Muro, Y., et al. (2016). Behavior of mouse spermatozoa in the female reproductive tract from soon after mating to the beginning of fertilization. *Biology of Reproduction*, *94*(4), 80.

Needham, K. B., et al. (2017). Repeated immune challenges affect testosterone but not sperm quality. *Journal of Experimental Zoology Part A: Ecological and Integrative Physiology*, *327*(6), 398–406.

Neubaum, D. M., & Wolfner, M. F. (1999). Mated *Drosophila melanogaster* females require a seminal fluid protein, Acp36DE, to store sperm efficiently. *Genetics*, *153*(2), 845–857.

Nishimura, H., & L'Hernault, S. W. (2017). Spermatogenesis. *Current Biology*, *27*(18), R988–R994.

O'Hanlon, D. E., Moench, T. R., & Cone, R. A. (2013). Vaginal pH and microbicidal lactic acid when lactobacilli dominate the microbiota. *PLoS One*, *8*(11), e80074.

Orr, T. J., & Brennan, P. L. R. (2015). Sperm storage: Distinguishing selective processes and evaluating criteria. *Trends in Ecology & Evolution*, *30*(5), 261–272.
Orr, T. J., & Zuk, M. (2012). Sperm storage. *Current Biology*, *22*(1), R8–R10.
Otti, O., McTighe, A. P., & Reinhardt, K. (2013). In vitro antimicrobial sperm protection by an ejaculate-like substance. *Functional Ecology*, *27*(1), 219–226.
Overstreet, J. W., & Cooper, G. W. (1978). Sperm transport in the reproductive tract of the female rabbit: II. The sustained phase of transport. *Biology of Reproduction*, *19*(1), 115–132.
Owen, D. H., & Katz, D. F. (2005). A review of the physical and chemical properties of human semen and the formulation of a semen simulant. *Journal of Andrology*, *26*(4), 459–469.
Parker, G. A. (1970). Sperm competition and its evolutionary consequences in the insects. *Biological Reviews*, *45*, 525–567.
Parker, G. A. (1990). Sperm competition games: Raffles and roles. *Proceedings of the Royal Society B: Biological Sciences*, *242*(1304), 120–126.
Parker, G. a., & Pizzari, T. (2010). Sperm competition and ejaculate economics. *Biological Reviews of the Cambridge Philosophical Society*, *85*, 897–934.
Parker, G. A., et al. (1996). Sperm competition games: Individual assessment of sperm competition intensity by group spawners. *Proc. R. Soc. London Ser. B*, *263*, 1291–1297.
Perry, J. C., Sirot, L., & Wigby, S. (2013). The seminal symphony: How to compose an ejaculate. *Trends in Ecology & Evolution*, *28*(7), 414–422.
Pizzari, T., & Foster, K. R. (2008). Sperm sociality: Cooperation, altruism, and spite. *PLoS Biology*. *6*(5), e130.
Pizzari, T., Løvlie, H., & Cornwallis, C. K. (2004). Sex-specific, counteracting responses to inbreeding in a bird. *Proceedings Biological sciences/The Royal Society*, *271*, 2115–2121.
Pompon, J., & Levashina, E. A. (2015). A new role of the mosquito complement-like Cascade in male fertility in *Anopheles gambiae*. D. S. Schneider, Ed. *PLoS Biology*, *13*(9), e1002255.
Racey, P. A., et al. (1987). Sperm-epithelium relationships in relation to the time of insemination in little brown bats (*Myotis lucifugus*). *Journal of Reproduction and Fertility*, *80*(2), 445–454.
Radhakrishnan, P., & Fedorka, K. M. (2012). Immune activation decreases sperm viability in both sexes and influences female sperm storage. *Proceedings of the Royal Society B: Biological Sciences*, *279*(1742), 3577–3583.
Reinhardt, K., Naylor, R., & Siva-Jothy, M. T. (2003). Reducing a cost of traumatic insemination: Female bedbugs evolve a unique organ. *Proceedings of the Biological Sciences*, *270*, 2371–2375.
Roberts, M. L., Buchanan, K. L., & Evans, M. R. (2004). Testing the immunocompetence handicap hypothesis: A review of the evidence. *Animal Behaviour*, *68*(2), 227–239.
Schmid-Hempel, P. (2005). Evolutionary ecology of insect immune defenses. *Annual Review of Entomology*, *50*, 529–551.
Schnakenberg, S. L., Siegal, M. L., & Bloch Qazi, M. C. (2012). Oh, the places they'll go: Female sperm storage and sperm precedence in *Drosophila melanogaster*. *Spermatogenesis*, *2*(3), 224–235.
Schulze, M., et al. (2018). Detection and characterization of Lactobacillus spp. in the porcine seminal plasma and their influence on boar semen quality. J. R. Drevet, Ed. *PLoS One*, *13*(9), e0202699.
Schuppe, H.-C., et al. (2017). Urogenital infection as a risk factor for male infertility. *Deutsches Aerzteblatt International*, *114*(19), 339–346.
Schwenke, R. A., Lazzaro, B. P., & Wolfner, M. F. (2016). Reproduction-immunity trade-offs in insects. *Annual Review of Entomology*, *61*, 239–256.

Sheldon, I. M., Owens, S.-E., & Turner, M. L. (2017). Innate immunity and the sensing of infection, damage and danger in the female genital tract. *Journal of Reproductive Immunology*, *119*, 67–73.

Sidaway, P. (2016). Infection: Sexual transmission of gonorrhoea is facilitated by other bacteria. *Nature Reviews Urology*, *13*(10), 562.

Simmons, L. W. (2011). Resource allocation trade-off between sperm quality and immunity in the field cricket, *Teleogryllus oceanicus*. *Behavioral Ecology*, *23*(1), 168–173.

Simmons, L. W., & Emlen, D. J. (2006). Evolutionary trade-off between weapons and testes. *Proceedings of the National Academy of Sciences of the United States of America*, *103*(44), 16346–16351.

Simmons, L. W., Lüpold, S., & Fitzpatrick, J. L. (2017). Evolutionary trade-off between secondary sexual traits and ejaculates. *Trends in Ecology & Evolution*, *32*(12), 964–976.

Simmons, L. W., & Roberts, B. (2005). Bacterial immunity traded for sperm viability in male crickets. *Science*, *309*(5743), 2031. 2031.

Sirot, L. K., Wolfner, M. F., & Wigby, S. (2011). Protein-specific manipulation of ejaculate composition in response to female mating status in *Drosophila melanogaster*. *Proceedings of the National Academy of Sciences of the United States of America*, *108*, 9922–9926.

Skau, P. A., & Folstad, I. (2005). Does immunity regulate ejaculate quality and fertility in humans? *Behavioral Ecology*, *16*(2), 410–416.

Sobrero, A. J., & Macleod, J. (1962). The immediate postcoital test. *Fertility and Sterility*, *13*, 184–189.

Soper, D. E. (2010). Pelvic inflammatory disease. *Obstetrics and Gynecology*, *116*(2 Pt. 1), 419–428.

Steele, M. G., & Wishart, G. J. (1992). Evidence for a species-specific barrier to sperm transport within the vagina of the chicken hen. *Theriogenology*, *38*(6), 1107–1114.

Stern, J. E., et al. (1992). Antisperm antibodies in women: Variability in antibody levels in serum, mucus, and peritoneal fluid. *Fertility and Sterility*, *58*(5), 950–958.

Suarez, S. S. (1987). Sperm transport and motility in the mouse oviduct: Observations in situ. *Biology of Reproduction*, *36*(1), 203–210.

Suarez, S. S. (2015). Gamete and zygote transport. In T. Plant & A. Zeleznich (Eds.), *Knobil and Neill's physiology of reproduction* (pp. 197–232). Oxford: Elsevier.

Suarez, S. S. (2016). Mammalian sperm interactions with the female reproductive tract. *Cell and Tissue Research*, *363*(1), 185–194.

Suarez, S. S., Brockman, K., & Lefebvre, R. (1997). Distribution of mucus and sperm in bovine oviducts after artificial insemination: The physical environment of the oviductal sperm reservoir. *Biology of Reproduction*, *56*(2), 447–453.

Suarez, S. S., & Pacey, A. A. (2006). Sperm transport in the female reproductive tract. *Human Reproduction Update*, *12*(1), 23–37.

Swallow, J. G., & Wilkinson, G. S. (2002). The long and short of sperm polymorphisms in insects. *Biological Reviews of the Cambridge Philosophical Society*, *77*(2), 153–182.

Taherali, F., Varum, F., & Basit, A. W. (2018). A slippery slope: On the origin, role and physiology of mucus. *Advanced Drug Delivery Reviews*, *124*, 16–33.

Taylor, M. L., Price, T. A. R., & Wedell, N. (2014). Polyandry in nature: A global analysis. *Trends in Ecology & Evolution*, *29*(7), 376–383.

Tokuhiro, K., et al. (2012). Protein disulfide isomerase homolog PDILT is required for quality control of sperm membrane protein ADAM3 and male fertility. *Proceedings of the National Academy of Sciences of the United States of America*, *109*(10), 3850–3855.

Troedsson, M. H., et al. (2001). Interaction between equine semen and the endometrium: The inflammatory response to semen. *Animal Reproduction Science*, *68*(3–4), 273–278.

Tung, C. K., et al. (2015a). Emergence of upstream swimming via a hydrodynamic transition. *Physical Review Letters*, *114*(10), 108102.

Tung, C. K., et al. (2015b). Microgrooves and fluid flows provide preferential passageways for sperm over pathogen *Tritrichomonas foetus*. *Proceedings of the National Academy of Sciences of the United States of America, 112*(17), 5431–5436.
Vujisić, S., et al. (2005). Antisperm antibodies in semen, sera and follicular fluids of infertile patients: Relation to reproductive outcome after in vitro fertilization. *American Journal of Reproductive Immunology, 54*(1), 13–20.
Wald, M. (2005). Male infertility: Causes and cures. *Sexuality, Reproduction and Menopause, 3*(2), 83–87.
Wardlow A.M. and Agrawal A.F., Sexual conflict and Sexually Transmitted Infections (STIs): Coevolution of sexually antagonistic host traits with an STI, The American Naturalist 193 (1), E1–E14. http://doi.org/10.1086/700564.
Wedekind, C., et al. (2004). MHC-linked susceptibility to a bacterial infection, but no MHC-linked cryptic female choice in whitefish. *Journal of Evolutionary Biology, 17*(1), 11–18.
Wedell, N. (1993). Spermatophore size in bushcrickets: Comparative evidence for nuptial gifts as a sperm protective device. *Evolution, 47*(4), 1203–1212.
Wedell, N., Gage, M. J. G. G., & Parker, G. A. (2002). Sperm competition, male prudence and sperm-limited females. *Trends in Ecology & Evolution, 17*, 313–320.
Wentworth, B. C., & Mellen, W. J. (1963). Effects of spermatozoal antibodies and method of insemination on fecundity of white leghorn females. In *Poultry science*. Savoy, IL: Poultry Science Associates Inc. 61874. p. 1317.
Wigby, S., & Chapman, T. (2004). Sperm competition. *Current Biology, 14*(3), R100–R103.
Wigby, S., et al. (2009). Seminal fluid protein allocation and male reproductive success. *Current Biology, 19*, 751–757.
Yaniz, J. L., et al. (2014). Three-dimensional architecture of the ovine oviductal mucosa. *Anatomia, Histologia, Embryologia, 43*(5), 331–340.
Yousef, M. S., et al. (2016). Sperm binding to oviduct epithelial cells enhances TGFB1 and IL10 expressions in epithelial cells as well as neutrophils in vitro: Prostaglandin E2 as a main regulator of anti-inflammatory response in the bovine oviduct. *PLoS One, 11*(9), e0162309.
Yudin, A. I., et al. (2005). Beta-defensin 126 on the cell surface protects sperm from immunorecognition and binding of anti-sperm antibodies. *Biology of Reproduction, 73*(6), 1243–1252.
Zambrano, F., et al. (2016). Leukocytes coincubated with human sperm trigger classic neutrophil extracellular traps formation, reducing sperm motility. *Fertility and Sterility, 106*(5), 1053–1060.e1.
Zera, A. J., & Harshman, L. G. (2001). The physiology of life history trade-offs in animals. *Annual Review of Ecology and Systematics, 32*, 95–126.
Zheng, W. M., et al. (2001). An in situ hybridization study of the effects of artificial insemination on the localization of cells expressing MHC class II mRNA in the chicken oviduct. *Reproduction, 122*(4), 581–586.
Ziegler, A., Kentenich, H., & Uchanska-Ziegler, B. (2005). Female choice and the MHC. *Trends in Immunology, 26*(9), 496–502.

CHAPTER NINE

A symbiont's guide to the germline

Shelbi L. Russell*, Laura Chappell, William Sullivan
Department of Molecular, Cell, and Developmental Biology, University of California Santa Cruz, Santa Cruz, CA, United States
*Corresponding author: e-mail address: shelbilrussell@gmail.com

Contents

1. Background: Symbiont transmission modes maintain symbiotic associations 316
2. Soma-to-soma strategies of vertical symbiont transmission 319
3. Vertical transmission through the germline part 1: Migration from the soma to oocytes/embryos 322
   3.1 Transmission without tissue types: Sponge symbiont transmission 323
   3.2 Extracellular storage of vertically transmitted intracellular symbionts 323
   3.3 Intracellular symbiont transmission from the soma to the germline involves cell-to-cell transfer 326
4. Vertical transmission through the germline part 2: Migration from the soma to germ stem cells (GSCs) 330
5. Vertical transmission through the germline part 3: Continuous germ cell association throughout development 331
   5.1 Segregation with host cell division: The first form of vertical transmission 331
   5.2 Symbiont segregation with the primordial germline 332
   5.3 Continuous germ line association is not the norm among strictly vertically transmitted symbionts 338
6. Parallels between symbiont transmission and the origin of the germline 339
7. Conclusions 342

References 343

Abstract

Microbial symbioses exhibit astounding adaptations, yet all symbionts face the problem of how to reliably associate with host offspring every generation. A common strategy is vertical transmission, in which symbionts are directly transmitted from the female to her offspring. The diversity of symbionts and vertical transmission mechanisms is as expansive as the diversity of eukaryotic host taxa that house them. However, there are several common themes among these mechanisms based on the degree to which symbionts associate with the host germline during transmission. In this review, we detail three distinct vertical transmission strategies, starting with associations that are transmitted from host somatic cells to offspring somatic cells, either due to lacking a germline or avoiding it. A second strategy involves somatically-localized symbionts that migrate into

Current Topics in Developmental Biology, Volume 135
ISSN 0070-2153
https://doi.org/10.1016/bs.ctdb.2019.04.007

the germline during host development. The third strategy we discuss is one in which the symbiont maintains continuous association with the germline throughout development. Unexpectedly, the vast majority of documented vertically inherited symbionts rely on the second strategy: soma-to-germline migration. Given that not all eukaryotes contain a sequestered germline and instead produce offspring from somatic stem cell lineages, this soma-to-germline migration is discussed in the context of multicellular evolution. Lastly, as recent genomics data have revealed an abundance of horizontal gene transfer events from symbiotic and non-symbiotic bacteria to host genomes, we discuss their impact on eukaryotic host evolution.

1. Background: Symbiont transmission modes maintain symbiotic associations

Symbiotic associations between microbes and eukaryotes are ubiquitous in nature and provide functions that enable their members to adopt novel niches relative to their ancestors. These associations range from highly integrated intracellular associations where bacteria reside in the host cytoplasm, often encompassed by a host-derived membrane, to extracellular associations where symbionts are housed in special structures or epithelial surfaces (Cavanaugh, McKiness, Newton, & Stewart, 2006; Moran, 2007; Moya, Peretó, Gil, & Latorre, 2008). Functionally, symbioses range from associations in which both host and symbiont benefit from the interaction, termed mutualism, to associations in which only one partner benefits, termed commensalism, to associations in which one member benefits at the expense of the other, termed parasitism. For this review, we will be focusing on interactions toward the mutualistic end of the spectrum (although, some are perhaps better described as addictive, Sullivan, 2017), as these situations select for symbiont association with the host germline.

Across the diversity of symbiotic associations, many reliable mechanisms have evolved to facilitate symbiont transmission. They range from horizontal transmission strategies, in which symbionts pass through an environmental intermediate to reach new hosts, to vertical strategies, in which symbionts are inherited directly through parental host tissues (Fig. 1). While paternal transmission via sperm occurs (Watanabe, Yukuhiro, Matsuura, Fukatsu, & Noda, 2014), it is rare, and vertically inherited symbionts are typically transmitted through the female germline or maternally-brooded embryos (Bright & Bulgheresi, 2010). Vertical transmission is associated with highly dependent associations (Fisher, Henry, Cornwallis, Kiers, & West, 2017) and is thought to evolve through non-additive (epistatic)

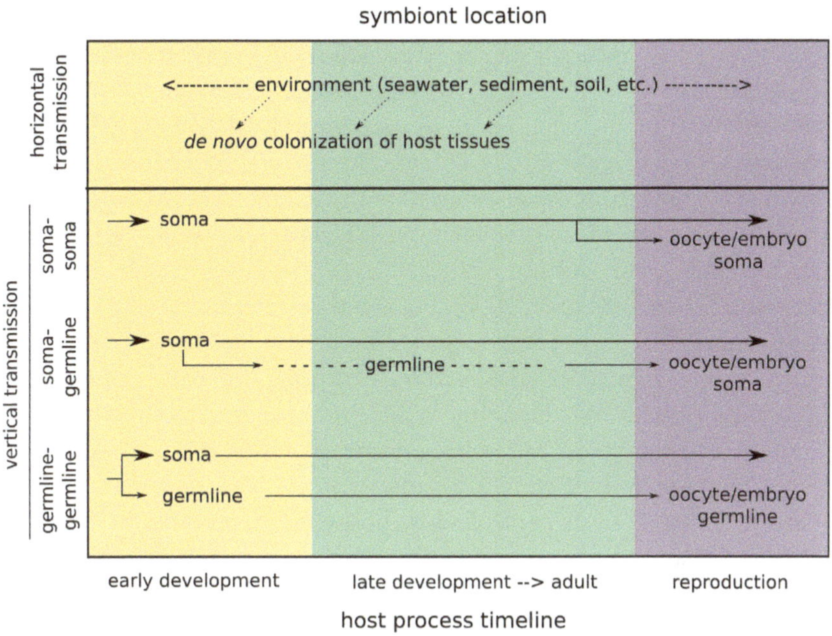

Fig. 1 General location patterns of symbionts during host development and reproduction in horizontally transmitted associations and vertically transmitted associations with three different strategies based upon when and how symbionts colonize the soma, germline, gametes, or offspring.

genetic interactions between host and symbiont that improve the fitness of both partners (Drown, Zee, Brandvain, & Wade, 2013). Thus, many associations have independently evolved elaborate routes and mechanisms for symbionts to reach host offspring (Bright & Bulgheresi, 2010).

Remarkably, many associations exhibit evidence of both vertical and horizontal transmission modes (Bright & Bulgheresi, 2010; Ebert, 2013; Russell, Corbett-Detig, & Cavanaugh, 2017), indicating that the mechanisms or constraints required by each mode do not necessarily preclude the other mode. The evolutionary reasons for the existence of these "mixed modes" that incorporate horizontal and vertical transmission are not well understood, but may involve selection for some amount of symbiont gene flow via horizontal transmission to maintain symbiont genome function (Russell et al., 2017). Mixed mode transmission certainly requires mechanisms for cell-to-cell transfer to be maintained for the horizontal component of transmission. As we discuss below, many vertical transmission strategies also utilize cell-to-cell transfer to enable symbiont migration from somatic host cells to the germline.

Regarding the evenness of symbiont distribution among host cells and tissues, there is an apparent tendency toward highly specific tissue tropism, or distribution, with increasing association obligacy/dependence. Mutualistic bacterial symbionts often reside in specialized host-derived cells termed bacteriocytes (or sometimes mycetocytes in insects). Bacteriocytes localize to specific host tissues or organs relevant to the symbiont function in the host, termed bacteriomes (Hentschel, Steinert, & Hacker, 2000; Hosokawa, Koga, Kikuchi, Meng, & Fukatsu, 2010; Moya et al., 2008; Pinto-Carbó, Gademann, Eberl, & Carlier, 2018). For example, chemosynthetic symbionts of bivalves reside in gill bacteriocytes where they have access to reduced chemicals for oxidation and carbon fixation (Stewart, Newton, & Cavanaugh, 2005), and the amino acid-synthesizing symbionts of aphids reside in bacteriocytes in paired bacteriome structures near the gut (Koga, Meng, Tsuchida, & Fukatsu, 2012). Interestingly, at least in insects, differentiation of these cells is controlled by host genes including the homeotic gene *ultrabithorax* (Matsuura, Kikuchi, Miura, & Fukatsu, 2015). More recently evolved or facultative associations exhibit less specific patterns of tissue tropism. For example, the more recently evolved *Rickettsia* symbionts of ticks exhibit nonspecific tissue tropism compared to their obligate *Coxiella* symbionts (Lalzar, Friedmann, & Gottlieb, 2014). Similarly, the secondary, facultative symbiont *Serratia* of aphids exhibits a far more disorganized and less specific tissue distribution than the primary, obligate symbiont *Buchnera* (Koga et al., 2012). Tissue distribution is an important consideration, as it often indicates how symbionts pass from one host generation to the next.

In this chapter, we examine vertical transmission strategies exhibited by endosymbionts associated with single-cell hosts to multicellular asexual and sexual hosts. By examining associations at these different levels of organismal complexity, we show how strategies scale across cellular and tissue complexity. As symbiotic associations exist across many eukaryotic and bacterial taxa, it is important to consider how strategies correlate with a particular taxonomic group or for some shared lifestyle, etc. For example, while intracellular symbioses are abundant in plants and invertebrate animals, excluding pathogens, they appear to be rare in vertebrates (Nyholm & Graf, 2012). The only example found to date is in the spotted salamander, *Ambystoma maculatum*, which was thought to exclusively host its algal symbiont, *Oophila amblystomatis*, extracellularly in its egg capsules. However, more detailed investigation showed that some of the algae invade salamander cells and tissues during embryogenesis (Kerney et al., 2011). Many other examples of vertebrate symbioses exist, such as gut-associated microbiomes (Shapira, 2016), but they are all extracellular associations permitted by both the innate and adaptive immune systems (Chu & Mazmanian, 2013). Thus, it appears that aspects of adaptive immunity prevent intracellular bacterial establishment in vertebrates (Chu & Mazmanian, 2013; Nyholm & Graf, 2012), limiting symbiont access to the germline.

As we will show, distinct mechanisms of vertical transmission exist across associations, but these mechanisms are united by common strategies for navigating between the soma and/or germline. For clarity, we use germline to refer to the specific lineages in the male and female reproductive organs that lead to the formation of sperm and eggs, respectively. We use the term germ stem cell (GSC) to refer to the stem cells that produce a self-renewing daughter cell and a daughter cell that produces the lineage leading to gamete production (Lehmann, 2012). Given that all known endosymbionts descended from free-living ancestors, the mechanisms for interacting with host somatic cells likely predated those for interacting with germ cells. Consequently, we find distinct differences in the continuity of symbiont association with the germline during vertical transmission. The following section describes the three major forms of vertical transmission: soma to soma, soma to germline, and germline to germline (Fig. 1, y-axis). Where possible, we also describe the underlying molecular and cellular mechanisms driving vertical transmission processes.

2. Soma-to-soma strategies of vertical symbiont transmission

There are many examples of transmission strategies that accomplish vertical transmission without symbionts directly associating with the germline. Naturally, this is the only option in host species that do not sequester a protected germline, such as basal metazoans and plants (Radzvilavicius, Hadjivasiliou, Pomiankowski, & Lane, 2016). However, strategies of soma-to-soma transmission also occur in host lineages with germlines, suggesting that germline association is either not necessary, not permitted, or has not yet evolved in these vertically transmitted associations.

An excellent example of vertical transmission exclusively through somatic lineages occurs in the obligate asexual catenulid flatworm, *Paracatenula galateia*, as illustrated in Fig. 2A. These platyhelminthes contain chemosynthetic bacterial symbionts and very basic body plans consisting of a limited number of cell types (Dirks et al., 2012). Reproduction is by asexual

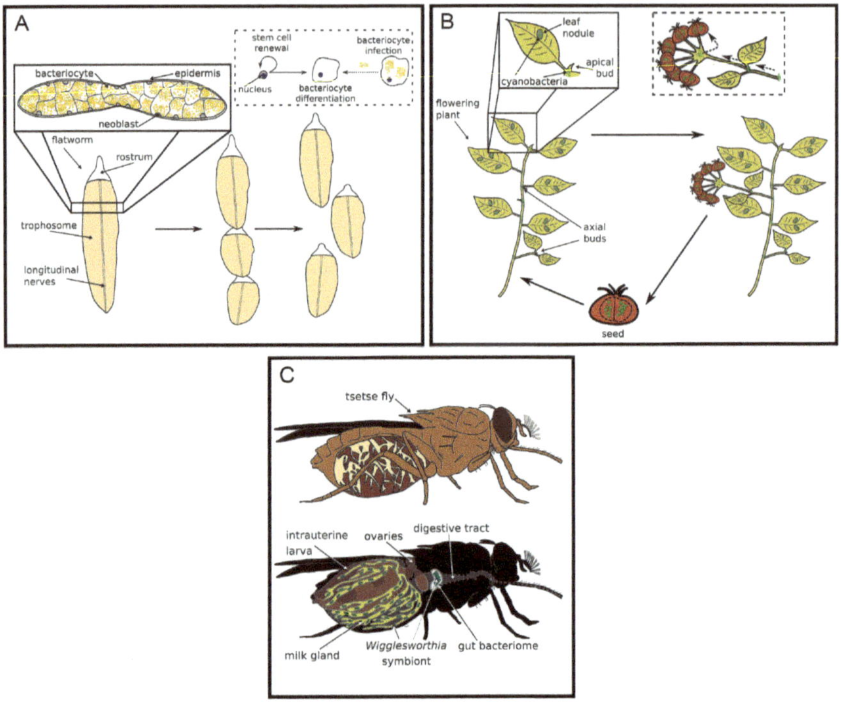

Fig. 2 Examples of soma-to-soma vertical transmission strategies. (A) Transmission of chemosynthetic alphaproteobacterial symbionts during asexual reproduction by fragmentation in *Paracatenula galateia* requires cell-to-cell transfer to bacteriocytes after they divide from neoblasts and differentiate (Dirks et al., 2012). (B) Cyanobacterial plant leaf nodule symbionts are transmitted vertically by colonizing the apical and axillary bud tissue after germination, from which they colonize either vegetative shoots or reproductive shoots (Pinto-Carbó et al., 2018). (C) The obligate intracellular gammaproteobacterial symbiont of tsetse flies, *Wigglesworthia glossinidia*, is housed in paired bacteriomes off the midgut. Symbionts are thought to colonize the milk gland through the digestive tract, where they are transmitted extracellularly via milk gland secretions to intrauterine larvae (Rio et al., 2012).

fragmentation and relies on stem-cell-like neoblast cells. Neoblasts are also responsible for the worm's regenerative abilities, and produce new symbiont-containing bacteriocytes restricted to the posterior of the worm. In natural reproduction, fragmentation begins along the anterior-posterior axis of the worm, splitting the symbiont population in half (Dirks et al., 2012). Interestingly, although the neoblasts can become bacteriocytes, they themselves are not infected (Dirks et al., 2012), and so symbionts must be acquired after differentiation by cell-to-cell transfer from the existing infected bacteriocytes (Dirks, 2011).

Also lacking a sequestered germline, plants in the families Rubiaceae, Primulaceae, and Dioscorea host endophytic extracellular *Burkholderia*-related bacterial symbionts that induce the formation of the leaf nodules they are housed in (Pinto-Carbó et al., 2018). The function of these symbionts is less well-understood than that for rhizobia root symbionts, but might involve cofactor metabolism and/or protection from reactive oxygen species. In these associations, leaf symbionts become associated with embryos when they are trapped after the axillary shoot meristems (i.e., stem cells) differentiate into inflorescences (see Fig. 2B). These leaf symbionts exhibit reduced genome sizes relative to their free-living relatives (2–6 vs 8 Mb) and have more non-coding elements (Pinto-Carbó et al., 2018), as expected for vertically transmitted symbionts (Toft & Andersson, 2010). Thus, while they are extracellular, these bacterial symbionts appear to have high fidelity mechanisms for localizing to specific plant tissues during embryogenesis. Transmission of leaf endophytes in *Dioscorea sansibarensis* occurs through the colonization of asexual reproductive tissues, tubers and bulbils (Pinto-Carbó et al., 2018), likely enabling far more symbionts to associate than is possible in small seeds. Interestingly, horizontally transmitted rhizobia bacteria localize to cells that respond to bacterially-induced mitotic signals, as opposed to existing dividing cells (Geurts, Xiao, & Reinhold-Hurek, 2016), which may partially explain why none of the rhizobia symbionts are found to be vertically transmitted (Bright & Bulgheresi, 2010; Jones, Kobayashi, Davies, Taga, & Walker, 2007).

In species with viviparous development, vertically transmitted symbionts have the option of colonizing offspring during development or during birth. In tsetse flies (Rio et al., 2012), and bat flies (Hosokawa et al., 2012), the gammaproteobacterial symbionts *Candidatus* Wigglesworthia glossinidia and *Candidatus* Aschnera chinzeii, respectively, migrate from the soma to colonize the milk gland. During embryogenesis, symbionts colonize offspring through milk secretions fed to the developing offspring (Rio et al., 2012), as shown in Fig. 2C. In the marine ascidian *Lissoclinum punctatum*, intracellular cyanobacterial symbionts are housed within tunic cells (a mesenchyme-like tissue that overlies the epidermis, Hirose, 2009). Instead of colonizing embryos during embryogenesis, symbionts transfer to offspring by direct tissue contact when larvae swim out of the mother's tunic (Kojima & Hirose, 2012). Similarly, genetic data on the host-associated epithelial microbiome communities in vertebrates indicates that many of these associations are transmitted from mother to offspring by contact (Funkhouser & Bordenstein, 2013).

Soma-to-soma vertical transmission strategies need not be contained within host tissues. This is best exemplified by the wide diversity of stink bug species that transfer their symbionts externally, in extracellular host-derived secretions. There are two main strategies for external transmission reported to date among stink bugs and relatives (Insecta: Heteroptera: Pentatomomorpha): egg-smearing with symbiont-containing secretions or deposition of symbiont-containing capsules. Species in the families Acanthosomatidae, Cydnidae, Pentatomidae, Pyrrhocoridae, and Scutelleridae apply symbiont-containing secretions to egg surfaces during oviposition (Hosokawa et al., 2013; Kaiwa et al., 2010; Kaiwa et al., 2011; Kikuchi et al., 2009; Prado & Almeida, 2009; Prado, Rubinoff, & Almeida, 2006). After hatching, nymphs probe the surface of the remaining egg mass and acquire symbionts (Hosokawa et al., 2013). In an amazing behavioral modification of egg-smearing, females of the subsocial stink bug *Parastrachia japonensis* wait until 5 min before egg hatching to apply symbiont-containing mucus to the eggs, conferring their infection (Hosokawa et al., 2012). A modification of egg-smearing is seen in stink bugs in the family Urostylididae, which produce a symbiont-infected jelly that coats eggs upon oviposition (Kaiwa et al., 2014). Production and deposition of symbiont-containing capsules can be seen in the Japanese common plataspid stink bug, *Megacopta punctatissima*. In this species, gammaproteobacterial symbionts are deposited in specialized capsules alongside egg masses. Upon hatching, nymphs consume these capsules, acquiring the symbionts needed for normal growth and development (Fukatsu & Hosokawa, 2002).

3. Vertical transmission through the germline part 1: Migration from the soma to oocytes/embryos

The vast majority of vertically transmitted symbioses exhibit transmission strategies involving the transfer of symbionts from somatic tissue directly to mature gametes or offspring (vertical transmission: soma-germline diagram in Fig. 1 and Supplemental Table 1 in the online version at https://doi.org/10.1016/bs.ctdb.2019.04.007). It is perhaps not surprising that the majority of endosymbionts exhibit very specific tissue distributions and only reside within bacteriocytes in adults, but it does suggest that many vertically transmitted symbionts cannot or have not evolved mechanisms to remain continuously associated with the germline. Below, we describe three general categories of this pattern based upon how and when symbionts become associated with host gametes or offspring, and the complexity of host tissues.

3.1 Transmission without tissue types: Sponge symbiont transmission

Basal animal taxa such as sponges present an excellent system to investigate basic regulation of symbiont localization and transmission, as they lack true tissues, and simply consist of multiple cell types (Cavalier-Smith, 2017). As depicted in Fig. 3A, sponge bodies consist of choanocyte (feeding cell) chambers suspended in mesohyl, lined with a layer of pinacocytes to serve as an external barrier. The mesohyl makes up most of the sponge by volume, and is an extracellular matrix consisting of connective tissue, sponge cells, and bacterial symbionts (Hentschel, Piel, Degnan, & Taylor, 2012). Oocytes are located in the mesohyl (Usher, Kuo, Fromont, & Sutton, 2001), and in cases of internal fertilization and viviparity, embryos are also brooded in the mesohyl, sometimes in specialized chambers (Kaye, 1991; Schmitt, Weisz, Lindquist, & Hentschel, 2007; Sharp, Eam, Faulkner, & Haygood, 2007; Vacelet, Fiala-Médioni, Fisher, & Boury-Esnault, 1996).

During sexual reproduction in sponges, four main strategies for vertical symbiont transmission have been observed (Ereskovsky, Gonobobleva, & Vishnyakov, 2005): (1) symbionts are phagocytosed into oocytes from their extracellular locations in the adult mesohyl (Maldonado, 2007), (2) bacteriocytes transfer symbionts to brooded embryos (Maldonado, 2007; Schmitt, Angermeier, Schiller, Lindquist, & Hentschel, 2008), (3) symbionts are transferred extracellularly via mucus (Kaye, 1991), and (4) symbionts are sequestered in the extracellular space between the egg and the follicle cells during oogenesis, with the symbionts entering the embryo during cleavage (Ereskovsky & Bouryesnault, 2002). Even in the simplest of multicellular body plans, there are a wide diversity of mechanisms and strategies for accomplishing proper localization patterns in the soma and germline. Sponges are also capable of asexual reproduction by fragmentation (Hentschel et al., 2002), further increasing the strategies available to them for symbiont transmission.

3.2 Extracellular storage of vertically transmitted intracellular symbionts

A distinct transmission strategy involves not only the extracellular transfer of endosymbionts during the vertical transmission process, but also their storage. For example, in the whiteflies *Aleurochiton aceris* and *Bemisia tabaci*, symbionts are positioned under the vitelline envelope, outside the oocyte plasma membrane. In these host species, the primary and secondary bacterial

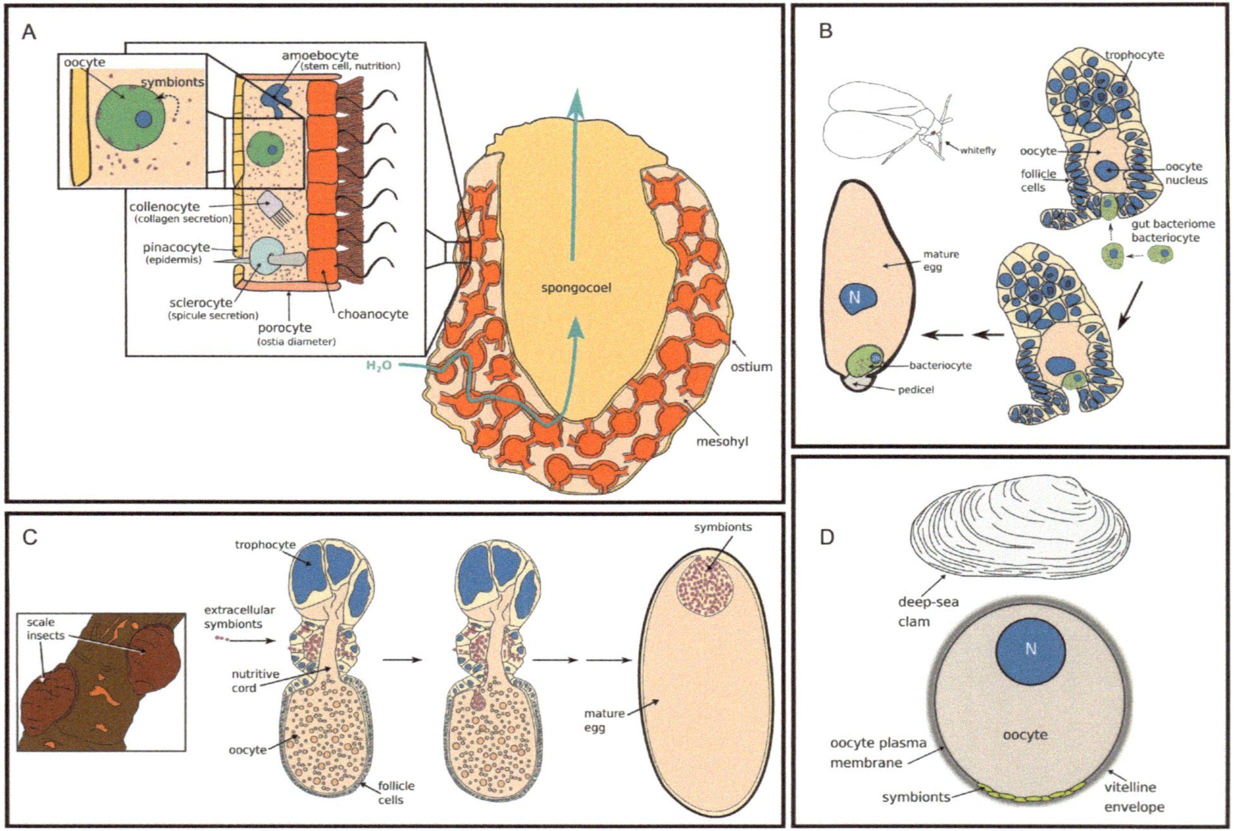

Fig. 3 See legend on opposite page.

symbionts travel together in host bacteriocytes from the body cavity to the ovary, beginning at the last larval instar stage. There, they pass between the posterior follicle cells to sit next to the oocyte plasma membrane, becoming enclosed between it and the vitelline envelope produced by the follicle cells (Costa et al., 1996; Szklarzewicz & Moskal, 2001) (Fig. 3B). Between five and seven bacteriocytes associate with each oocyte, which contain dense aggregations of actin. The bacteriocytes appear to maintain an extracellular position in both *A. aceris* (Szklarzewicz & Moskal, 2001) and *B. tabaci* (Costa et al., 1996) through the end of oogenesis. In a similar strategy, the gammaproteobacterial symbionts of *Greenisca brachypodii* scale insects, *Candidatus* Kotejella greeniscae and *Arsenophonus*, migrate between and through follicle cells to cluster at the connection between the oocyte and supporting cells. Once there, they cluster at the anterior of the developing oocyte in a deep, but extracellular depression of the oocyte plasma membrane, and are enclosed by the egg envelope (Michalik et al., 2018) (Fig. 3C).

While extracellular routes of oocyte-mediated symbiont transmission are relatively rare in insects, they may be common in marine molluscs, as the vesicomyid and solemyid species examined to date exhibit evidence consistent with this trend. In the deep-sea chemosynthetic clam *Calyptogena okutanii*, the chemosynthetic gammaproteobacterial symbionts were shown to be located under the vitelline envelope at the vegetal/posterior pole of mature oocytes (Ikuta et al., 2016) (Fig. 3D). Similarly, in *Solemya velum*, symbionts may be associated with the oocyte perimeter (Russell, McCartney, & Cavanaugh, 2018). It is not clear why this strategy is used

Fig. 3 Examples of soma-to-germline vertical transmission strategies that use extracellular routes. (A) Illustration of a demosponge body plan and route of symbiont transfer to oocytes/embryos from extracellular populations in the mesohyl. (B) In whiteflies, bacteriocytes containing bacterial symbionts migrate from the gut bacteriome to the ovary and become associated with the perivitelline space between the oocyte plasma membrane and the follicle cells prior to vitelline envelope formation (Szklarzewicz & Moskal, 2001). This results in the bacteriocyte being located under the shell, but outside the oocyte, at the posterior of mature eggs (Costa, Toscano, & Henneberry, 1996), as illustrated. (C) The two gammaproteobacterial symbionts of scale insects migrate extracellularly from dissociated bacteriocytes in the gut bacteriome to the ovary, where they cross between or through follicle cells to become associated with the perivitelline space between oocyte and follicle cell plasma membranes. As illustrated, this collection of extracellular symbionts becomes enclosed in the perivitelline space in the mature oocyte (Michalik et al., 2018). (D) Similarly, the vertically transmitted symbionts of deep-sea chemosynthetic clams localize to the perivitelline space of broadcast spawned eggs (Ikuta et al., 2016), as illustrated.

in these species, as both broadcast spawn their gametes directly into the oceanic water column, exposing them to environmental dangers. Perhaps these related chemosynthetic symbionts, which are more closely related to free-living colorless sulfur-oxidizing bacteria than other endosymbionts, do not possess the mechanisms needed to colonize the oocyte directly, or colonization is prohibitive to host development.

3.3 Intracellular symbiont transmission from the soma to the germline involves cell-to-cell transfer

Obligate endosymbiosis is at a very high frequency among hemipteran insect taxa, such as aphids, mealybugs, whiteflies, and planthoppers, due to an exclusive diet of nutrient-poor plant fluids (Hansen & Moran, 2014). In these associations, symbionts provide metabolites such as amino acids to supplement the host's nitrogen-poor diet (Douglas, 2016). The majority of these taxa host one primary symbiont, which is always present and co-speciates with the host, and one or more secondary symbionts, which are more facultative and have their own evolutionary histories independent of the host. Both primary and secondary symbionts are vertically transmitted in these associations (Douglas, 2016), but they may have different cellular routes of inheritance (Koga et al., 2012). In many of these associations, intracellular symbionts pass from adult bacteriocyte cells to gametes or embryos through cell-to-cell transfer mechanisms. Below, we describe what is known about the best studied of these associations.

In pea aphids, *Acyrthosiphon pisum*, females reproduce by either sexual or asexual reproduction depending on the season. Over the summer months, many asexual generations are produced via telescoping viviparous parthenogenesis (in which adult aphids contain embryonic aphids, containing embryonic aphids), and sexual eggs are produced for overwintering. In aphid sexual reproduction, the gammaproteobacterial endosymbiont *Candidatus* Buchnera aphidicola is delivered directly to the posterior pole of the oocyte in the final stages of oogenesis via cell-to-cell transfer from the follicle cells (Miura et al., 2003) (Fig. 4). During parthenogenetic embryogenesis, *Buchnera* are transported in host bacteriocytes to the posterior pole of the blastula (Miura et al., 2003). There, *Buchnera* are exocytosed from the bacteriocytes and endocytosed by the blastula membrane, and incorporated into the syncytial cytoplasm (Koga et al., 2012). These bacteria localize near host nuclei in the mesodermal syncytium and become enclosed in individual cells during cellularization, producing a new generation of bacteriocytes (Fig. 4). Interestingly, the secondary symbiont *Serratia* is taken

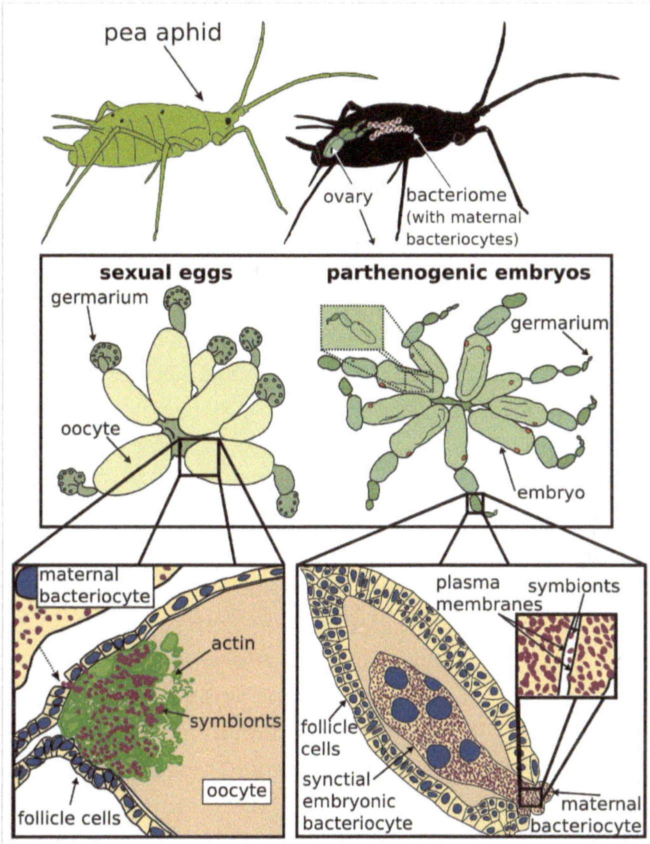

Fig. 4 Intracellular soma-to-germline vertical transmission strategies illustrated for sexual and parthenogenic pea aphids infected with the primary gammaproteobacterial symbiont, *Buchnera aphidicola*. In both modes of host reproduction symbionts are transmitted from maternal bacteriocytes, that originate in the gut bacteriome, to the posterior of developing oocytes or embryos through cell-to-cell transfer. During development, the bacteriome remains in close contact with the germband, ultimately residing near both the gut and the ovary (Braendle et al., 2003; Koga et al., 2012; Miura et al., 2003).

up by the blastula from the hemolymph, and sorts separately from *Buchnera* during cellularization (Koga et al., 2012). During later stages of embryogenesis, bacteriocytes cluster together, forming a paired bacteriome organ that remains in close proximity to the germ cells throughout development, maintaining this position in the adult (Koga et al., 2012; Miura et al., 2003). As a limited number of *Buchnera* are transferred in either reproduction mode, the vertical transmission process imposes a fairly harsh bottleneck on within-host symbiont population sizes (Mira & Moran, 2002).

Appropriation of symbionts to the germline or embryo from somatic tissues late in development is a common strategy for vertical transmission, and routes between tissues can be complex. For example, the *Arsenophonus*-like bacterial endosymbiont of human lice, *Candidatus* Riesia pediculicola (Allen, Reed, Perotti, & Braig, 2007), completes a complex pattern of migration across host cells and tissues during development, crossing extracellularly from germline in embryogenesis to the soma during the nymphal stages, and back to the ovary in the adult. There, symbionts pass into fully developed eggs through hydropyle structures in the shell (Perotti, Allen, Reed, & Braig, 2007). Similarly, during embryogenesis the fat-body bacteroidetes symbionts of cockroaches and the related basal termites (genus *Mastotermes*) migrate extracellularly from the embryonic bacteriome to pre-bacteriocyte cells in the fat body (Bandi et al., 1995; Lambiase, Grigolo, Laudani, Sacchi, & Baccetti, 1997), and then migrate from there to the ovary in the second nymphal instar (Sacchi et al., 1988). During the third and fourth instars, symbionts exit bacteriocytes and migrate extracellularly across the ovariole sheath, between the follicle cells, and to the plasma membrane of the oocyte. There they are surrounded by microvilli until after vitellogenesis when these bacteria are taken up by oocytes via pseudopod-like extensions (Sacchi et al., 1988). In *Camponotus floridanus* carpenter ants, endosymbionts also have a dynamic pattern of migration during development, ending up in the midgut prior to metamorphosis. They are thought to migrate from this tissue to the ovary (Stoll, Feldhaar, Fraunholz, & Gross, 2010), colonizing oocytes shortly after division from the stem cell (Kupper, Stigloher, Feldhaar, & Gross, 2016). Lastly, in one of the more bizarre localization patterns reported, the gammaproteobacterial symbionts of mealybugs reside within a second betaproteobacterial endosymbiont, and are transported to oocytes in this configuration. The nested symbiont cells are transported within bacteriocytes from the symbiont-housing organ (bacteriome) in the abdomen to the ovary. There, they are released from host cells, cluster around the connection between the oocyte and supporting cells (similar to scale insects, Michalik et al., 2018), and are taken into the germline at this point (von Dohlen, Kohler, Alsop, & McManus, 2001).

Interestingly, several associations demonstrate the phenomenon of the "symbiont ball," where symbionts cluster together in a clumped, ball-shaped structure during transmission and embryogenesis. While some symbiont balls are bound by the oocyte membrane (Michalik et al., 2018), the examples below are not. In brown planthoppers, yeast-like symbionts appear to migrate from the adult fat body to the ovary, passing between follicle cells to enter the posterior oocyte cytoplasm in late oogenesis. There, they form a ball of symbionts that migrates, and ultimately colonizes the fat body in the embryonic abdomen (Nan et al., 2016). Similarly, the *Carsonella* and *Profftella* symbionts of the Asian citrus psyllid, *Diaphorina citri*, migrate extracellularly from the abdomen to the oocyte where they pass between the follicle cells and are incorporated in the oocyte as a ball (Dan, Ikeda, Fujikami, & Nakabachi, 2017). In the stink bugs *Nysius ericae*, *Nysius plebius*, and *Nithecus jacobaeae*, bacteriocytes containing gammaproteobacterial symbionts exist within membranes adjoining those of previtellogenic oocytes, and transfer symbionts across the membranes. As oocytes mature, symbionts form a ball at the oocyte anterior (Matsuura et al., 2012; Swiatoniowska, Ogorzalek, Golas, Michalik, & Szklarzewicz, 2013). Although it is not its normal distribution, a symbiont ball can also be seen in wMel-infected *Drosophila melanogaster* when symbiont transport via host microtubules is increased (Russell, Lemseffer, & Sullivan, 2018).

Passage of symbionts between follicle cells for direct uptake by vitellogenic host oocytes is another common theme in ovarial symbiont transmission strategies, as described above for scale insects, termites/cockroaches, planthoppers, and psyllids, and has been shown for the adelgid aphid *Adelges viridis* and its betaproteobacterial symbionts (Michalik, Gołas, Kot, Wieczorek, & Szklarzewicz, 2013). Fortunately, this is also how the *Spiroplasma* symbionts of the model organism *Drosophila melanogaster* are transmitted, enabling experiments to determine the underlying molecular and cellular mechanisms. Work by Herren et al. has shown that in infected *D. melanogaster*, *Spiroplasma* colonize the oocyte following extracellular transport from the hemolymph. After passing between the follicle cells of vitellogenic oocytes (stages 8–10), symbionts are endocytosed with yolk granules and use the Yolkless receptor involved in normal yolk uptake from follicle cells (Herren, Paredes, Schupfer, & Lemaitre, 2013). Given the diversity of symbionts that enter the oocyte through the peri-follicular space (Dan et al., 2017; Miura et al., 2003; Nan et al., 2016), the high yolk content of insect embryos (Izumi, Yano, Yamamoto, & Takahashi, 1994), and that intracellular pathogens have also been found to co-opt the yolk machinery for ovarial transmission (Herren et al., 2013), this is a potential mechanism for other endosymbionts.

Infection of oocytes in pre-vitellogenic stages of oogenesis is observed in other associations in addition to those in stink bugs. In the bulrush bug, *Chilacis typhae*, symbionts are housed in bacteriocyte-like cells in the midgut epithelium as well as in the ovary germaria. Symbionts enter oocytes from the surrounding cells near the posterior of the germarium (Kuechler, Dettner, & Kehl, 2011). Other examples of transfer to pre-vitellogenic oocytes may exist, however, resolving the position of symbionts in the germarium's dense tissue structure may limit the detection of this transmission route.

The full and precise details about how symbionts colonize host tissues during host development are not known for many associations, but much can be inferred from their localization patterns in adults. For example, *Rhipicephalus* spp. ticks host *Coxiella* sp. symbionts, which are present in the malpighian tubules of both males and females, as well as in the female gonad. While it is unclear when in development the *Coxiella* symbionts migrate to the ovary, Lalzar et al. showed that they are at high concentration in the oviduct and interstitial ovary cells, and associate with host oocytes beginning in mid-oogenesis. Interestingly, *Coxiella* concentrate at opposite poles during mid-oogenesis (stage 3) and become restricted to one pole by late oogenesis (stages 4–5) (Lalzar et al., 2014). Based on studies in *Drosophila*, this suggests that these symbionts may rely on the host actin and microtubule cytoskeleton and microtubule motor proteins (Ferree et al., 2005; Russell, Lemseffer, & Sullivan, 2018; Serbus & Sullivan, 2007).

4. Vertical transmission through the germline part 2: Migration from the soma to germ stem cells (GSCs)

While the examples above demonstrate that many symbionts colonize the germline late in gametogenesis, or even after fertilization during embryonic development, some symbionts have evolved strategies for colonizing the primordial germ stem cells. There are a number of reasons that a symbiont might be selected to colonize the germ stem cell. These include: (1) sequestration with the GSC to prevent the symbiont genome from accumulating mutations during cell division and to protect the symbionts from host cellular defenses (Radzvilavicius et al., 2016; Russell & Cavanaugh, 2017), (2) manipulation of host reproduction (Fast et al., 2011; Foray, Perez-Jimenez, Fattouh, & Landmann, 2018), and (3) insurance that the symbiont is present in all host offspring.

Transfer from somatic cells to the germ stem cell is well documented in *Wolbachia*, a widespread bacterial endosymbiont of insects and filarial nematodes (Correa & Ballard, 2016; Werren, Baldo, & Clark, 2008). During early embryogenesis in filarial nematodes, *Wolbachia* preferentially concentrate in the blastomere leading to lineages that produce the germline and hypodermal chords, a somatic tissue that runs the length of the body. However, when this blastomere divides, the bacteria are excluded from the daughter blastomere destined to form the germline, and instead concentrate in the daughter blastomere destined to form the hypodermis (Landmann et al., 2012). *Wolbachia* remains there until the fourth larval instar stage when a portion of the population migrates extracellularly from the hypodermis to the neighboring distal somatic sheath cells of the gonad, and from there colonize the germline stem cells. This process occurs in the female germline, but not the male. This migration is associated with a disruption in cortical actin, consistent with a *Wolbachia*-induced endocytic event. The bacteria then concentrate in the ovary syncytium, possibly using the host's actin-rich rachis structure for motility, or replication (Foray et al., 2018; Landmann et al., 2012). Interestingly, *Wolbachia's* requirement for host fertility may be partially explained by its localization to the germline stem cell. In *B. malayi*, the wBm strain of *Wolbachia* confers mitotic quiescence in the stem cells, but then activates transit-amplifying mitotic replication when cells divide away from the niche. In uninfected hosts, this process is incorrectly regulated, resulting in apoptosis and sterile oocytes (Foray et al., 2018; Landmann et al., 2012).

5. Vertical transmission through the germline part 3: Continuous germ cell association throughout development

Continuous association with the germline presents the most basic form of linkage with host reproduction for symbionts and appears quite important for bacterial taxa such as *Wolbachia*. While it is likely that other manipulative bacteria such as *Rickettsia* may exhibit similar patterns, the data are limited, so we present what is known currently about continuous germline association strategies, starting with the most basic form in unicellular hosts.

5.1 Segregation with host cell division: The first form of vertical transmission

The most rudimentary form of vertical transmission is symbiont segregation to host daughter cells during mitosis in unicellular hosts. Conceptually, this can be accomplished two ways: (1) via active coordination with the host cell machinery or (2) by having a high concentration of symbionts passively distributed around the dividing cell. Mitochondria, which arose via an ancient endosymbiotic event in the single-celled eukaryotic ancestor, appear to use both strategies in different organisms and cell types.

Active mechanisms that shuttle mitochondria along microtubules to daughter cells are used during asymmetric cell division in budding yeast and during *Drosophila* oogenesis (Mishra & Chan, 2014). Passive mechanisms involve mitosis-induced fission of multiple, fused mitochondria, which then are subdivided into daughter cells by cytokinesis (Kowald & Kirkwood, 2011).

In more recently evolved unicellular endosymbioses, both active and passive inheritance strategies are observed. For example, in *Hartmannella* species of amoeba, the alphaproteobacterial symbiont *Nucleicultrix* localizes to the host nucleus and segregates with the daughter nuclei during binary fission (i.e., cell division) (Schulz et al., 2014). In contrast, the methanotrophic archaeal symbionts of *Nyctotherus* spp. ciliates are randomly distributed among daughter cells following host cell division (van Hoek et al., 2000), suggesting they likely utilize a passive strategy.

These mechanisms for vertical transmission in single-celled hosts form the basis for intracellular interactions that not only enable transmission to the next generation, but facilitate the proper colonization of the somatic cells in multicellular hosts. For example, as discussed above, the wBm strain of *Wolbachia* segregates with particular host cell lineages during early embryogenesis in filarial nematodes (Landmann et al., 2012). If mechanisms exist to precisely control patterns of symbiont segregation to host daughter cells during cell division, these can be used to ensure proper tissue tropism during development. An example of this was discussed earlier for the pea aphid, *Acyrthosiphon pisum*, in which *Buchnera* and *Serratia* symbionts exhibit independent, tightly coordinated movements with host syncytial nuclei and cells during embryogenesis (Koga et al., 2012) (Fig. 4). Similarly, in *Drosophila*, *Wolbachia* segregate asymmetrically during the larval neuroblast stem cell divisions such that they fate map to specific regions of the adult brain (Albertson, Casper-Lindley, Cao, Tram, & Sullivan, 2009). In this way, vertically transmitted symbionts can achieve specific tissue localization patterns via faithful segregation with particular cell lineages during development.

5.2 Symbiont segregation with the primordial germline

Given their affinity for the germline, it not surprising that many *Wolbachia* strains that infect insect species appear to localize to the germline continuously throughout development. Patterns in continuous germline localization are variable from strain to strain, but commonalities exist among strategies in a somewhat modular fashion, indicating that different strains share different combinations of mechanisms. An overview of what is known about *Wolbachia* localization to the germ line during development is presented below (see Fig. 5).

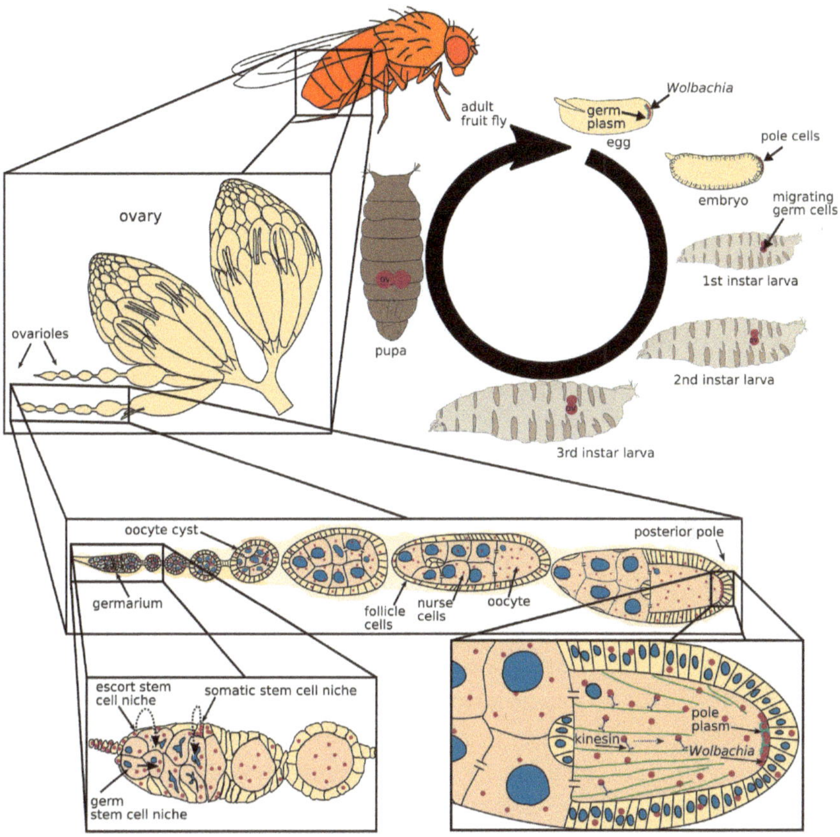

Fig. 5 Continuous germline association, as exemplified for *Drosophila* fruit flies infected with strains of *Wolbachia* bacteria. Some strains of *Wolbachia*, such as wWil in *D. willistoni*, localize to the germline throughout development by first localizing to the germline stem cell and somatic stem cell niches in the adult ovary. During oogenesis, *Wolbachia* localize to the germ plasm at the posterior pole of the oocyte. The localization patterns established during oogenesis persist in the embryo, and presence at the posterior pole of the oocyte/embryo enables the bacteria to be enclosed in host pole cells during cellularization. These cells then migrate during embryogenesis to form the primordial germline and future gonad (Serbus & Sullivan, 2007; Toomey, Panaram, Fast, Beatty, & Frydman, 2013; Veneti, Clark, Karr, Savakis, & Bourtzis, 2004).

In adult *Drosophila* flies, the wMel, wWil, and wAu strains of *Wolbachia* are found within the germline stem cells at the end of the ovariole terminal filaments (Ferree et al., 2005; Miller & Riegler, 2006). The wMel strain has been shown to segregate with both the regenerating germline stem cell and the differentiating daughter cell in *Drosophila melanogaster*, ensuring high vertical transmission fidelity (Albertson et al., 2009). Interestingly, while wMel segregate equally between germline stem cells and their daughters, they segregate unequally between neuroblast stem cells and daughter cells (Albertson et al., 2009), suggesting that *Wolbachia* are able to manipulate these processes depending on the cell type.

As oogenesis progresses, wMel replicate and are transported on host microtubules to colonize the nurse cells and developing *D. melanogaster* oocyte. In early oogenesis (stages 3–6), *Wolbachia* concentrate at the anterior end of the oocyte (Ferree et al., 2005; Serbus & Sullivan, 2007). This localization pattern is dependent on the host minus-end directed microtubule motor protein dynein, as both the depolymerization of microtubules and knockdown of dynein heavy chain or its dynactin linker protein disrupt anterior localization (Ferree et al., 2005). Following microtubule reorganization at stage 7 of oogenesis, *Wolbachia* become evenly distributed throughout the oocyte cytoplasm (Serbus & Sullivan, 2007). Starting at stage 9, the germ plasm begins to form in the cytoplasm at the posterior pole of the oocyte, which will form the germline in embryogenesis (Lu et al., 2018; Sinsimer, Lee, Thiberge, & Gavis, 2013). Slightly after germ plasm assembly begins, wMel use the host plus-end directed microtubule-dependent motor protein kinesin heavy chain to move to the posterior pole and colocalize with the germ plasm, as evidenced by inhibition of this process via microtubule depolymerization and kinesin heavy chain knockdown (Serbus & Sullivan, 2007). Importantly, wMel achieve kinesin-mediated transport without interfering in host development by being a poor competitor for this motor protein (see Box 1). Following stage 10a of oogenesis, the remaining *Wolbachia* contained in the nurse cells are dumped into the oocyte during nurse cell dumping (Russell, Lemseffer, & Sullivan, 2018; Serbus & Sullivan, 2007), which also appears to occur in the *Westeberhardia* gammaproteobacterial symbionts of *Cardiocondyla obscurior* ants (Klein et al., 2016).

Germ plasm colocalization during oogenesis is also reported for wMel in *Drosophila simulans*, wWil in *Drosophila willistoni*, wSty in *Drosophila teissieri, Drosophila yakuba, and Drosophila santomea*, wCle in *Cimex lectularius* bed bugs, wAtab in *Asobara tabida*, the strain in *Aphytis*, the strain in *Trichogramma, and* the strain in *Nasonia* wasps (Breeuwer & Werren, 1990; Dedeine et al., 2001; Hosokawa et al., 2010; Miller & Riegler, 2006; Pintureau et al., 2000; Veneti et al., 2004; Zchori-Fein, Roush, & Rosen, 1998). Interestingly, other strains exhibit alternative localization patterns that may mimic other patterning factors. *Wolbachia* strains wMo, wKa, and wKi in *Drosophila simulans* exhibit anterior localization, similar to the *bicoid* mRNA gradient during embryogenesis. In a range of *Drosophila* species, the wRi strain achieves an even distribution across late oocytes and embryos (Albertson et al., 2009). Compared to wMel, wRi colonizes *D. simulans*

BOX 1 Wolbachia use but do not abuse the host transport system.

Wolbachia's migration through the developing oocyte to the posterior pole plasm is coincident with recruitment of factors required for germline formation as well as for anterior/posterior and dorsal/ventral axis formation (Serbus & Sullivan, 2007). Thus, *Wolbachia* must navigate and use the host transport system without disrupting transport of vital host components. In oocytes in which *Wolbachia* titer is too high, dorsal/ventral axis determination is disrupted (Serbus et al., 2011). Insight into the mechanisms by which *Wolbachia* ensures its localization to the posterior pole without disrupting germline establishment comes from a recent study by Russell et al. (2018) demonstrating that *Wolbachia* is a weak competitor for the plus-end directed motor protein kinesin heavy chain. Knocking down a key kinesin linker protein, kinesin light chain (KLC) that associates with a number of host components required for pole plasm formation, surprisingly results in a dramatic increase of *Wolbachia* at the posterior pole. One interpretation of this result is that knocking down KLC results in a greater concentration of kinesin for *Wolbachia's* poleward transport. That is, kinesin is limiting for *Wolbachia* but not host components. Experimental support for this idea comes from finding that overexpressing kinesin in the oocyte results in a dramatic increase of *Wolbachia* at the posterior pole, similar to KLC knockdown. Thus, *Wolbachia* may have evolved to weakly compete with host cargo for association with motor proteins, ensuring that germline formation is not disrupted. A key next aim is to identify the kinesin linker protein used by *Wolbachia*. Whether this is a host protein or a protein encoded by *Wolbachia* is unknown.

and *D. melanogaster* oocytes at higher titers, ensuring that at least some *Wolbachia* are included in the posterior-located germ plasm (Serbus & Sullivan, 2007; Veneti et al., 2004).

In the embryo, wMel within the germ plasm become cellularized with the pole cells during cycle 11 of embryogenesis (Serbus & Sullivan, 2007; Veneti et al., 2004), and from within these primordial germ cells, undergo migration to the future gonad as in normal fly development (Asaoka, 2004). While strains such as wAu and wMel colonize the somatic embryo in addition to the pole cells (Albertson et al., 2009; Miller & Riegler, 2006; Serbus & Sullivan, 2007), wWil from *D. willistoni* exhibits strict tropism for the pole cells and the resulting germline stem cells (Miller & Riegler, 2006). Thus, several strains of *Wolbachia* are able to target the germline continuously throughout development, and some do so to the exclusion of the somatic tissue. See Boxes 2 and 3 for further discussions about how *Wolbachia* interacts with the germline to influence host reproduction.

BOX 2 Manipulation of host reproduction by *Wolbachia*-induced cytoplasmic incompatibility.

In addition to navigating the germline, *Wolbachia* interacts and dramatically influences a diversity of germline functions including transcription, translation, and the cell cycle (Correa & Ballard, 2016). Because *Wolbachia* is exclusively transmitted through the female germline, manipulations that serve to promote infected female fecundity and provide a selective advantage to infected females serve to increase *Wolbachia* infection frequencies in the population. These include *Wolbachia*-induced male killing, feminization, parthenogenesis and, most famously, cytoplasmic incompatibility (aka. CI).

Wolbachia-induced CI is a conditional form of male sterility. Matings between *Wolbachia*-infected males and uninfected females produce dramatically reduced hatch rates. However, if the females are infected, normal hatch rates occur whether she mates with infected or uninfected males. Thus, in an infected population, infected females are at a great selective advantage over uninfected females. Both in the laboratory and field settings, CI results in rapid sweeps of *Wolbachia* through insect populations (Kriesner, Hoffmann, Lee, Turelli, & Weeks, 2013; Turelli & Hoffmann, 1991). Cellular analysis of embryos derived from the CI-cross exhibit a failure of the paternal chromosome complement to condense and properly align on the metaphase plate. Consequently, there is either a partial or complete failure of paternal chromosome segregation during the first zygotic division. Subsequent analysis revealed the proximal cause of these defects in paternal chromosome dynamics is a delay in the protamine-histone exchange that occurs as the sperm transitions into a pronucleus. Likely as a consequence of these delays, DNA replication, nuclear envelope breakdown, CDK1 activation, and entry of the paternal chromosome set into anaphase are delayed. Interestingly, in crosses between infected females and infected males, condensation and segregation of the paternal chromosome set is normal. Furthermore, neither CDK1 activation nor mitosis is delayed (Landmann, Orsi, Loppin, & Sullivan, 2009), as would be expected if the infected female simply had matching modifications on its chromosomes. Instead, some *Wolbachia*-generated product in the female germline may reverse the male modification.

Recent insight into the molecular basis of CI may help resolve these two models. Proteomic studies of sperm derived from infected and uninfected male mosquitoes identified a wPip *Wolbachia* protein only present in the former. This protein, currently named CidA, is encoded in an operon containing a second gene, *cidB*, which encodes a deubiquitylating enzyme. Transgenic male *Drosophila* expressing CidA and CidB produce paternal chromosome segregation defects strikingly similar to those observed in CI crosses (Beckmann, Ronau, & Hochstrasser, 2017). Biochemical studies showed that CidA binds and inhibits CidB, inhibiting its deubiquitylating activity. These results support a toxin-antitoxin model for how sperm and oocyte modifications induced by *Wolbachia* function in cytoplasmic incompatibility. A parallel study by LePage et al. using wMel *Wolbachia* relied on bioinformatics approaches to identify bacterial genes that correlate with CI induction, returning CifA and CifB, homologs of the *cidAB* genes (LePage et al., 2017). These genes are encoded by a prophage integrated

BOX 2 Manipulation of host reproduction by *Wolbachia*-induced cytoplasmic incompatibility.—cont'd

into the wMel genome. A number of lines of evidence strongly suggest CifA and CifB induce CI. First, these genes are present in CI-inducing strains but absent in non-CI inducing strains. Second, as with the Beckman et al. (2017) study, expression of *cifA* and *cifB* in the germline of males expressing CI resulted in chromosome segregation defects equivalent to those observed in CI crosses and reduced hatch rates. Furthermore, normal hatch rates were recovered when these transgenic males were mated to *Wolbachia*-infected females.

These exciting findings raise a number of questions, most significantly what are the targets of the CidB deubiquitylating enzyme? What is the relationship between CidAB and the cell cycle and chromosome defects observed in CI crosses? How is rescue achieved by infected female oocytes? It is an exciting era for *Wolbachia* functional and genomic research.

BOX 3 *Wolbachia* control of germline differentiation.

Accumulating evidence from different strains of *Wolbachia* suggests that these intracellular symbionts have the ability to modulate host cell differentiation by controlling host gene expression. In *Drosophila melanogaster*, the wMel strain is able to rescue loss of the *sex-lethal* (*sxl*) and *bag-of-marbles* (*bam*) genes, both of which are involved in controlling germ stem cell maintenance and daughter cell differentiation in early oogenesis (Flores, Bubnell, Aquadro, & Barbash, 2015; Sun & Cline, 2009). Recently, Ote et al. (2016) found that expression of the *Wolbachia* protein Toxic manipulator of oogenesis (TomO) is able to rescue the germline stem cell maintenance function of Sxl. However, another *Wolbachia*-encoded factor must be involved in full *Wolbachia*-based *sxl* rescue, as TomO does not restore female fertility. The mechanism by which TomO rescues Sxl function appears to be through the disruption of mRNA complexes, which causes increased Nos expression (Ote, Ueyama, & Yamamoto, 2016). TomO has also been found to bind and disrupt other mRNA complexes at later stages of oogenesis, such as those bound to orb mRNA (Ote & Yamamoto, 2018), suggesting that this bacterial protein may broadly regulate host translation.

In support of the idea that *Wolbachia* is a general regulator of host cellular differentiation, *Wolbachia* control over this process has been shown in other strains and host taxa. Recently, it has been shown that the presence of wBm in *Brugia malayi* filarial nematode ovaries controls germ stem cell quiescence. If wBm are removed by antibiotic treatment, then the germ stem cells differentiate and are lost. Further along in the ovary, this inappropriate loss of quiescence manifests as oocyte apoptosis and polarity defects, culminating in a loss of fertility (Foray et al., 2018; Landmann, Voronin, Sullivan, & Taylor, 2011). *Wolbachia* have also been reported to be required for normal oogenesis in *Asobara tabida* wasps because they inhibit apoptosis (Pannebakker, Loppin, Elemans, Humblot, & Vavre, 2007). However, the results from filarial nematodes suggest that a mechanism also involving control of host cell differentiation may be involved upstream of the apoptotic effects.

Over the evolutionary history of *Wolbachia*, many strains have switched hosts (Riegler, Sidhu, Miller, & O'Neill, 2005; Turelli et al., 2018) through horizontal transmission events (Ahmed, Breinholt, & Kawahara, 2016; Le Clec'h et al., 2013; Li et al., 2017). Host switching requires mechanisms of cell-to-cell transfer and germline targeting to restart the vertical transmission process. Processes for both the release from and uptake into host cells happen readily for wMel in cell culture (White et al., 2017). *Wolbachia's* route from the soma to the germline is observed in experiments on whole female flies injected with wMel-infected hemolymph (Frydman, Li, Robson, & Wieschaus, 2006). After injection, *Wolbachia* migrate through the hemolymph to the ovary. There, they enter the oocytes through the germarium's somatic stem cell niche, as opposed to localizing to the germ stem cells or their niche directly. The normal pattern of germ stem cell localization is, however, resumed in the next vertically transmitted generation (Frydman et al., 2006). Given that the somatic stem cell niche is a route to the oocytes in other associations (discussed above for stink bugs and bulrush bugs), it may be the ancestral route for *Wolbachia*, used before these bacteria gained access to the germ stem cell.

One other association which may prioritize appropriating symbionts to the germline over the functional somatic tissue is the leafhopper *Scaphoideus titanus*. In *S. titanus*, *Cardinium* symbionts (Bacteroidetes) are transmitted to oocytes from adjacent bacteriocyte cells in the ovary. Interestingly, during nymphal development, symbionts are continuously associated only with the ovary, as the fat body and salivary gland are not colonized until adulthood (Sacchi et al., 2008), suggesting that the bacteria associated with the germline form a discrete population from the bacteria in the soma, with little need for exchange between the two.

5.3 Continuous germ line association is not the norm among strictly vertically transmitted symbionts

While continuous association with the germline may conceptually be the most intimate/integrated form of vertical transmission, it clearly cannot be in reality because few endosymbionts exhibit this pattern of inheritance. According to patterns of genome evolution, the obligate endosymbionts of plant-feeding insects are some of the most extreme examples of evolutionarily derived symbionts (relative to their free-living relatives) known. Compared to free-living taxa with genome sizes of 3–7 Mb, vertically transmitted insect symbionts possess genomes down to 0.2–0.6 Mb, often only containing genes for their symbiotic functions and even lacking genes for cellular maintenance

(Toft & Andersson, 2010). However, few of these associations remain associated with the germline continuously, and instead exhibit one of the various forms of soma-to-germline routes of vertical transmission.

In contrast, the genomes of many *Wolbachia* endosymbionts exhibit evidence that they experience environments other than the host cell, suggesting that they are less integrated with their hosts than some of the endosymbionts that have to migrate to the ovary from the soma. The genomes of the various *Wolbachia* strains are around 1.3 Mb and contain abundant mobile elements (Klasson et al., 2009; Lindsey, Werren, Richards, & Stouthamer, 2016; Wu et al., 2004). As the cut and paste activities of mobile elements across the genome are generally deleterious, the elements become inactivated and lost over time. Thus, finding active copies in the genome indicates that these bacteria experience regular exposure to the environment, during which time they acquire new, functional mobile elements (Toft & Andersson, 2010). Interestingly, the strains in supergroups C and D do exhibit signs of genome degradation and stasis, consistent with their obligate nature and strict vertical transmission strategy (Comandatore et al., 2015). So, while highly integrated endosymbionts may continuously associate with the germline, they do not seem to require it.

6. Parallels between symbiont transmission and the origin of the germline

Similar to the evolution of multicellularity and the germline itself, a range of solutions have evolved for handling and distributing inherited endosymbionts among cells and tissues in organisms of differing cellular and structural complexity. Multicellularity has independently evolved at least 25 times in organisms ranging from bacteria to single-celled eukaryotes (Rokas, 2008). While all of these make some form of reproductive cell type, not all have a dedicated and sequestered tissue type for it, i.e., a germline. For example, no discrete germline evolved in plants and basal animal taxa, such as cnidaria, and platyhelminthes. Instead, gametes are generated in the adult from pluripotent stem cells in the soma (Radzvilavicius et al., 2016). Thus, a germline is clearly not necessary for sexual reproduction and multicellular development. Similarly, despite vertically-transmitted endosymbionts' ties to host reproduction for their own reproduction, continuous localization within the germline or pluripotent stem cells is not generally observed (see Sections 2 and 3 above, and Supplemental Table 1 in the online version at https://doi.org/10.1016/bs.ctdb.2019.04.007).

It is interesting to consider how the population of host tissues and sequestration of a germline affects mitochondria, the intracellular relict of the endosymbiont that enabled oxidative phosphorylation in the ancestral eukaryote. Oogamy and uniparental mitochondrial inheritance evolved long before germline sequestration, and cannot be implicated in its evolution via older hypotheses such as to avoid competition among mitochondrial genotypes (Cosmides & Tooby, 1981; Eberhard, 1980) or to enable tissue complexity (Ispolatov, Ackermann, & Doebeli, 2012). Instead, it has been proposed that mitochondrial loss of function mutations drove the sequestration of the germline in animal early development. When per-generation mitochondrial mutation rates are high, either due to a high per-division mutation rate (e.g., due to metabolically-induced damage (Goldsby, Knoester, Ofria, & Kerr, 2014)) or a high number of genome/cellular replications, or both, sequestration of the germline serves to limit the number of mutations acquired by the mitochondrial genome. Over-replication of mitochondria during oogamy further enables effective selection for the best phenotype/genotype, restoring the functions served by selecting mitochondria from a large somatic population (Radzvilavicius et al., 2016).

During multicellular development endosymbionts are not distributed to every cell type in the body like mitochondria organelles, but are restricted to particular tissues like differentiated cell types (Cheng & Aksoy, 1999; Dirks et al., 2012; Wentrup, Wendeberg, Huang, Borowski, & Dubilier, 2013). Assuming the same population-level principles apply to endosymbionts as mitochondria, the Radzvilavicius et al. 2016 theory suggests that mutation accumulation is not an issue for most endosymbionts. This could potentially be due to low symbiont mutation rates and/or the inability to ramp up symbiont copy number during oogamy to enable within-individual selection for fit genotypes (Radzvilavicius et al., 2016). In these cases, the large variance in mutational loads of symbionts across host cells may help ensure a functional symbiont is available for transmission. Many endosymbiont populations number in the millions to billions of cells in somatic tissue (Duperron, Quiles, Szafranski, Léger, & Shillito, 2016; Komaki & Ishikawa, 2000; Mitchell & Cavanaugh, 1983; Sender, Fuchs, & Milo, 2016; Wollenberg & Ruby, 2009) and are typically at much lower abundances in the germline (Ferree et al., 2005; Ikuta et al., 2016; Mira & Moran, 2002; Russell, McCartney, & Cavanaugh, 2018). Thus, mechanisms may be selected that enable the choice of symbionts from the much larger somatic pool of genotypes to occupy the germline. Perhaps relevant to this topic, population sizes of the human head lice endosymbiont (Perotti et al., 2007) and the carpenter ant endosymbionts (Stoll et al., 2010) change radically throughout development.

Another popular theory for the evolution of multicellularity and germline sequestration proposes that cellular processes required for animal life cause mutational damage, thus somatic cells have been relegated to the job, protecting germ cells from mutations that could be passed to offspring (Goldsby et al., 2014). While this is unlikely to be the driver for endosymbionts, which are generally thought to have evolved associations with their hosts for their functions in somatic tissues (Moran, 2007), it may offer insights into influential processes. For example, if symbionts are unable to regulate their metabolic functions on a tissue-level basis, then they could cause more harm than benefit if they were to continuously associate with the germline. By restricting symbionts to very specific cell types (i.e., bacteriocytes) in host tissues, often in large numbers, the mutational damage can be kept isolated, and a large symbiont pool is left available to be selected from to occupy the germline. See Box 4 for a further discussion of the evolutionary pressures facing somatic and germline-associated symbiont populations.

BOX 4 Endosymbionts: case-studies in kin selection.

As a whole, endosymbionts with somatic populations that function to better host fitness and transmit a subset of their population to host offspring present a case study in kin selection (Griffin & West, 2002). Next to nothing is known about how or when symbionts are selected for transmission or relegation to a lifetime in the soma, with no individually-gained fitness. Maintaining low intrahost genetic diversity likely helps, as there is nothing on which to select (Russell & Cavanaugh, 2017). Alternatively, selecting symbionts from the somatic population, as many associations do, may help keep the population honest and prevent "cheaters," i.e., symbionts receiving benefit from, but not providing services to the host. Thus, as long as they are genetically related, symbionts in the soma increase their fitness by enabling symbionts in the germline to reproduce and be transmitted to host offspring. This is analogous to how sibling reproduction increases the inclusive fitness of sexual eukaryotes (Griffin & West, 2002).

Endosymbionts with more general tissue tropisms and less well-resolved contributions to host fitness, such as *Wolbachia*, present an interesting thought experiment on how these situations evolve. Somatic support for germline symbionts can either evolve as primary factors, when symbionts start out as tissue infections, or secondary factors, when germline-associated bacteria colonize the germline to improve host fitness. As there are few reasons besides selfish ones for a symbiont to exclusively colonize the germline, the second situation likely represents an initially pathogenic/parasitic one. *Wolbachia's* fairly unspecific somatic tissue distribution, and its lack of a need for specific bacteriocyte cells

Continued

BOX 4 Endosymbionts: case-studies in kin selection.—cont'd

(Pietri, DeBruhl, & Sullivan, 2016) suggests that *Wolbachia* began as a general somatic infection, like some *Rickettsia* species (Lalzar et al., 2014), and has evolved highly specific germline-association mechanisms from there. However, intriguingly, in bed bugs, *Cimex lectularius, Wolbachia* occupies bacteriocytes and produces vitamin B12 for the host (Hosokawa et al., 2010). The alternative is also possible, that it was originally exclusively germline associated, which implies that a pathogenic relationship turned commensal and then beneficial, or at least addictive (Sullivan, 2017), sometime in the history of its relationship with insects. In either case, once symbiont and host reproductive interests have become linked, symbiont populations should be selected to be well mixed and have minimal diversity so that symbiont reproduction in the germline fulfills the fitness interests of the symbionts in the soma.

7. Conclusions

Our review of the literature reveals that the majority of microbial symbionts across a diversity of host and symbiont taxa are transmitted to the host germline from somatic cells or tissues, making association with the host germline rare for endosymbionts. It is not clear why this pattern prevails, but it could be due to either a lack of need for continuous association or a constraint involving its evolution. For example, it is conceivable that the presence of symbionts at particular stages of primordial germline formation could disrupt the process.

Bacterial access to the germline has impacted host genome evolution and has prepared hosts for interacting with microbial symbionts. It has been repeatedly observed that insect genomes contain horizontally transferred genes from bacterial genomes. As any horizontally transferred gene must make its way to the germline to be inherited with the rest of the genome, transfers from germline-associated symbionts have high likelihoods of inheritance. While some of these transfers are from recent endosymbionts, such as the new sex chromosome in the isopod *Armadillidium vulgare* (Leclercq et al., 2016), others are from unassociated, or not currently associated bacteria. For example, mealybug genomes contain genes from diverse bacterial lineages that likely supplement functions lost from their endosymbionts' reduced genomes (Husnik et al., 2013). Psyllid genomes contain metabolic genes from many different bacteria that complement symbiont pathways (Sloan et al., 2014), and aphid genomes contain non-*Buchnera* genes whose products are transferred to the *Buchnera* symbionts (Nakabachi, Ishida, Hongoh, Ohkuma, & Miyagishima, 2014). Furthermore, in a recent preprint, Blondel et al. propose that the *oskar* gene, which is necessary and sufficient to induce the formation of the germline in holometabolous insects, is the product of a horizontal gene transfer event combined with a fusion event between a bacterial gene and a eukaryotic gene (Blondel, Jones, & Extavour, 2018). Thus, germline-associated endosymbionts may even be implicated in the formation of the germline itself.

While these studies have provided significant insights regarding the patterns and mechanisms of vertical endosymbiont transmission, much remains unknown. Key outstanding questions include: Why is continuous association with the germline rare in vertically inherited endosymbionts? Are there events during germline development that are particularly susceptible to interference by the presence of an endosymbiont? What are the molecular and cellular mechanisms by which different endosymbionts target and associate with the host germline? What are the host and endosymbiont factors that determine endosymbiont abundance in the germline? How genetically diverse are within-host symbiont populations, and how does that diversity impact the inclusive fitness of symbionts relegated to the soma (and thus destined never to reproduce)? And similarly, how do vertical transmission strategies impact the size and genetic diversity (i.e., genetic bottleneck) of inherited symbiont populations? As more knowledge is gained about host and symbiont genomics, and we learn how symbiont and host genes function, answers to these questions will be forthcoming.

References

Ahmed, M. Z., Breinholt, J. W., & Kawahara, A. Y. (2016). Evidence for common horizontal transmission of *Wolbachia* among butterflies and moths. *BMC Evolutionary Biology*, *16*, 118.

Albertson, R., Casper-Lindley, C., Cao, J., Tram, U., & Sullivan, W. (2009). Symmetric and asymmetric mitotic segregation patterns influence *Wolbachia* distribution in host somatic tissue. *Journal of Cell Science*, *122*, 4570–4583.

Allen, J. M., Reed, D. L., Perotti, M. A., & Braig, H. R. (2007). Evolutionary relationships of "Candidatus Riesia spp.," endosymbiotic enterobacteriaceae living within hematophagous primate lice. *Applied and Environmental Microbiology*, *73*, 1659–1664.

Asaoka, M. (2004). Germline stem cells in the Drosophila ovary descend from pole cells in the anterior region of the embryonic gonad. *Development*, *131*, 5079–5089.

Bandi, C., Sironi, M., Damiani, G., Magrassi, L., Nalepa, C. A., Laudani, U., et al. (1995). The establishment of intracellular symbiosis in an ancestor of cockroaches and termites. *Proceedings of the Biological Sciences*, *259*, 293–299.

Beckmann, J. F., Ronau, J. A., & Hochstrasser, M. (2017). A *Wolbachia* deubiquitylating enzyme induces cytoplasmic incompatibility. *Nature Microbiology*, *2*, 17007.

Blondel, L., Jones, T. E. M., & Extavour, C. G. (2018). Bacterial contribution to genesis of the novel germ line determinant oskar. *bioRxiv*, 453514. https://doi.org/10.1101/453514.

Braendle, C., Miura, T., Bickel, R., Shingleton, A. W., Kambhampati, S., & Stern, D. L. (2003). Developmental origin and evolution of bacteriocytes in the aphid–*Buchnera* symbiosis. *PLoS Biology*, *1*, e21.

Breeuwer, A. J., & Werren, J. H. (1990). Microorganisms associated with chromosome destruction and reproductive isolation between two insect species. *Nature*, *346*, 558–560.

Bright, M., & Bulgheresi, S. (2010). A complex journey: Transmission of microbial symbionts. *Nature Reviews Microbiology*, *8*, 218–230.

Cavalier-Smith, T. (2017). Origin of animal multicellularity: Precursors, causes, consequences—the choanoflagellate/sponge transition, neurogenesis and the Cambrian explosion. *Philosophical Transactions of the Royal Society, B: Biological Sciences*, *372*, 20150476.

Cavanaugh, C. M., McKiness, Z. P., Newton, I. L. G., & Stewart, F. J. (2006). Marine chemosynthetic symbioses. In M. Dworkin, S. Falkow, E. Rosenberg, K.-H. Schleifer, & E. Stackebrandt (Eds.), *The prokaryotes* (pp. 475–507). New York, NY: Springer.

Cheng, Q., & Aksoy, S. (1999). Tissue tropism, transmission and expression of foreign genes in vivo in midgut symbionts of tsetse flies. *Insect Molecular Biology*, *8*, 125–132.

Chu, H., & Mazmanian, S. K. (2013). Innate immune recognition of the microbiota promotes host-microbial symbiosis. *Nature Immunology*, *14*, 668–675.

Comandatore, F., Cordaux, R., Bandi, C., Blaxter, M., Darby, A., Makepeace, B. L., et al. (2015). Supergroup C *Wolbachia*, mutualist symbionts of filarial nematodes, have a distinct genome structure. *Open Biology*, *5*, 150099.

Correa, C. C., & Ballard, J. W. O. (2016). Wolbachia associations with insects: Winning or losing against a master manipulator. *Frontiers in Ecology and Evolution*, *3*, 153.

Cosmides, L. M., & Tooby, J. (1981). Cytoplasmic inheritance and intragenomic conflict. *Journal of Theoretical Biology*, *89*, 83–129.

Costa, H. S., Toscano, N. C., & Henneberry, T. J. (1996). Mycetocyte inclusion in the oocytes of Bemisia argentifolii (Homoptera: Aleyrodidae). *Annals of the Entomological Society of America*, *89*, 694–699.

Dan, H., Ikeda, N., Fujikami, M., & Nakabachi, A. (2017). Behavior of bacteriome symbionts during transovarial transmission and development of the Asian citrus psyllid. *PLoS One*, *12*, e0189779.

Dedeine, F., Vavre, F., Fleury, F., Loppin, B., Hochberg, M. E., & Boulétreau, M. (2001). Removing symbiotic Wolbachia bacteria specifically inhibits oogenesis in a parasitic wasp. *Proceedings of the National Academy of Sciences*, *98*, 6247–6252.

Dirks, U. (2011). *Novel strategies of maintaining and transmitting specific microbial symbionts in marine flatworms, uniwien.* Dissertation. Fakultät für Lebenswissenschaften: University of Vienna.

Dirks, U., Gruber-Vodicka, H. R., Leisch, N., Bulgheresi, S., Egger, B., Ladurner, P., et al. (2012). Bacterial symbiosis maintenance in the asexually reproducing and regenerating flatworm paracatenula galateia. *PLoS One*, 7, e34709.

Douglas, A. E. (2016). How multi-partner endosymbioses function. *Nature Reviews. Microbiology*, *14*, 731–743.

Drown, D. M., Zee, P. C., Brandvain, Y., & Wade, M. J. (2013). Evolution of transmission mode in obligate symbionts. *Evolutionary Ecology Research*, *15*, 43.

Duperron, S., Quiles, A., Szafranski, K. M., Léger, N., & Shillito, B. (2016). Estimating symbiont abundances and gill surface areas in specimens of the hydrothermal vent mussel bathymodiolus puteoserpentis maintained in pressure vessels. *Frontiers in Marine Science*, *3*, 16.

Eberhard, W. G. (1980). Evolutionary consequences of intracellular organelle competition. *The Quarterly Review of Biology*, *55*, 231–249.
Ebert, D. (2013). The epidemiology and evolution of symbionts with mixed-mode transmission. *Annual Review of Ecology, Evolution, and Systematics*, *44*, 623–643.
Ereskovsky, A. V., & Bouryesnault, N. (2002). Cleavage pattern in Oscarella species (Porifera, Demospongiae, Homoscleromorpha): Transmission of maternal cells and symbiotic bacteria. *Journal of Natural History*, *36*, 1761–1775.
Ereskovsky, A. V., Gonobobleva, E., & Vishnyakov, A. (2005). Morphological evidence for vertical transmission of symbiotic bacteria in the viviparous sponge Halisarca dujardini Johnston (Porifera, Demospongiae, Halisarcida). *Marine Biology*, *146*, 869–875.
Fast, E. M., Toomey, M. E., Panaram, K., Desjardins, D., Kolaczyk, E. D., & Frydman, H. M. (2011). Wolbachia enhance drosophila stem cell proliferation and target the germline stem cell niche. *Science*, *334*, 990–992.
Ferree, P. M., Frydman, H. M., Li, J. M., Cao, J., Wieschaus, E., & Sullivan, W. (2005). Wolbachia utilizes host microtubules and dynein for anterior localization in the Drosophila oocyte. *PLoS Pathogens*, *1*, e14.
Fisher, R. M., Henry, L. M., Cornwallis, C. K., Kiers, E. T., & West, S. A. (2017). The evolution of host-symbiont dependence. *Nature Communications*, *8*, 15973.
Flores, H. A., Bubnell, J. E., Aquadro, C. F., & Barbash, D. A. (2015). The *Drosophila* bag of marbles gene interacts genetically with *Wolbachia* and shows female-specific effects of divergence. *PLoS Genetics*, *11*, e1005453.
Foray, V., Perez-Jimenez, M. M., Fattouh, N., & Landmann, F. (2018). Wolbachia control stem cell behavior and stimulate germline proliferation in filarial nematodes. *Developmental Cell*, *45*, 198–211.
Frydman, H. M., Li, J. M., Robson, D. N., & Wieschaus, E. (2006). Somatic stem cell niche tropism in Wolbachia. *Nature*, *441*, 509–512.
Fukatsu, T., & Hosokawa, T. (2002). Capsule-transmitted gut symbiotic bacterium of the Japanese common plataspid stinkbug, Megacopta punctatissima. *Applied and Environmental Microbiology*, *68*, 389–396.
Funkhouser, L. J., & Bordenstein, S. R. (2013). Mom knows best: The universality of maternal microbial transmission. *PLoS Biology*, *11*, e1001631.
Geurts, R., Xiao, T. T., & Reinhold-Hurek, B. (2016). What does it take to evolve a nitrogen-fixing endosymbiosis? *Trends in Plant Science*, *21*, 199–208.
Goldsby, H. J., Knoester, D. B., Ofria, C., & Kerr, B. (2014). The evolutionary origin of somatic cells under the dirty work hypothesis. *PLoS Biology*, *12*, e1001858.
Griffin, A. S., & West, S. A. (2002). Kin selection: Fact and fiction. *Trends in Ecology & Evolution*, *17*, 15–21.
Hansen, A. K., & Moran, N. A. (2014). The impact of microbial symbionts on host plant utilization by herbivorous insects. *Molecular Ecology*, *23*, 1473–1496.
Hentschel, U., Hopke, J., Horn, M., Friedrich, A. B., Wagner, M., Hacker, J., et al. (2002). Molecular evidence for a uniform microbial community in sponges from different oceans. *Applied and Environmental Microbiology*, *68*, 4431–4440.
Hentschel, U., Piel, J., Degnan, S. M., & Taylor, M. W. (2012). Genomic insights into the marine sponge microbiome. *Nature Reviews. Microbiology*, *10*, 641–654.
Hentschel, U., Steinert, M., & Hacker, J. (2000). Common molecular mechanisms of symbiosis and pathogenesis. *Trends in Microbiology*, *8*, 226–231.
Herren, J. K., Paredes, J. C., Schupfer, F., & Lemaitre, B. (2013). Vertical transmission of a Drosophila endosymbiont via cooption of the yolk transport and internalization machinery. *MBio*, *4*, e00532-12.
Hirose, E. (2009). Ascidian tunic cells: Morphology and functional diversity of free cells outside the epidermis. *Invertebrate Biology*, *128*, 83–96.

Hosokawa, T., Hironaka, M., Inadomi, K., Mukai, H., Nikoh, N., & Fukatsu, T. (2013). Diverse strategies for vertical symbiont transmission among subsocial stinkbugs. *PLoS One*, *8*, e65081.

Hosokawa, T., Hironaka, M., Mukai, H., Inadomi, K., Suzuki, N., & Fukatsu, T. (2012). Mothers never miss the moment: A fine-tuned mechanism for vertical symbiont transmission in a subsocial insect. *Animal Behaviour*, *83*, 293–300.

Hosokawa, T., Koga, R., Kikuchi, Y., Meng, X.-Y., & Fukatsu, T. (2010). Wolbachia as a bacteriocyte-associated nutritional mutualist. *Proceedings of the National Academy of Sciences*, *107*, 769–774.

Hosokawa, T., Nikoh, N., Koga, R., Satô, M., Tanahashi, M., Meng, X.-Y., et al. (2012). Reductive genome evolution, host–symbiont co-speciation and uterine transmission of endosymbiotic bacteria in bat flies. *The ISME Journal*, *6*, 577–587.

Husnik, F., Nikoh, N., Koga, R., Ross, L., Duncan, R. P., Fujie, M., et al. (2013). Horizontal gene transfer from diverse bacteria to an Insect genome enables a tripartite nested mealybug symbiosis. *Cell*, *153*, 1567–1578.

Ikuta, T., Igawa, K., Tame, A., Kuroiwa, T., Kuroiwa, H., Aoki, Y., et al. (2016). Surfing the vegetal pole in a small population: Extracellular vertical transmission of an "intracellular" deep-sea clam symbiont. *Royal Society Open Science*, *3*, 160130.

Ispolatov, I., Ackermann, M., & Doebeli, M. (2012). Division of labour and the evolution of multicellularity. *Proceedings of the Royal Society B: Biological Sciences*, *279*, 1768–1776.

Izumi, S., Yano, K., Yamamoto, Y., & Takahashi, S. Y. (1994). Yolk proteins from insect eggs: Structure, biosynthesis and programmed degradation during embryogenesis. *Journal of Insect Physiology*, *40*, 735–746.

Jones, K. M., Kobayashi, H., Davies, B. W., Taga, M. E., & Walker, G. C. (2007). How rhizobial symbionts invade plants: The Sinorhizobium–Medicago model. *Nature Reviews Microbiology*, *5*, 619–633.

Kaiwa, N., Hosokawa, T., Kikuchi, Y., Nikoh, N., Meng, X. Y., Kimura, N., et al. (2010). Primary gut symbiont and secondary, sodalis-allied symbiont of the scutellerid stinkbug Cantao ocellatus. *Applied and Environmental Microbiology*, *76*, 3486–3494.

Kaiwa, N., Hosokawa, T., Kikuchi, Y., Nikoh, N., Meng, X. Y., Kimura, N., et al. (2011). Bacterial symbionts of the giant jewel stinkbug *Eucorysses grandis* (Hemiptera: Scutelleridae). *Zoological Science*, *28*, 169–174.

Kaiwa, N., Hosokawa, T., Nikoh, N., Tanahashi, M., Moriyama, M., Meng, X.-Y., et al. (2014). Symbiont-supplemented maternal investment underpinning host's ecological adaptation. *Current Biology*, *24*, 2465–2470.

Kaye, H. R. (1991). Sexual reproduction in four Caribbean commercial sponges. II. Oogenesis and transfer of bacterial symbionts. *Invertebrate Reproduction and Development*, *19*, 13–24.

Kerney, R., Kim, E., Hangarter, R. P., Heiss, A. A., Bishop, C. D., & Hall, B. K. (2011). Intracellular invasion of green algae in a salamander host. *Proceedings of the National Academy of Sciences of the United States of America*, *108*, 6497–6502.

Kikuchi, Y., Hosokawa, T., Nikoh, N., Meng, X.-Y., Kamagata, Y., & Fukatsu, T. (2009). Host-symbiont co-speciation and reductive genome evolution in gut symbiotic bacteria of acanthosomatid stinkbugs. *BMC Biology*, *7*, 2.

Klasson, L., Westberg, J., Sapountzis, P., Näslund, K., Lutnaes, Y., Darby, A. C., et al. (2009). The mosaic genome structure of the Wolbachia wRi strain infecting Drosophila simulans. *Proceedings of the National Academy of Sciences of the United States of America*, *106*, 5725–5730.

Klein, A., Schrader, L., Gil, R., Manzano-Marín, A., Flórez, L., Wheeler, D., et al. (2016). A novel intracellular mutualistic bacterium in the invasive ant Cardiocondyla obscurior. *The ISME Journal*, *10*, 376–388.

Koga, R., Meng, X.-Y., Tsuchida, T., & Fukatsu, T. (2012). Cellular mechanism for selective vertical transmission of an obligate insect symbiont at the bacteriocyte-embryo interface. *Proceedings of the National Academy of Sciences of the United States of America, 109*, E1230–E1237.

Kojima, A., & Hirose, E. (2012). Transmission of cyanobacterial symbionts during embryogenesis in the coral reef ascidians Trididemnum nubilum and T. clinides (Didemnidae, Ascidiacea, Chordata). *The Biological Bulletin, 222*, 63–73.

Komaki, K., & Ishikawa, H. (2000). Genomic copy number of intracellular bacterial symbionts of aphids varies in response to developmental stage and morph of their host. *Insect Biochemistry and Molecular Biology, 30*, 253–258.

Kowald, A., & Kirkwood, T. B. (2011). Evolution of the mitochondrial fusion–fission cycle and its role in aging. *Proceedings of the National Academy of Sciences of the United States of America, 108*, 10237–10242.

Kriesner, P., Hoffmann, A. A., Lee, S. F., Turelli, M., & Weeks, A. R. (2013). Rapid sequential spread of two *Wolbachia* variants in *Drosophila simulans*. *PLoS Pathogens, 9*, e1003607.

Kuechler, S. M., Dettner, K., & Kehl, S. (2011). Characterization of an obligate intracellular bacterium in the midgut epithelium of the bulrush bug Chilacis typhae (Heteroptera, Lygaeidae, Artheneinae). *Applied and Environmental Microbiology, 77*, 2869–2876.

Kupper, M., Stigloher, C., Feldhaar, H., & Gross, R. (2016). Distribution of the obligate endosymbiont Blochmannia floridanus and expression analysis of putative immune genes in ovaries of the carpenter ant Camponotus floridanus. *Arthropod Structure & Development, 45*, 475–487.

Lalzar, I., Friedmann, Y., & Gottlieb, Y. (2014). Tissue tropism and vertical transmission of *Coxiella* in *Rhipicephalus sanguineus* and *Rhipicephalus turanicus* ticks: Tissue tropism of tick symbionts. *Environmental Microbiology, 16*, 3657–3668.

Lambiase, S., Grigolo, A., Laudani, U., Sacchi, L., & Baccetti, B. (1997). Pattern of bacteriocyte formation in Periplaneta americana (L.) (Blattaria: Blattidae). *International Journal of Insect Morphology and Embryology, 26*, 9–19.

Landmann, F., Bain, O., Martin, C., Uni, S., Taylor, M. J., & Sullivan, W. (2012). Both asymmetric mitotic segregation and cell-to-cell invasion are required for stable germline transmission of Wolbachia in filarial nematodes. *Biology Open, 1*, 536–547.

Landmann, F., Orsi, G. A., Loppin, B., & Sullivan, W. (2009). Wolbachia-mediated cytoplasmic incompatibility is associated with impaired histone deposition in the male pronucleus. *PLoS Pathogens, 5*, e1000343.

Landmann, F., Voronin, D., Sullivan, W., & Taylor, M. J. (2011). Anti-filarial activity of antibiotic therapy is due to extensive apoptosis after wolbachia depletion from filarial nematodes. *PLoS Pathogens, 7*, e1002351.

Le Clec'h, W., Chevalier, F. D., Genty, L., Bertaux, J., Bouchon, D., & Sicard, M. (2013). Cannibalism and predation as paths for horizontal passage of Wolbachia between terrestrial isopods. *PLoS One, 8*, e60232.

Leclercq, S., Thézé, J., Chebbi, M. A., Giraud, I., Moumen, B., Ernenwein, L., et al. (2016). Birth of a W sex chromosome by horizontal transfer of *Wolbachia* bacterial symbiont genome. *Proceedings of the National Academy of Sciences of the United States of America, 113*, 15036–15041.

Lehmann, R. (2012). Germline stem cells: Origin and destiny. *Cell Stem Cell, 10*, 729–739.

LePage, D. P., Metcalf, J. A., Bordenstein, S. R., On, J., Perlmutter, J. I., Shropshire, J. D., et al. (2017). Prophage WO genes recapitulate and enhance Wolbachia-induced cytoplasmic incompatibility. *Nature, 543*, 243–247.

Li, S.-J., Ahmed, M. Z., Lv, N., Shi, P.-Q., Wang, X.-M., Huang, J.-L., et al. (2017). Plantmediated horizontal transmission of Wolbachia between whiteflies. *The ISME Journal, 11*, 1019–1028.

Lindsey, A. R. I., Werren, J. H., Richards, S., & Stouthamer, R. (2016). Comparative genomics of a parthenogenesis-inducing wolbachia symbiont. *Genes|Genomes|Genetics*, *6*, 2113–2123.

Lu, W., Lakonishok, M., Serpinskaya, A. S., Kirchenbüechler, D., Ling, S.-C., & Gelfand, V. I. (2018). Ooplasmic flow cooperates with transport and anchorage in *Drosophila* oocyte posterior determination. *The Journal of Cell Biology*, *217*, 3497–3511.

Maldonado, M. (2007). Intergenerational transmission of symbiotic bacteria in oviparous and viviparous demosponges, with emphasis on intracytoplasmically-compartmented bacterial types. *Journal of the Marine Biological Association of the UK*, *87*, 1701–1713.

Matsuura, Y., Kikuchi, Y., Hosokawa, T., Koga, R., Meng, X.-Y., Kamagata, Y., et al. (2012). Evolution of symbiotic organs and endosymbionts in lygaeid stinkbugs. *The ISME Journal*, *6*, 397–409.

Matsuura, Y., Kikuchi, Y., Miura, T., & Fukatsu, T. (2015). *Ultrabithorax* is essential for bacteriocyte development. *Proceedings of the National Academy of Sciences of the United States of America*, *112*, 9376–9381.

Michalik, A., Gołas, A., Kot, M., Wieczorek, K., & Szklarzewicz, T. (2013). Endosymbiotic microorganisms in Adelges (Sacchiphantes) viridis (Insecta, Hemiptera, Adelgoidea: Adelgidae): Molecular characterization, ultrastructure and transovarial transmission. *Arthropod Structure & Development*, *42*, 531–538.

Michalik, A., Schulz, F., Michalik, K., Wascher, F., Horn, M., & Szklarzewicz, T. (2018). Coexistence of novel gammaproteobacterial and *Arsenophonus* symbionts in the scale insect *Greenisca brachypodii* (Hemiptera, Coccomorpha: Eriococcidae): Symbionts of *Greenisca brachypodii*. *Environmental Microbiology*, *20*, 1148–1157.

Miller, W. J., & Riegler, M. (2006). Evolutionary dynamics of wAu-like wolbachia variants in neotropical Drosophila spp. *Applied and Environmental Microbiology*, *72*, 826–835.

Mira, A., & Moran, N. A. (2002). Estimating population size and transmission bottlenecks in maternally transmitted endosymbiotic Bacteria. *Microbial Ecology*, *44*, 137–143.

Mishra, P., & Chan, D. C. (2014). Mitochondrial dynamics and inheritance during cell division, development and disease. *Nature Reviews. Molecular Cell Biology*, *15*, 634–646.

Mitchell, T., & Cavanaugh, C. (1983). Numbers of symbiotic bacteria in the gill tissue of the bivalve Solemya velum Say. *The Biological Bulletin*, *165*, 521.

Miura, T., Braendle, C., Shingleton, A., Sisk, G., Kambhampati, S., & Stern, D. L. (2003). A comparison of parthenogenetic and sexual embryogenesis of the pea aphid Acyrthosiphon pisum (Hemiptera: Aphidoidea). *Journal of Experimental Zoology. Part B, Molecular and Developmental Evolution*, *295*, 59–81.

Moran, N. A. (2007). Symbiosis as an adaptive process and source of phenotypic complexity. *Proceedings of the National Academy of Sciences of the United States of America*, *104*, 8627–8633.

Moya, A., Peretó, J., Gil, R., & Latorre, A. (2008). Learning how to live together: Genomic insights into prokaryote–animal symbioses. *Nature Reviews. Genetics*, *9*, 218–229.

Nakabachi, A., Ishida, K., Hongoh, Y., Ohkuma, M., & Miyagishima, S. (2014). Aphid gene of bacterial origin encodes a protein transported to an obligate endosymbiont. *Current Biology*, *24*, R640–R641.

Nan, G.-H., Xu, Y.-P., Yu, Y.-W., Zhao, C.-X., Zhang, C.-X., & Yu, X.-P. (2016). Oocyte vitellogenesis triggers the entry of yeast-like symbionts into the oocyte of brown planthopper (Hemiptera: Delphacidae). *Annals of the Entomological Society of America*, *109*, 753–758.

Nyholm, S. V., & Graf, J. (2012). Knowing your friends: Invertebrate innate immunity fosters beneficial bacterial symbioses. *Nature Reviews. Microbiology*, *10*, 815–827.

Ote, M., Ueyama, M., & Yamamoto, D. (2016). *Wolbachia* protein TomO targets nanos mRNA and restores germ stem cells in *Drosophila* sex-lethal mutants. *Current Biology*, *26*, 2223–2232.

Ote, M., & Yamamoto, D. (2018). The *Wolbachia* protein TomO interacts with a host RNA to induce polarization defects in *Drosophila* oocytes. *Archives of Insect Biochemistry and Physiology*, *99*, e21475.
Pannebakker, B. A., Loppin, B., Elemans, C. P., Humblot, L., & Vavre, F. (2007). Parasitic inhibition of cell death facilitates symbiosis. *Proceedings of the National Academy of Sciences of the United States of America*, *104*, 213–215.
Perotti, M. A., Allen, J. M., Reed, D. L., & Braig, H. R. (2007). Host-symbiont interactions of the primary endosymbiont of human head and body lice. *The FASEB Journal*, *21*, 1058–1066.
Pietri, J. E., DeBruhl, H., & Sullivan, W. (2016). The rich somatic life of *Wolbachia*. *MicrobiologyOpen*, *5*, 923–936.
Pinto-Carbó, M., Gademann, K., Eberl, L., & Carlier, A. (2018). Leaf nodule symbiosis: Function and transmission of obligate bacterial endophytes. *Current Opinion in Plant Biology*, *44*, 23–31.
Pintureau, B., Grenier, S., Boléat, B., Lassablière, F., Heddi, A., & Khatchadourian, C. (2000). Dynamics of Wolbachia populations in transfected lines of Trichogramma. *Journal of Invertebrate Pathology*, *76*, 20–25.
Prado, S. S., & Almeida, R. P. P. (2009). Role of symbiotic gut bacteria in the development of *Acrosternum hilare* and *Murgantia histrionica*. *Entomologia Experimentalis et Applicata*, *132*, 21–29.
Prado, S. S., Rubinoff, D., & Almeida, R. P. P. (2006). Vertical transmission of a pentatomid caeca-associated symbiont. *Annals of the Entomological Society of America*, *99*, 577–585.
Radzvilavicius, A. L., Hadjivasiliou, Z., Pomiankowski, A., & Lane, N. (2016). Selection for mitochondrial quality drives evolution of the germline. *PLoS Biology*, *14*, e2000410.
Riegler, M., Sidhu, M., Miller, W. J., & O'Neill, S. L. (2005). Evidence for a global *Wolbachia* replacement in *Drosophila melanogaster*. *Current Biology*, *15*, 1428–1433.
Rio, R. V. M., Symula, R. E., Wang, J., Lohs, C., Wu, Y.-N., Snyder, A. K., et al. (2012). Insight into the transmission biology and species-specific functional capabilities of tsetse (Diptera: Glossinidae) obligate symbiont *Wigglesworthia*. *MBio*, *3*, e00240-11.
Rokas, A. (2008). The origins of multicellularity and the early history of the genetic toolkit for animal development. *Annual Review of Genetics*, *42*, 235–251.
Russell, S. L., & Cavanaugh, C. M. (2017). Intrahost genetic diversity of bacterial symbionts exhibits evidence of mixed infections and recombinant haplotypes. *Molecular Biology and Evolution*, *34*, 2747–2761.
Russell, S. L., Corbett-Detig, R. B., & Cavanaugh, C. M. (2017). Mixed transmission modes and dynamic genome evolution in an obligate animal–bacterial symbiosis. *The ISME Journal*, 1359–1371.
Russell, S. L., Lemseffer, N., & Sullivan, W. T. (2018). Wolbachia and host germline components compete for kinesin-mediated transport to the posterior pole of the Drosophila oocyte. *PLoS Pathogens*, *14*, e1007216.
Russell, S. L., McCartney, E., & Cavanaugh, C. M. (2018). Transmission strategies in a chemosynthetic symbiosis: Detection and quantification of symbionts in host tissues and their environment. *Proceedings of the Royal Society B: Biological Sciences*, *285*, 9.
Sacchi, L., Genchi, M., Clementi, E., Bigliardi, E., Avanzati, A. M., Pajoro, M., et al. (2008). Multiple symbiosis in the leafhopper Scaphoideus titanus (Hemiptera: Cicadellidae): Details of transovarial transmission of Cardinium sp. and yeast-like endosymbionts. *Tissue & Cell*, *40*, 231–242.
Sacchi, L., Grigolo, A., Mazzinit, M., Bigliardi, E., Baccetti, B., & Laudani, U. (1988). Symbionts in the oocytes of Blattella germanica (L.) (Dictyoptera: Blattellidae): Their mode of transmission. *International Journal of Insect Morphology and Embryology*, *17*, 437–446.

Schmitt, S., Angermeier, H., Schiller, R., Lindquist, N., & Hentschel, U. (2008). Molecular microbial diversity survey of sponge reproductive stages and mechanistic insights into vertical transmission of microbial symbionts. *Applied and Environmental Microbiology*, *74*, 7694–7708.

Schmitt, S., Weisz, J. B., Lindquist, N., & Hentschel, U. (2007). Vertical transmission of a phylogenetically complex microbial consortium in the viviparous sponge *Ircinia felix*. *Applied and Environmental Microbiology*, *73*, 2067–2078.

Schulz, F., Lagkouvardos, I., Wascher, F., Aistleitner, K., Kostanjšek, R., & Horn, M. (2014). Life in an unusual intracellular niche: A bacterial symbiont infecting the nucleus of amoebae. *The ISME Journal*, *8*, 1634–1644.

Sender, R., Fuchs, S., & Milo, R. (2016). Revised estimates for the number of human and bacteria cells in the body. *PLoS Biology*, *14*, e1002533.

Serbus, L. R., Ferreccio, A., Zhukova, M., McMorris, C. L., Kiseleva, E., & Sullivan, W. (2011). A feedback loop between Wolbachia and the Drosophila gurken mRNP complex influences Wolbachia titer. *Journal of Cell Science*, *124*, 4299–4308.

Serbus, L. R., & Sullivan, W. (2007). A cellular basis for *Wolbachia* recruitment to the host germline. *PLoS Pathogens*, *3*, e190.

Shapira, M. (2016). Gut microbiotas and host evolution: Scaling up symbiosis. *Trends in Ecology & Evolution*, *31*, 539–549.

Sharp, K. H., Eam, B., Faulkner, D. J., & Haygood, M. G. (2007). Vertical transmission of diverse microbes in the tropical sponge corticium sp. *Applied and Environmental Microbiology*, *73*, 622–629.

Sinsimer, K. S., Lee, J. J., Thiberge, S. Y., & Gavis, E. R. (2013). Germ plasm anchoring is a dynamic state that requires persistent trafficking. *Cell Reports*, *5*, 1169–1177.

Sloan, D. B., Nakabachi, A., Richards, S., Qu, J., Murali, S. C., Gibbs, R. A., et al. (2014). Parallel histories of horizontal gene transfer facilitated extreme reduction of endosymbiont genomes in sap-feeding insects. *Molecular Biology and Evolution*, *31*, 857–871.

Stewart, F. J., Newton, I. L. G., & Cavanaugh, C. M. (2005). Chemosynthetic endosymbioses: Adaptations to oxic–anoxic interfaces. *Trends in Microbiology*, *13*, 439–448.

Stoll, S., Feldhaar, H., Fraunholz, M. J., & Gross, R. (2010). Bacteriocyte dynamics during development of a holometabolous insect, the carpenter ant Camponotus floridanus. *BMC Microbiology*, *10*, 1.

Sullivan, W. (2017). *Wolbachia*, bottled water, and the dark side of symbiosis. *Molecular Biology of the Cell*, *28*, 2343–2346.

Sun, S., & Cline, T. W. (2009). Effects of Wolbachia infection and ovarian tumor mutations on sex-lethal germline functioning in drosophila. *Genetics*, *181*, 1291–1301.

Swiatoniowska, M., Ogorzalek, A., Golas, A., Michalik, A., & Szklarzewicz, T. (2013). Ultrastructure, distribution, and transovarial transmission of symbiotic microorganisms in Nysius ericae and Nithecus jacobaeae (Heteroptera: Lygaeidae: Orsillinae). *Protoplasma*, *250*, 325–332.

Szklarzewicz, T., & Moskal, A. (2001). Ultrastructure, distribution, and transmission of endosymbionts in the whitefly Aleurochiton aceris Modeer (Insecta, Hemiptera, Aleyrodinea). *Protoplasma*, *218*, 45–53.

Toft, C., & Andersson, S. G. E. (2010). Evolutionary microbial genomics: Insights into bacterial host adaptation. *Nature Reviews. Genetics*, *11*, 465–475.

Toomey, M. E., Panaram, K., Fast, E. M., Beatty, C., & Frydman, H. M. (2013). Evolutionarily conserved *Wolbachia*-encoded factors control pattern of stem-cell niche tropism in Drosophila ovaries and favor infection. *Proceedings of the National Academy of Sciences of the United States of America*, *110*, 10788–10793.

Turelli, M., Cooper, B. S., Richardson, K. M., Ginsberg, P. S., Peckenpaugh, B., Antelope, C. X., et al. (2018). Rapid global spread of w Ri-like *Wolbachia* across multiple *Drosophila*. *Current Biology*, *28*, 963–971.e8.

Turelli, M., & Hoffmann, A. A. (1991). Rapid spread of an inherited incompatibility factor in California Drosophila. *Nature, 353*, 440–442.

Usher, K. M., Kuo, J., Fromont, J., & Sutton, D. C. (2001). Vertical transmission of cyanobacterial symbionts in the marine sponge Chondrilla australiensis (Demospongiae). *Hydrobiologia, 461*, 9–13.

Vacelet, J., Fiala-Médioni, A., Fisher, C., & Boury-Esnault, N. (1996). Symbiosis between methane-oxidizing bacteria and a deep-sea carnivorous cladorhizid sponge. *Marine Ecology: Progress Series, 145*, 77–85.

van Hoek, A. H. A. M., van Alen, T. A., Sprakel, V. S. I., Leunissen, J. A. M., Brigge, T., Vogels, G. D., et al. (2000). Multiple acquisition of methanogenic archaeal symbionts by anaerobic ciliates. *Molecular Biology and Evolution, 17*, 251–258.

Veneti, Z., Clark, M. E., Karr, T. L., Savakis, C., & Bourtzis, K. (2004). Heads or tails: Host-parasite interactions in the *Drosophila-Wolbachia* system. *Applied and Environmental Microbiology, 70*, 5366–5372.

von Dohlen, C., Kohler, S., Alsop, S. T., & McManus, W. R. (2001). Mealybug beta-proteobacterial endosymbionts contain gamma-proteobacterial symbionts. *412*, 433–436.

Watanabe, K., Yukuhiro, F., Matsuura, Y., Fukatsu, T., & Noda, H. (2014). Intrasperm vertical symbiont transmission. *Proceedings of the National Academy of Sciences of the United States of America, 111*, 7433–7437.

Wentrup, C., Wendeberg, A., Huang, J. Y., Borowski, C., & Dubilier, N. (2013). Shift from widespread symbiont infection of host tissues to specific colonization of gills in juvenile deep-sea mussels. *The ISME Journal, 7*, 1244–1247.

Werren, J. H., Baldo, L., & Clark, M. E. (2008). *Wolbachia*: Master manipulators of invertebrate biology. *Nature Reviews Microbiology, 6*, 741–751.

White, P. M., Pietri, J. E., Debec, A., Russell, S. L., Patel, B., & Sullivan, W. (2017). Mechanisms of horizontal cell-to-cell transfer of *Wolbachia* spp. in *Drosophila melanogaster*. *Applied and Environmental Microbiology, 83*, e03425-16.

Wollenberg, M. S., & Ruby, E. G. (2009). Population structure of *Vibrio fischeri* within the light organs of *Euprymna scolopes* squid from two Oahu (Hawaii) populations. *Applied and Environmental Microbiology, 75*, 193–202.

Wu, M., Sun, L. V., Vamathevan, J., Riegler, M., Deboy, R., Brownlie, J. C., et al. (2004). Phylogenomics of the reproductive parasite *Wolbachia pipientis* wMel: A streamlined genome overrun by mobile genetic elements. *PLoS Biology, 2*, e69.

Zchori-Fein, E., Roush, R. T., & Rosen, D. (1998). Distribution of parthenogenesis-inducing symbionts in ovaries and eggs of aphytis (Hymenoptera: Aphelinidae). *Current Microbiology, 36*, 1–8.

Kolibri 360 GmbH
Ferdinand-Jühlke-Straße 7
99095 Erfurt, Deutschland
produktsicherheit@kolibri360.de